Toxicity of Nanoparticles - Recent Advances and New Perspectives

Edited by Mohammed Muzibur Rahman, Jamal Uddin, Abdullah Mohamed Asiri and Md Rezaur Rahman

Published in London, United Kingdom

Toxicity of Nanoparticles - Recent Advances and New Perspectives
http://dx.doi.org/10.5772/intechopen.111007
Edited by Mohammed Muzibur Rahman, Jamal Uddin, Abdullah Mohamed Asiri and Md Rezaur Rahman

Contributors
Krishna Vaddadi, Nagimani Naidu, Nookaraju Muralasetti, Subash Adhikari, Iman M. Alfagih, Krishna Gudikandula, Trimothi Dasari, Reeja Sundaram, Divya Bijukumar, Neeraja Revi, Oluwatosin D. Oladejo, Dalia Chávez-García, Karla Juárez-Moreno, Anahi Rodríguez-López, Gerardo González De La Cruz, Patricia Rodríguez-Fragoso, Lourdes Rodríguez-Fragoso, Akeem Omolaja Akinfenwa, Ahmed Abdelfattah Hussein, Hemant Sarin, V. Sandhiya, K. Dhunmati, Senthilkumar Balakrishnan, C.N. Nalini, Ganesh Mani, Ubaidulla Uthumansha, Anna Permanasari

First published in London, United Kingdom, 2024 by IntechOpen
IntechOpen is the global imprint of INTECHOPEN LIMITED, registered in England and Wales, registration number: 11086078, 5 Princes Gate Court, London, SW7 2QJ, United Kingdom

British Library Cataloguing-in-Publication Data
A catalogue record for this book is available from the British Library

Additional hard and PDF copies can be obtained from orders@intechopen.com

Toxicity of Nanoparticles - Recent Advances and New Perspectives
Edited by Mohammed Muzibur Rahman, Jamal Uddin, Abdullah Mohamed Asiri and Md Rezaur Rahman
p. cm.
Print ISBN 978-1-83768-452-6
Online ISBN 978-1-83768-604-9
eBook (PDF) ISBN 978-1-83768-605-6

For EU product safety concerns:
IN TECH d.o.o., Prolaz Marije Krucifikse Kozulić 3, 51000 Rijeka, Croatia,
info@intechopen.com or visit our website at intechopen.com.

Meet the editors

Mohammed Muzibur Rahman received his BSc and MSc from Shahjalal University of Science and Technology, Bangladesh, in 1999 and 2001, respectively. He received his Ph.D. from Chonbuk National University, South Korea in 2007. After his Ph.D., he worked as a postdoctoral fellow and assistant professor at pioneer research centers and universities in South Korea, Japan, and Saudi Arabia (2007–2011). Presently, he is a professor at the Center of Excellence for Advanced Materials Research (CEAMR) and Chemistry Department at King Abdulaziz University, Saudi Arabia. His research interests include carbon nanotubes, sensors, nanotechnology, nanocomposites, nanomaterials, carbon nanofibers, photocatalysis, semiconductors, nanoparticles, electrocatalysis, ionic liquids, surface chemistry, and electrochemistry.

Dr. Jamal Uddin is a professor in the Department of Natural Sciences, Coppin State University (CSU), Baltimore, MD, USA, where he teaches teaching chemistry, physical science, environmental science, and nanotechnology. He is also the founder and director of the Center for Nanotechnology at the same university. He is also a visiting scientist at the Center for Fluorescence Spectroscopy, Department of Biochemistry and Molecular Biology, School of Medicine, University of Maryland, USA. Dr. Uddin's research interests include solar energy, laser photochemistry, nanomaterials, quantum dots, carbon nanotubes, graphene, dendrimers, fluorescence spectroscopy, and metal-enhanced fluorescence. Some of his significant publications are on electron transfer chemistry in donor–acceptor complexes. He is a member of the American Chemical Society (ACS) and corresponding secretary of the American Association of University Professors (AAUP), CSU, Maryland chapter. Dr. Uddin is the recipient of the 2012 and 2016 Daily Record Innovator of the Year for Maryland individuals who have had a positive effect and tremendous impact on the state.

Abdullah Mohamed Asiri obtained a Ph.D. from the University of Wales, College of Cardiff, UK, in 1995. He is the head of the Chemistry Department at King Abdul Aziz University, Saudi Arabia, and the founder and director of the Center of Excellence for Advanced Materials Research (CEAMR). He is a Professor of Organic Photochemistry. His research interests include color chemistry, synthesis of novel photochromics, thermochromic systems, synthesis of novel coloring matters and dyeing of textiles, materials chemistry, nanochemistry, nanotechnology, polymers, and plastics. He is the editor-in-chief of King Abdulaziz University Journal of Science. He is also a member of the editorial board of Pigments and Resin Technology, Organic Chemistry Insights, and Recent Patents on Materials Science. He is the vice president of the Saudi Chemical Society (Western Province Branch).

Dr. Md Rezaur Rahman Obtained a Ph.D. in Polymer Nanocomposites from the University of Malaysia Sarawak (UNIMAS) in 2011. He is a member of the Institution of Chemical Engineers, UK. He is currently a senior lecturer in the Department of Chemical Engineering and Energy Sustainability, Faculty of Engineering, UNIMAS. His research interests include bio-plastic research and development, cellulose-based micro and nanocomposite technology, industrial biomaterials and biocomposites manufacturing, biomass technology, value-added utilization of residual biomass and biochemicals, polymer solar cells, polymeric scaffolds for soft tissue engineering, and pulp and paper technology. He has collaborated with the Institute of Technology and Science, University of Tokushima, Japan, and Swinburne University of Technology, Malaysia. He is currently supervising five MSc and three Ph.D. students. He leads three research projects and is a co-researcher for five other projects. He has published eighty-three international journal articles and two books.

Contents

Preface

Toxicity of Nanoparticles – Recent Advances and New Perspectives focuses on the development of nanoparticles for advanced applications. This book reviews the state of the art of various nanoparticles, their facile synthesis, detailed characterizations, and potential applications in toxicological purposes including drug delivery, bioengineering, microbiology, seafood and aquacultures, phytometallics, nanotoxicological assessment, and immunotoxicity. It covers nanoscale aspects of the synthesis, growth, and development of spherical nanoparticles as well as new paths and emerging frontiers in nanoparticle toxicity. The discussion includes fundamentals and conventional applied experimental routes of toxicity, the interaction of nanoparticles in biological and chemical aspects, and the interface of nanoscience and nanotechnology.

This book seeks to bridge the gap between undergraduate, graduate, and research study in advanced nanoscale materials, to introduce scientists to the opportunities offered by applied science and technology. I hope that this contribution will bring new entrants into the fields of applied material science and nanotechnology and help scientists to develop their specializations.

Mohammed Muzibur Rahman and Abdullah Mohamed Asiri
Center of Exellence for Advanced Materials Research (CEAMR),
Department of Chemistry,
King Abdulaziz University,
Jeddah, Saudi Arabia

Jamal Uddin
Department of Chemistry,
Coppin State University,
Baltimore, United States

Md Rezaur Rahman
Faculty of Engineering,
Department of Chemical Engineering and Energy Sustainability,
University Malaysia Sarawak,
Kota Samarahan, Malaysia

Section 1

State-of-Art Nanotechnology

Chapter 1

Toxicity of Quantum Dots

Gerardo González De la Cruz, Lourdes Rodríguez-Fragoso, Patricia Rodríguez-Fragoso and Anahi Rodríguez-López

Abstract

Quantum dots (QD) have been deeply studied due to their physicochemical and optical properties with important advantages of a wide range biomedical applications. Nevertheless, concern prevails about its toxic effects, mainly in those QD whose core contains cadmium. Therefore, there are reports about the toxicity caused by the release of ions of cadmium and the effects related to its tiny nanometric size. The aim of this chapter is to show the evaluations about the toxicity of QD, which include studies on viability, proliferation, uptake, and distribution *in vitro* and *in vivo* models. What are the worrying toxic effects of QD? There are reports about some mechanisms of toxicity caused by QD, such as immunological toxicity, cell death (apoptosis and necrosis), genotoxicity, among others. In addition, we discuss how coating QD with passivating agents that improve their biocompatibility. Likewise, this coating modifies their size and surface charge, which are fundamental aspects of the interaction with other biomolecules. We consider highlighting information about more precise techniques and methodologies that help us to understand how QD induce damage in several biological systems.

Keywords: quantum dots, cytotoxicity, cadmium, nanotoxicity, biocompatibility

1. Introduction

In recent decades, there have been countless publications on the use of nanomaterials, particularly in the biomedical area. The main use of semiconductor nanoparticles (NPs) lies in the development of formulations for the delivery of anticancer therapies, specifically targeting diseased tissues and organs. Moreover, quantum dots (QDs) provide remarkable specificity while avoiding damage to surrounding healthy cells and thus avoiding the dreaded side and adverse effects of current treatments. However, among the great applications and their attractive physicochemical and optical properties are a myriad of toxicological effects in biological systems [1]. QDs are inorganic semiconductors with a size range of 1–10 nm. Unlike other types of nanomaterials (NMs), QDs possess a unique and exceptional luminescent property. QDs have become the focus of a study by many researchers [2]. So far, QDs are the most promising option that have exhibited potential for applications in bioimaging (luminescence detection) [3, 4].

Quantum dots have properties, such as luminescent intensity, broad emission spectrum, tight size control, and selectivity, based on their composition. In addition, quantum dots have high resistance to photobleaching, physicochemical robustness, and better half-life than other conventional fluorochromes [5–9]. These nanomaterials are constituted by central semiconductor core consisting of elements from groups II, VI, III-V, or IV-VI of the periodic table and mostly can be composed of heavy metals and toxic materials (e.g., Cd, Te and Hg, CdS, CdTe, CdSe, among others) [10, 11]. Because their main component is cadmium and because of their tiny size they imply a potential hazard, especially for medical applications. There are different types of cadmium-free quantum dots, such as InP/ZnS, CuInS2/ZnS, AgInS2/ZnS, silicon, and graphene. Although they are cadmium-free in their composition, they are still subject to rigorous toxicological studies [12].

In order to reduce the cytotoxicity of quantum dots, there are some strategies such as the use of some shells composed of ZnS, CdS, ZnSe, or even CdS/ZnS multishells. By covering the core not only improved luminescent effect but also reducing the toxicity by avoiding the release of heavy metal ions [13, 14]. Achieving functionalization of the QD core shell with a polymeric shell can give the desired biocompatibility and decrease its cytotoxic effects [15, 16]. Among some functionalized QDs, there are those coated with polymers such as dextrin or maltodextrin, which make the semiconductor able to target organs and can even be taken up by cellular organelles [17–19]. This advantage allows QD to be more specific and selective for applications for disease diagnosis and treatment purposes. However, the negative effects that QDs may have on cells are difficult to assess. QDs have higher fluorescence intensity, prolonged lifetime, specificity, and possess optical stability compared to conventional fluorochromes. In addition, the wavelength at which they emit is given by tight control of the core size. **Figure 1** shows the characteristic image of QDs emitting photoluminescence.

The characteristics of QDs include size, which is what determines the wavelength at which they emit, although in some cases it does not depend on their composition. Thus, QDs of smaller size (2 nm) emit in blue, QDs of 3–5 nm in green, 6–8 nm in orange, and sizes of approximately 8–10 nm in red [10]. The controversial mechanisms by which QDs are introduced into cells are of great interest among the scientific community and thus the molecular and physiological basis of cytotoxicity. These cytotoxic effects have been classified into *in vivo* and *in vitro*. Thus, cell culture-based tests have become the first choice for bioassessment of QD toxicity [20]. However, *in vitro* studies include assessments of cell membrane integrity, morphological changes, organelle dysfunction, and in some cases quantification of viable cells. Nevertheless,

Figure 1.
Fluorescence image of cadmium QD. La emisión de fluorescencia es dependiente del tamaño de los QD. Por lo que, la fotoluminiscencia va del Azul Para aquellos QD más pequeños y hasta el rojo Para los de un núcleo mayor.

information on the behavior of quantum dots in a biological system is still scarce and does not emphasize the cell type-dependent toxicity induced by quantum dots. In this review, we have summarized the efforts in achieving a less toxic design, its advantages and disadvantages in the synthesis of single and bioconjugated quantum dots for application as nanovehicles.

2. Cytotoxic effects of quantum dots on diverse cell lines

The cell membrane is the first barrier that divides intracellular from extracellular mechanisms. The process by which QDs enter the cell is not well defined, although it includes anchoring of QDs to the cell membrane, transmembrane transport, distribution and localization within subcellular compartments, and intracellular accumulation. All these processes are linked to their future application, their potential toxic effects, and the adverse effects induced in a dose-time-dependent manner [21]. Tests such as *in vitro* cytotoxicity are important because of the significant morphological changes caused by QDs at the cellular and subcellular levels. In recent years, a huge variety of *in vitro* studies suggest that QDs have toxic effects on cells at different levels [22, 23]. In addition, the passage of QDs across the cell membrane has been demonstrated, the effects are oxidative stress, direct damage to membrane, morphological alterations, and various types of cell death.

In vitro models are necessary for safety assessment in preclinical testing of nanomaterials for diagnostic purposes. Although some models for cytotoxicity are not sufficient due to lack of human cells available for culture or even lack of reproducibility in assays. Therefore, the predictability about the safety of a nanodrug is a difficult task for nanotoxicology researchers [24]. However, there are *in vitro* models considered as standard patterns for toxicological studies of nanomedicines such as the use of human renal Hek293 cells [25]. Over a decade, our research group has focused its interest on the study of dextrin-coated 3.5 nm sized cadmium sulfide QDs (CdS-dex) [26] and their potential biomedical application as is the case of doxorubicin-conjugated CdS-dex QDs (CdS-dex/dox) [27]. Therefore, we have established several *in vitro* tests using Hek293, HeLa (cervix adenocarcinoma), and HepG2 (hepatic cells) cells for preclinical studies on CdS-dextrin quantum dots and with maltodextrin. Therefore, our results demonstrate that CdS-dex QDs and CdS-dex/dox QDs induce exposure to dose-dependent cytotoxic effects. In addition to this, we consider that one of the main evaluations to be performed on QDs is the monitoring of their cellular uptake and distribution. We observed that Hek293, HeLa and HepG2 cells when being treated with concentrations of 0.01 and 1 μg/mL, CdS-dex QDs cross the cell membrane, induce morphological changes, and distribute uniformly at different cellular level. Due to their nanometer size, QDs caused cytotoxicity in the three different cell types by crossing the cell membrane. However, morphological changes varied significantly between Hek293, HepG2, and HeLa cells and the concentration of CdS-dex QDs (**Figure 2**). When QDs have contact with the extracellular membrane, they interact with components of the plasma membrane which allows them to somehow enter the cell by some mechanism such as endocytosis. Endocytosis engulfs the QDs by invagination of the membrane to form endocytic vesicles, which transport the QDs to subcellular compartments. Depending on the cell type, as well as some biomolecules involved in the process, endocytosis can occur in different types [28, 29]. Some authors refer to the uncertainty about the toxic effect that quantum dots may cause as they are transported through the bloodstream and leach into the kidneys.

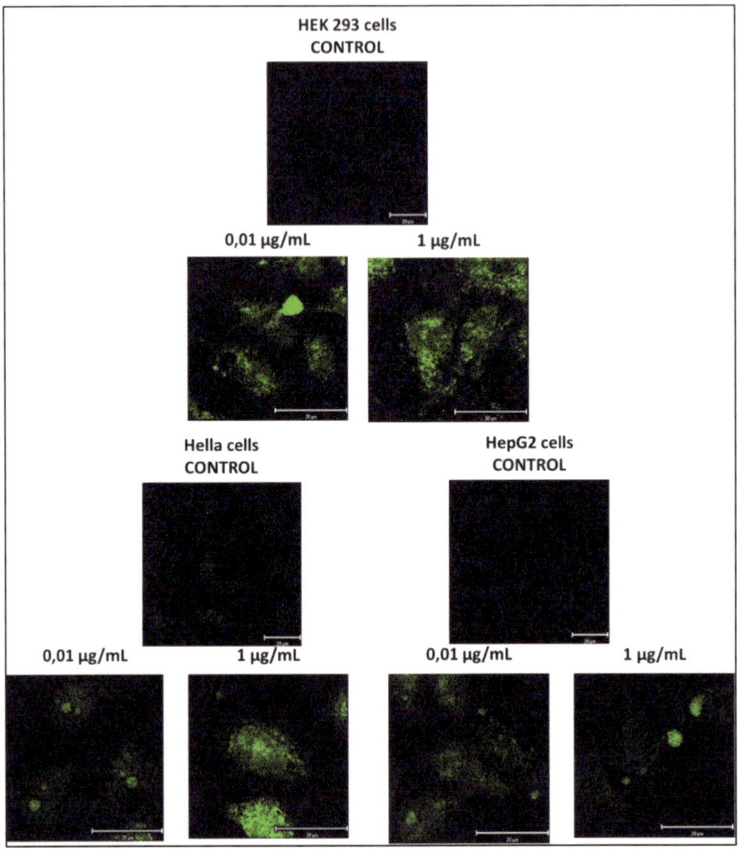

Figure 2.
Fluorescent microscopic visualization of CdS-dex QD in human cell lines. Cells were treated for 24 h with CdS-dex QD (0,01–1 µg/mL). Cells were seeded on slides by smearing and allowed to dry, then analyzed using confocal epifluorescence microscope. Green fluorescence shows the presence of QD surrounding the cytoplasm of Hek293, HeLa, and HepG2 cells. Scale bar 20 µm.

However, there is no information on the nephrotoxic effects of quantum dots both *in vitro* and *in vivo*. Nevertheless, some studies aim to understand the cytotoxic effect on renal cells caused by quantum dots. Therefore, quantum dots, such as titanium oxide (TiO_2), zinc oxide (ZnO), and cadmium sulfide (CdS), have been evaluated in tubular cells (HK-2) in which the cellular and molecular mechanism through oxidative stress induced by quantum dots was demonstrated. In which it was observed that the cytotoxicity of quantum dots was size and solubility dependent. Furthermore, quantum dots that were soluble such as CdS and ZnO were found to cause dose-dependent cell death and degradation/discharge of their ions, respectively [30].

In another investigation, carboxylated CdTe QDs were used and the induced cytotoxicity was evaluated in HeLa cells treated at concentrations from 0.1 to 1000 ng/mL during different exposure times. The effect of CdTe QDs on cell death type, genotoxic effect, and cellular uptake was also evaluated. In this study, they demonstrated that carboxylated QDs did not prove to be less cytotoxic compared to CdTe alone in a concentration-dependent manner. Furthermore, they concluded that CdTe-COOH QDs have genotoxic properties and antiproliferative effects in HeLa cells [31].

Although CdS-dex quantum dots produced different cytotoxic effects on human tumor cells, these effects are not necessarily benign. In fact, our study showed that these nanoparticles had the ability to enter even subcellular compartments. Thus, their biological behavior could trigger pathophysiological effects in a concentration-dependent intrinsic manner. Our CdS quantum dots are coated with a polymeric layer of dextrin. However, many nanomaterials are known to have an inorganic or polymer layer protecting the core to prevent degradation. Even so, heavy metal ions such as cadmium can be released through low stability [32–34]. Studies are needed to know if the cadmium core degrades and releases metal ions and what effects are related to this degradation.

Despite the remarkable effects caused by CdS-dex quantum dots, we clearly need to reinforce the studies and strategies that allow us to learn more about their toxicity. We are getting closer and closer to obtaining biocompatible semiconductor nanoparticles with useful capabilities in diagnosis, treatment, and monitoring of pathologies such as cancer.

Evidently, QDs have physicochemical properties and capabilities and characteristics similar to biological molecules that allow them to be used in biodiagnostics, bioimaging, and targeted drug delivery. For a drug to be effectively delivered using nanocarriers such as QDs, the core component of the QD, the drug or molecule with which it will bioconjugate, and the core shell must be considered. That is, this set of components must be carefully selected to have therapeutic efficiency and optimal safety for use in a biological system [35, 36]. Currently, QDs are considered a tool with promising uses and applications in nanomedicine. However, their cytotoxic effects remain among the main challenges regarding their biocompatibility. The QDs with the highest capacity to emit luminescence and with the highest efficiency in carrying molecules with active principle are those containing cadmium (Cd). However, one of the limitations for the use of Cd QDs in nanomedicine and clinical research is that it is suggested that the core disintegrates and is potentially toxic. That is, it has been considered that it is the core of the QD that largely determines the cytotoxic response and pathophysiological effects [37–39].

Some authors refer that the safety assessment of QDs alone or conjugated is of vital importance since it will allow predicting the effects when interacting with a biological system. They suggest that a nanomaterial is small enough to enter a cell and its cellular compartments, regardless of the route of administration [40–42]. For systemic drug delivery, the intravenous (IV) route is used, which is a major challenge in the development of nanotherapies [43]. The US Food and Drugs Administration (FDA) has approved NMs that have been studied in rigorous preclinical studies combining therapeutic and biological targets as drug delivery agents [44–46].

Our working group has been given the task of synthesizing colloidal CdS-dex/dox QD and evaluating on HeLa cell. We treated HeLa cells with CdS-dex and CdS-dex/dox to compare the selectividad of uptake alone as well as bioconjugated (1 μg/mL) in both cases and with doxorubicin at the same concentration. After 24 h of incubation and in order to investigate the cellular absorption of QD, cells were fixed on slides for visualization by confocal fluorescence microscopy. Through visualization of fluorescence and cellular uptake, we can observe that in cells treated with CdS-dex QDs without bioconjugation, there was a higher distribution in cytoplasm, nucleus, and nucleoli of the cell. However, this cellular uptake and distribution were not the same in the case of HeLa cells treated with doxorubicin and CdS-dex/dox. Nevertheless, in cells treated with doxorubicin and CdS-dex/dox, a significant increase in cell size was observed compared to cells treated with QDs alone. Although, QDs did not appear homogeneous throughout the cytoplasm and with lower fluorescence intensity in the nucleus (**Figure 3**). They can also induce not only cytotoxic but also genotoxic effects in both normal and cancer cells [47–50].

Although, it has been shown that the effect after cellular uptake of various QDs depends on their size, shape, concentration, and cell type. The cytotoxic effect and mechanisms of nanotoxicity by the interaction of QDs with cells remain complex to assess and far from fully understood. However, this nanotoxicity has been shown to occur intracellularly or extracellularly [51]. QDs can even interact directly with biomolecules once inside the cell, due to their minute size. As a result of this interaction, an alteration in cellular equilibrium coexists, as well as irreversible morphological

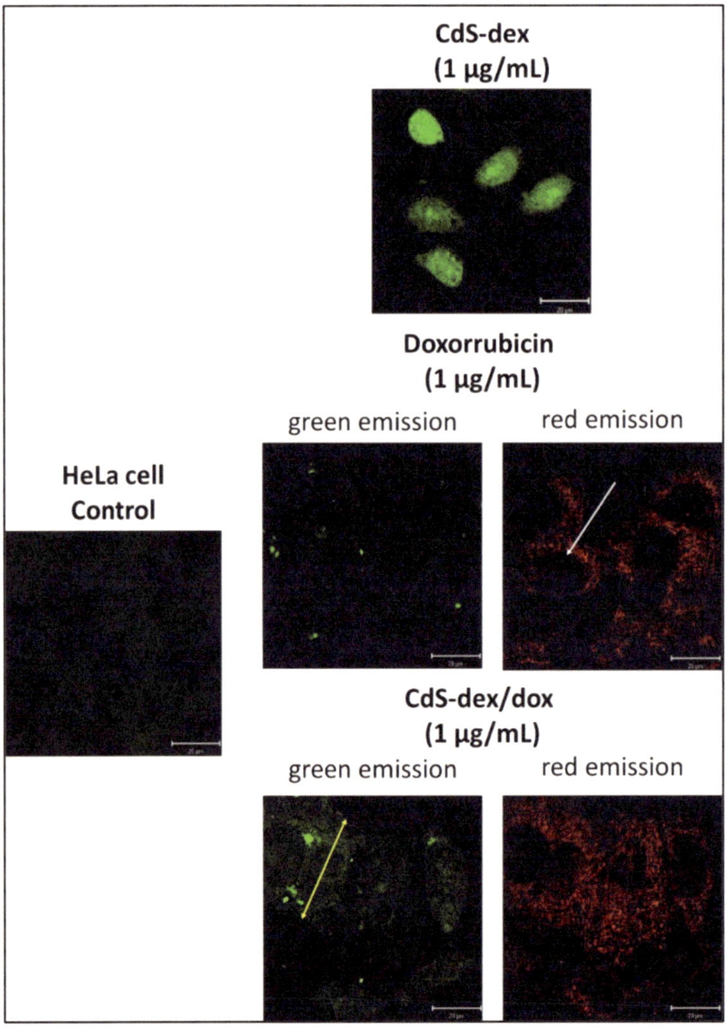

Figure 3.
Fluorescent microscopic visualization of doxorubicin, CdS-dex, and CdS-dex/dox QD in HeLa cell. Cells were treated for 24 h at 0,01–1 µg/mL concentration of doxorubicin, CdS-dex, and CdS-dex/dox QD. Cells were seeded on slides by smearing and allowed to dry, then analyzed using a confocal epifluorescence microscope. Green fluorescence shows the presence of CdS-dex QD. Red emission shows fluorescence in the presence of doxorubicin and CdS-dex/dox QD. The yellow arrow represents the increase in size and the white arrow indicates the absence of QD.

and functional damage [51]. Even if indirectly the outside of the interacts with QDs through membrane receptors that cause activation and inhibition of different signaling pathways, causing toxic reactions or cell death [52].

Therefore, the cytotoxicity of QDs is more complex than we can imagine, it can cause not only the interaction with heavy metals contained in QDs but the disintegration of the core and the release of Cd ions, which increases their toxic potential. Under this condition, researchers have expressed concern about the use of NM and the parameters to be evaluated for future medical applications. This question arises from the association of adverse effects derived from the ability of QDs to enter cells and lodge in various subcellular compartments. This implies that they could evade the defense mechanisms of the human body, cross biological barriers and even interact with components of blood circulation [53]. Moreover, the blood circulation is the primary passage of NMs to the distribution of target organs. Thus, vascular endothelial cells serve as the first barrier and are tasked with maintaining vascular integrity [54]. In a study with ZnO nanoparticles, it has been shown that they are capable of causing cytotoxicity in HUVEC cells due to the increase of intracellular reactive oxygen species (ROS) in a dose-dependent manner [55]. Our studies have shown that at concentrations of 0.01 μg/mL, CdS-dex QDs already cause cytotoxic effects in HUVEC cells. The QDs are distributed around the cytoplasm, producing an increase in cell size and completely changing the characteristic morphology of the endothelial cell (**Figure 4**). Although it does not penetrate into the nucleus and nucleoli, cellular uptake occurs in a dose-dependent manner. In addition, endosome formation is observed, suggesting that cell deformation and toxicity are caused by cellular stress following the passage of the QD into the cell. The cytotoxicity produced by QDs is the

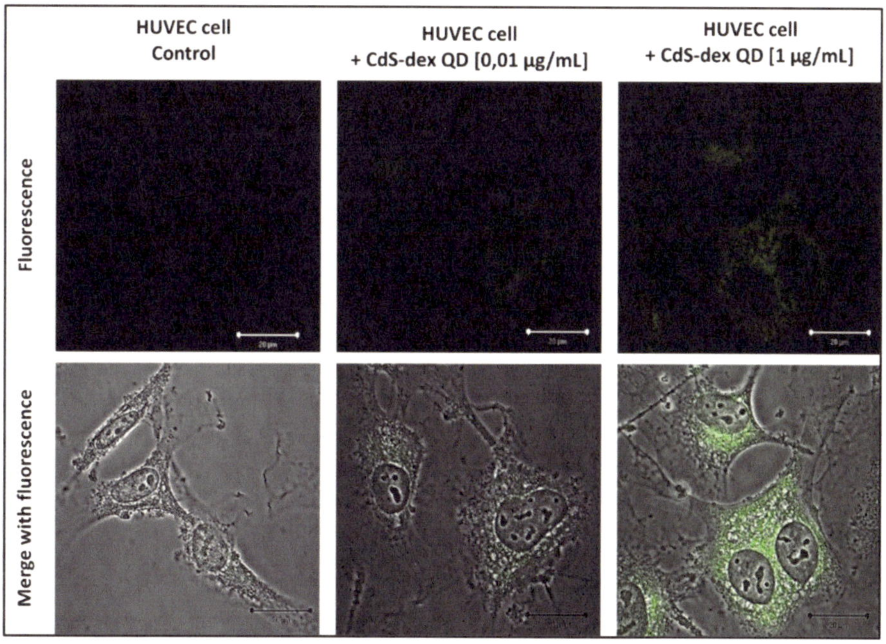

Figure 4.
Fluorescent microscopic visualization of HUVEC cells treated with CdS-dex QD at 0,01–1 μg/mL concentration and 24 h time exposure. Cells were seeded on slides by smearing and allowed to dry, then analyzed using a confocal epifluorescence microscope. Green fluorescence shows the presence of CdS-dex QD.

main parameter limiting their use in bioimaging research. The idea of applying QDs that produce morphological changes and ultimately cell death is a determining factor. Currently, joint efforts are being made for the development of innovative QDs capable of meeting the needs in healthcare areas. This progress in QD design and synthesis has resulted in improved safety *in vitro* studies. However, a myriad of factors that lead to cytotoxicity of QDs in normal, cancer, and endothelial cells remain in question. It has also been demonstrated that when QDs come into contact with organisms, they produce toxicity that is size-dependent, concentration threshold-dependent, and varies according to cytosensitivity [56]. However, factors such as concentration range are responsible for the intracellular distribution, which necessitates storage and bioaccumulation and thus increases cytotoxicity [57]. There is still a long way to go to achieve an accurate understanding and standardized parameters on safety for the use of quantum dots in the field of biomedicine.

In a whole decade, we have been dedicated to the design, synthesis, and nanotoxicological evaluation of quantum dots so we are very clear that, quantum dots can be improved in their design and composition. In addition, the nanoparticle size must be strictly controlled as it is one of the main factors influencing the toxicological effects of quantum dots [53]. The idea of having a complete profile of a type of nanomaterial is not unrealistic. However, it is necessary to demonstrate with studies on its preclinical evaluation. These evaluations include physicochemical characterization, *in vitro* evaluations with different types of human tumor and healthy cells, biodistribution, bioaccumulation, and pharmacokinetic studies. In addition, to perform exhaustive evaluations on its hemocompatibility as a starting point to rule out the toxic effect of a nanomaterial.

3. Conclusion

The development of newer drug delivery systems based on the use of quantum dots is one of the advantages for various disease treatments, such as cancer and gene therapy, as noted above. This modality allows for site-specific drug therapies and a higher safety profile. However, the pharmaceutical industry is far from knowing everything about the toxicological profile of all nanomaterials. However, nanotechnological challenges are evolving and it is necessary to focus our attention on the standardization of parameters for the evaluation of the cytotoxicity of nanomaterials such as quantum dots in order to broaden their safety range and thus ensure lower toxic effects. In the meantime, let us not forget that the key to the toxicity caused by quantum dots is given by the interaction of the elements that compose them and the biomolecules of the biological system. In the very near future, we can include scientific bases that tell us about physicochemical perspectives of quantum dots, better experimental conditions already standardized and reliable comparative analyses (*in vitro* and *in vivo*).

Appendices and nomenclature

NP	nanoparticles
QD	quantum dots
NM	nanomaterials
Cd	cadmium

Te	tellurium
Hg	mercury
CdS	cadmium sulfide
CdTe	cadmium telluride
CdSe	cadmium selenide
CdS-dex	sulfuro de cadmio core/capped dextrina
CdS-dex/dox	sulfuro de cadmio core/capped dextrina, with doxorubicin
HUVEC	umbilical cordon human cells
HepG2	hepatic cells
Hek293	kidney human cells
HeLa	adenocarcinoma of cervix
µg	micrograms
mL	milliliters

Author details

Gerardo González De la Cruz[1], Lourdes Rodríguez-Fragoso[2], Patricia Rodríguez-Fragoso[1] and Anahi Rodríguez-López[1]*

1 Departamento de Física, CINVESTAV – I.P.N, México, México

2 Facultad de Farmacia, Universidad Autónoma del Estado de Morelos, Cuernavaca, México

*Address all correspondence to: yanethanahi@gmail.com

References

[1] Chiu HI, Samad NA, Fang L, Lim V. Cytotoxicity of targeted PLGA nanoparticles: A systematic review. RSC Advances. 2021;**11**(16):9433-9449. DOI: 10.1039/d1ra00074h

[2] Huang X, Tang M. Research advance on cell imaging and cytotoxicity of different types of quantum dots. Journal of Applied Toxicology: JAT. 2021;**41**(3):342-361. DOI: 10.1002/jat.4083

[3] Wang L, Xu D, Gao J, Chen X, Duo Y, Zhang H. Semiconducting quantum dots: Modification and applications in biomedical science. Science China Materials. 2020;**63**:1631-1650. DOI: 10.1007/s40843-020-1330-7

[4] Zhu C, Chen Z, Gao S, Leng GB, Bin SI, Wen LK, et al. Recent advances in non-toxic quantum dots and their biomedical applications. Progress in Natural Science: Materials International. 2020;**29**:628-640. DOI: 10.1016/j.pnsc.2019.11.007

[5] Huang Z, Gao Y, Huang Z, Chen D, Sun J, Zhou L. Sulfur quantum dots: A novel fluorescent probe for sensitive and selective detection of Fe3+ and Phytic acid. Microchemical Journal. 2021;**170**:106656. DOI: 10.1016/j.microc.2021.106656

[6] Sheng Y, Huang Z, Zhong Q, Deng H, Lai M, Yang Y, et al. Sizefocusing results in highly Photoluminescent sulfur quantum dots with a stable emission wavelength. Nanoscale. 2021;**13**:2519-2526. DOI: 10.1039/D0NR07251F

[7] Zhang C, Zhang P, Ji X, Wang H, Kuang H, Cao W, et al. Ultrasonication-promoted synthesis of luminescent sulfur Nano-dots for cellular imaging applications. Chemical Communications. 2019;**55**:13004-13007. DOI: 10.1039/C9CC06586E

[8] Arshad F, Sk MP, Maurya SK, Siddique HR. Mechanochemical synthesis of sulfur quantum dots for cellular imaging. ACS Applied Nano Materials. 2021;**4**:3339-3344. DOI: 10.1021/acsanm.1c00509

[9] Reshma VG, Mohanan PV. Quantum dots: Applications and safety consequences. Journal of Luminescence. 2018;**205**:287-298. DOI: 10.1016/j.jlumin.2018.09.015

[10] Mansur HS. Quantum dots and nanocomposites. Nanomedicine and Nanobiotechnology. 2010;**2**(2):113-129. DOI: 10.1002/wnan.78

[11] Hu L, Zhong H, He Z. The cytotoxicities in prokaryote and eukaryote varied for CdSe and CdSe/ZnS quantum dots and differed from cadmium ions. Ecotoxicology and Environmental Safety. 2019;**181**:336-344. DOI: 10.1016/j.ecoenv.2019.06.027

[12] Xu G, Zeng S, Zhang B, Swihart MT, Yong KT, Prasad PN. New generation cadmium-free quantum dots for biophotonics and Nanomedicine. Chemical Reviews. 2016;**116**(19):12234-12327. DOI: 10.1021/acs.chemrev.6b00290

[13] Moulick A, Milosavljevic V, Vlachova J, Podgajny R, Hynek D, Kopel P, et al. Using CdTe/ZnSe core/shell quantum dots to detect DNA and damage to DNA. International Journal of Nanomedicine. 2017;**12**:1277-1291. DOI: 10.2147/ijn.s121840

[14] Zhang F, Liu B, Zhang Y, Wang J, Lu Y, Deng J, et al. Application of CdTe/

CdS/ZnS quantum dot in immunoassay for aflatoxin B1 and molecular modeling of antibody recognition. Analytica Chimica Acta. 2019;**1047**:139-149. DOI: 10.1016/j.aca.2018.09.058

[15] Ahmad R, Kaus NHM, Hamid S. Synthesis and characterization of PLGA-PEG Thymoquinone nanoparticles and its cytotoxicity effects in tamoxifen-resistant breast cancer cells. Advances in Experimental Medicine and Biology. 2020;**1292**:65-82. DOI: 10.1007/5584_2018_302

[16] Manoochehri S, Darvishi B, Kamalinia G, Amini M, Fallah M, Ostad SN, et al. Surface modification of PLGA nanoparticles via human serum albumin conjugation for controlled delivery of docetaxel. Daru. 2013;**21**(1):58. DOI: 10.1186/2008-2231-21-58

[17] Rodríguez-Fragoso P, Reyes-Esparza J, León-Buitimea A, Rodríguez-Fragoso L. Synthesis, characterization and toxicological evaluation of maltodextrin capped cadmium sulfide nanoparticles in human cell lines and chicken embryos. Journal of Nanbiotechnology. 2012;**10**(47):1-11. DOI: 10.1186/1477-3155-10-47

[18] Gutiérrez-Sancha I, Reyes-Esparza J, Rodríguez-Fragoso P, García-Vázquez F, Rodríguez-Fragoso L. Bright green emitting maltodextrin coated cadmium sulfide quantum dots as contrast agents for bioimaging: A biocompatibility study. International Journal of Nanomedicine and Nanosurgery. 2015;**1**(2):1-10. DOI: 10.16966/2470-3206.107

[19] Gutiérrez-Sancha I, Reyes-Esparza J, Rodríguez-Fragoso P, Rodríguez-Fragoso L. Pharmacokinetic of maltodextrin coated cadmium sulfide quantum dots in rats. Journal of Nanomedicine & Biotherapeutic Discovery. 2016;**6**:139. DOI: 10.4172/2155-983X

[20] Mahmoudi M, Hofmann H, Rothen-Rutishauser B, Petri-Fink A. Assessing the in vitro and in vivo toxicity of superparamagnetic iron oxide nanoparticles. Chemical Reviews. 2012;**112**(4):2323-2338. DOI: 10.1021/cr2002596

[21] Manshian BB, Abdelmonem AM, Kantner K, Pelaz B, Klapper M, Nardi Tironi C, et al. Evaluation of quantum dot cytotoxicity: Interpretation of nanoparticle concentrations versus intracellular nanoparticle numbers. Nanotoxicology. 2016;**10**:1318-1328. DOI: 10.1080/17435390.2016.1210691

[22] Paesano L, Perotti A, Buschini A, Carubbi C, Marmiroli M, Maestri E, et al. Markers for toxicity to HepG2 exposed to cadmium sulphide quantum dots; damage to mitochondria. Toxicology. 2016;**374**:18-28. DOI: 10.1016/j.tox.2016.11.012

[23] Kuznetsova VA, Visheratina AK, Ryan A, Martynenko IV, Loudon A, Maguire CM, et al. Enantioselective cytotoxicity of ZnS: Mn quantum dots in A549 cells. Chirality. 2017;**29**(8):403-408. DOI: 10.1002/chir.22713

[24] Shinde V, Sureshkumar P, Sotiriadou I, Hescheler J, Sachinidis A. Human embryonic and induced pluripotent stem cell based toxicity testing models: Future applications in new drug discovery. Current Medicinal Chemistry. 2016;**23**(30):3495-3509. DOI: 10.2174/0929867323666160627113436

[25] Zhao J, Qi X, Dai Q, He X, Dweep H, Guo M, et al. Toxicity study of ochratoxin a using HEK293 and HepG2 cell lines based on microRNA profiling. Human & Experimental Toxicology. 2017;**36**(1):8-22. DOI: 10.1177/0960327116632048

[26] Reyes-Esparza J, Martinez-Mena A, Gutierrez-Sancha I, Rodriguez-Fragoso P,

de la Gonzalez Cruz G, Mondragón R, et al. Synthesis, characterization and biocompatibility of cadmium sulfide nanoparticles capped with dextrin for in vivo and in vitro imaging application. Journal of Nanobiotechnology. 2015;**13**:83. DOI: 10.1186/s12951-015-0145-x

[27] Gonzalez de la Cruz G, Rodriguez-Fragoso P, Mastache-Juarez A, Rodriguez-Fragoso L. Doxorubicin-bioconjugated cadmium sulfide dextrin quantum dots for imaging cells. Indian Journal of Pharmaceutical Sciences. 2020;**82**:230-241

[28] Doherty GJ, McMahon HT. Mechanisms of endocytosis. Annual Review of Biochemistry. 2009;**78**(1):857-902. DOI: 10.1146/annurev.biochem.78.081307.110540

[29] Kumari S, MG S, Mayor S. Endocytosis unplugged: Multiple ways to enter the cell. Cell Research. 2010;**20**(3):256-275. DOI: 10.1038/cr.2010.19

[30] Pujalte I et al. Cytotoxic effects and cellular oxidative mechanisms of metallic nanoparticles on renal tubular cells: Impact of particle solubility. Toxicology Research. 2015;**4**:409-422. DOI: 10.1039/c4tx00184b

[31] Rodríguez-López A, Agarwal V, Reyes Esparza J, Rodríguez-Fragoso L. Cellular localization and toxicity assessment of Cdte-COOH quantum dots in Hela cells. International Journal of Nanomedicine and Nanosurgery. 2017;**3**(2):1-6. DOI: 10.16966/2470-3206.122

[32] Hu L, Zhong H, He Z. Toxicity evaluation of cadmium-containing quantum dots: A review of optimizing physicochemical properties to diminish toxicity. Colloids and Surfaces. B, Biointerfaces. 2021;**200**:111609. DOI: 10.1016/j.colsurfb.2021.111609

[33] Rizeq BR, Younes NN, Rasool K, Nasrallah GK. Synthesis, bioapplications, and toxicity evaluation of chitosan-based nanoparticles. International Journal of Molecular Sciences. 2019;**20**(22):5776. DOI: 10.3390/ijms20225776

[34] Wu D, Ma Y, Cao Y, Zhang T. Mitochondrial toxicity of nanomaterials. The Science of the Total Environment. 2020;**702**:134994. DOI: 10.1016/j.scitotenv.2019.134994

[35] Wong SY, Han L, Timachova K, Veselinovic J, Hyder MN, Ortiz C, et al. Drastically lowered protein adsorption on microbicidal hydrophobic/hydrophilic polyelectrolyte multilayers. Biomacromolecules. 2012;**13**(3):719-726

[36] Gao X, Cui Y, Levenson RM, Chung LW, Nie S. In vivo cancer targeting and imaging with semiconductor quantum dots. Nature Biotechnology. 2004;**22**(8):969-976. DOI: 10.1038/nbt994

[37] Mansur A, Mansur H, Carvalho S, Lobato Z, Guedes MI, Leite MDF. Surface biofunctionalized CdS and ZnS quantum dot nanoconjugates for nanomedicine and oncology: To be or not to be nanotoxic? International Journal of Nanomedicine. 2016;**11**:4669-4690. DOI: 10.2147/IJN.S115208

[38] Bell IR, Ives JA, Jonas WB. Nonlinear effects of nanoparticles: Biological variability from hormetic doses, small particle sizes, and dynamic adaptive interactions. Dose Response. 2013;**12**(2):202-232. DOI: 10.2203/dose-response.13-025.bell

[39] Chang Y-N, Zhang M, Xia L, Zhang J, Xing G. The toxic effects and mechanisms of CuO and ZnO nanoparticles. Materials. 2012;**5**(12):2850-2871. DOI: 10.3390/ma5122850

[40] Kim KB, Kim YW, Lim SK, Roh TH, Bang DY, Choi SM, et al. Risk assessment of zinc oxide, a cosmetic ingredient used as a UV filter of sunscreens. Journal of Toxicology and Environmental Health. Part B, Critical Reviews. 2017;**20**(3):155-182. DOI: 10.1080/10937404.2017.1290516

[41] Mohammed YH, Holmes A, Haridass IN, Sanchez WY, Studier H, Grice JE, et al. Support for the safe use of zinc oxide nanoparticle sunscreens: Lack of skin penetration or cellular toxicity after repeated application in volunteers. The Journal of Investigative Dermatology. 2019;**139**(2):308-315. DOI: 10.1016/j.jid.2018.08.024

[42] Fornaguera C, Calderó G, Mitjans M, Vinardell MP, Solans C, amp; Vauthier, C. Interactions of PLGA nanoparticles with blood components: Protein adsorption, coagulation, activation of the complement system and hemolysis studies. Nanoscale. 2015;**7**(14):6045-6058. DOI: 10.1039/c5nr00733j

[43] O'Brien ME, Wigler N, Inbar M, Rosso R, Grischke E, Santoro A, et al. Reduced cardiotoxicity and comparable efficacy in a phase III trial of pegylated liposomal doxorubicin HCl (CAELYX/Doxil) versus conventional doxorubicin for first-line treatment of metastatic breast cancer. Annals of Oncology. 2004;**15**(3):440-449

[44] Lim WT, Tan EH, Toh CK, Hee SW, Leong SS, Ang PC, et al. Phase I pharmacokinetic study of a weekly liposomal paclitaxel formulation (Genexol-PM) in patients with solid tumors. Annals of Oncology. 2010;**21**(2):382-388

[45] Winter PM, Morawski AM, Caruthers SD, Fuhrhop RW, Zhang H, Williams TA, et al. Molecular imaging of angiogenesis in early-stage atherosclerosis with alpha(v) beta3-integrintargeted nanoparticles. Circulation. 2003;**108**(18):2270-2274

[46] Davis ME, Zuckerman JE, Choi CH, Seligson D, Tolcher A, Alabi CA, et al. Evidence of RNAi in humans from systemically administered siRNA via targeted nanoparticles. Nature. 2010;**464**(7291):1067-1070

[47] Yan M, Zhang Y, Xu K, Fu T, Qin H, Zheng X. An in vitro study of vascular endothelial toxicity of CdTe quantum dots. Toxicology. 2011;**282**(3):94-103. DOI: 10.1016/j.tox.2011.01.015

[48] Yan M, Zhang Y, Qin H, Liu K, Guo M, Ge Y, et al. Cytotoxicity of CdTe quantum dots in human umbilical vein endothelial cells: The involvement of cellular uptake and induction of pro-apoptotic endoplasmic reticulum stress. International Journal of Nanomedicine. 2016;**11**:529-542. DOI: 10.2147/IJN.S93591

[49] Tian J, Hu J, Liu G, Yin H, Chen M, Miao P, et al. Altered gene expression of ABC transporters, nuclear receptors and oxidative stress signaling in zebrafish embryos exposed to CdTe quantum dots. Environmental Pollution. 2019, 1987;**244**:588-599. DOI: 10.1016/j.envpol.2018.10.092

[50] Wang L, Zhang J, Zheng Y, Yang J, Zhang Q, Zhu X. Bioeffects of CdTe quantum dots on human umbilical vein endothelial cells. Journal of Nanoscience and Nanotechnology. 2010;**10**(12):8591-8596. DOI: 10.1166/jnn.2010.2681

[51] Zhao LN, Zong WS, Zhang H, Liu RT. Kidney toxicity and response of selenium containing protein-glutathione peroxidase (Gpx3) to CdTe QDs on different levels. Toxicological Sciences. 2019;**168**(1):201-208. DOI: 10.1093/toxsci/kfy297

[52] Chang E, Thekkek N, Yu WW, Colvin VL, Drezek R. Evaluation of quantum dot cytotoxicity based on intracellular uptake. Small. 2006;**2**(12):1412-1417. DOI: 10.1002/smll.200600218

[53] Cruz GGDL, Rodríguez-Fragoso P, Reyes-Esparza J, Rodríguez-López A, Gómez-Cansino R, Rodriguez-Fragoso L. Interaction of Nanoparticles with Blood Components and Associated Pathophysiological Effects. London, UK: InTech; 2018. DOI: 10.5772/intechopen.69386

[54] Cao Y, Gong Y, Liu L, Zhou Y, Fang X, Zhang C, et al. The use of human umbilical vein endothelial cells (HUVECs) as an in vitro model to assess the toxicity of nanoparticles to endothelium: A review. Journal of Applied Toxicology: JAT. 2017;**37**(12):1359-1369. DOI: 10.1002/jat.3470

[55] Yuxiu G, Cheng S, Chen G, Shen Y, Li X, Jiang Q, et al. The effects of endoplasmic reticulum stress inducer thapsigargin on the toxicity of ZnO or TiO2 nanoparticles to human endothelial cells. Toxicology Mechanisms and Methods. 2017;**27**(3):191-200. DOI: 10.1080/15376516.2016.1273429

[56] Wang Y, Tang M. Review of in vitro toxicological research of quantum dot and potentially involved mechanisms. Science of the Total Environment. 2018;**625**:940-962. DOI: 10.1016/j.scitotenv.2017.12.334

[57] Liang X, Tang M. Research advances on cytotoxicity of cadmium containing quantum dots. Journal of Nanoscience and Nanotechnology. 2019;**19**(9):5375-5387. DOI: 10.1166/jnn.2019.16783

Chapter 2

Nanotoxicological Assessments of Upconversion Nanoparticles

Dalia Chávez-García and Karla Juarez-Moreno

Abstract

Upconversion nanoparticles (UCNPs) are highly efficient luminescent nanomaterials with emission in the visible spectra while being excited by near-infrared region light (NIR). With their unique properties such as high luminescence intensity, sharp emission peaks with narrow bandwidth, large anti-Stokes' shift, and sizes smaller than 100 nm, UCNPs have emerged as promising candidates for diverse biomedical applications such as cancer detection and therapy, fluorescence imaging, magnetic resonance imaging (MRI), and drug delivery. The UCNPs are composed of a crystalline matrix doped with lanthanide ions that can absorb NIR light (~980 nm) and upconvert it to visible light. However, to achieve successful biomedical applications, proper functionalization, target-specific cell interaction, and biocompatibility are critical factors that must be considered. Additionally, a comprehensive nanotoxicological assessment is necessary to ensure that UCNPs are not cytotoxic or genotoxic. This assessment is particularly important for long-term studies of nanoparticles' tracking *in vivo*. Therefore, this chapter aims to provide an in-depth evaluation of the nanotoxicological issues related to nanoparticles (NPs) and UCNPs in biomedical applications, and ensure their safety and efficacy as bioimaging and chemotherapeutic delivery tools.

Keywords: cytotoxicity, nanoparticles, upconversion, nanotoxicological, luminescent

1. Introduction

The toxicity assessment of nanoparticles (NPs) is a relevant issue since many researchers are using, specially, luminescent nanoparticles for various applications, such as bioimaging or drug delivery for *in vivo* and *in vitro* applications [1–3]. In this chapter, we will analyze how the approach in this analysis has been carried out for upconversion luminescent nanoparticles (UCNPs), which are a special type of NPs since they can receive energy in the near-infrared region (NIR) and emit in the visible or NIR spectrum. These NPs are composed of a matrix cell that can be made of oxides, oxysulfides, oxyhalides, phosphates, molybdates, tungstates, gallates, vanadates, and fluorides. The UCNPs are doped with lanthanide elements such as: Yb^{3+}, Er^{3+}, Tm^{3+}, and Ho^{3+}, among others. It is common in the upconversion process to have lanthanide elements co-doped to bring about a photon transfer between energy levels. For example, for the doping of Yb/Er, the Yb^{3+} absorbs NIR radiation at 970–980 nm of wavelength in its base state ($^2F_{7/2}$–$^2F_{5/2}$), then this energy is transferred to Er^{3+} and the electron is populated to level $^4I_{11/2}$, then, a second photon is absorbed and by Yb^{3+} and it is transferred to Er^{3+}, so the electron is raised to level $^4F_{7/2}$. From this state, it decays rapidly to $^4S_{3/2}$, and the green emission happens ($^4S_{3/2}$–$^4I_{15/2}$), and this process is called the APTE (Addition de photons par transfert d'énergie, i.e., photon energies by adding transfers), as can be seen in **Figure 1**. There are more upconversion processes with different doping combinations and concentrations of the ions, where the percentage of doping directly affects the color of emission [4].

The UCNPs have emerged as a promising nanomaterial for identifying specific cells and for drug delivery. Unlike other dyes, UCNPs exhibit stable emission if the source of excitation is maintained, making them more reliable. There are other types of upconversion processes such as: two-step absorption, cooperative sensitization, cooperative luminescence, the second harmonic generation, and two-photon absorption [4].

One crucial aspect of using UCNPs in biomedical applications lies in ensuring their biocompatibility on cells and or organisms. To achieve this, UCNPs must be functionalized with different ligands that specifically target the desired cells and organs. Several chemical groups, including polyethylene glycol (PEG) [5], polyethyleneimine (PEI) [6], polyvinylpyrrolidone (PVP) [7], polyacrylic acid (PAA) [8], and silica [6], have been used for this purpose. However, it is important to highlight that the toxicology of UCNPs depends on their physicochemical and physiological properties. Physicochemical properties include size, shape, surface area, and chemical composition, while physiological properties refer to the disease conditions, genetics, and other factors [9]. The recommended size for optimal penetration of NPs is below 100 nm. However, this size may also pose a risk of toxicity due to their potential to penetrate cellular structures and organs via the circulatory system. Moreover, UCNPs may generate reactive oxygen species (ROS) that can induce DNA damage, which not only affects the cell growth by means of protein oxidation, but also impacts mitochondrial respiration [10].

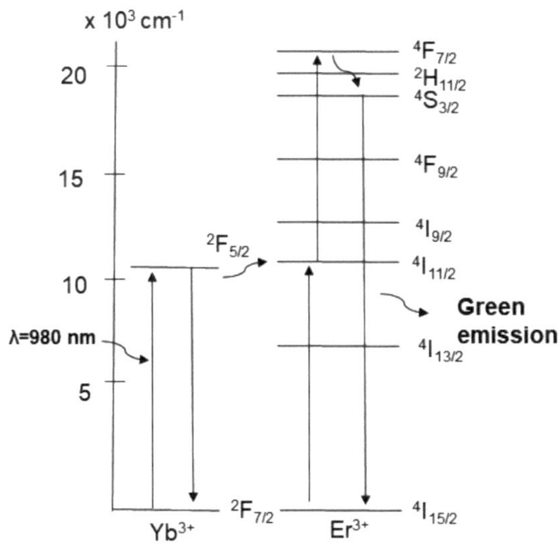

Figure 1.
Upconversion process between Yb^{3+} and Er^{3+} ions.

Several toxicological studies have been conducted on both *in vivo* and *in vitro* human cell lines and organs to assess the potential harmful effects of UCNPs. These studies have evaluated the effects of gene expression, growth, and reproduction of the organisms. It is crucial to continue monitoring and evaluating the toxicity of UCNPs as their use becomes more prevalent in biomedical applications.

2. Biocompatibility of nanoparticles

This section will provide an overview of different methods that researchers use to achieve biocompatibility of UCNPs. **Figure 2** depicts the common way to coat and functionalize UCNPs for several researches, as generally, the UCNPs or NPs need to be coated to ensure biocompatibility and they need functional groups to attach to several types of ligands that can bind to the surface of the targeted cells, as depicted.

2.1 Polyethylene glycol

The most used method to achieve biocompatibility is through PEGylation, which is both effective and straightforward. Although the specific approach may vary among different authors, PEGylation generally refers to the covalent conjugation of PEG to other molecules. This process enhances the physicochemical properties of the molecules, leading to reduce the immunogenicity and improve solubility, electrostatic binding, and hydrophobicity of a given biomolecule [11]. Overall, PEGylation represents a valuable tool for improving the biocompatibility of drugs and biomolecules, allowing for safer and more effective biomedical applications.

The first polymer conjugation was developed by Abuchowski et al. in 1977 [12], and various authors have developed different PEGylation methods for diverse applications, ranging from biocompatibility to trimodal fluorescence. For instance, Zeng et al. [13] developed PEG-modified $BaGdF_5$:Yb/Er UCNPs for multimodal fluorescence/CT (computed X-ray tomography)/magnetic bioimaging applications, which exhibited low cytotoxicity and long circulation time. Similarly, Maldiney et al. [14]

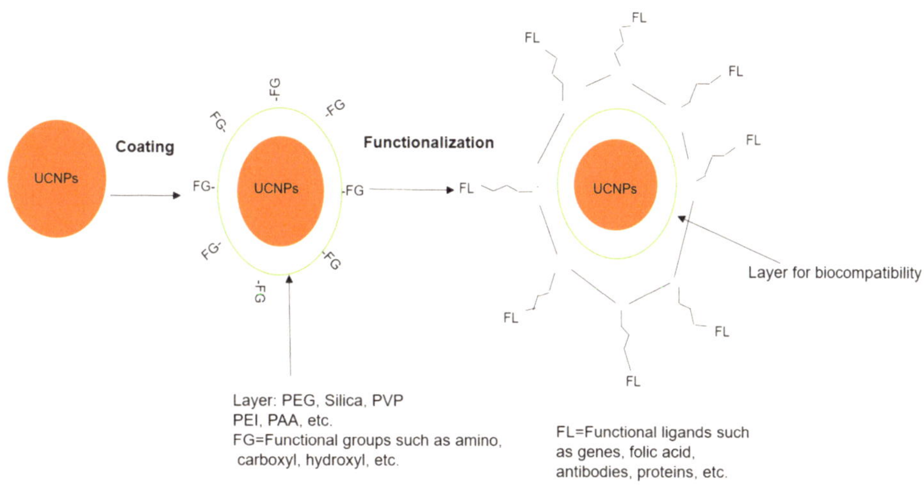

Figure 2.
Biocompatibility and functionalization of several types of UCNPs.

utilized luminescent NPs emitting in the near-infrared spectra, with two types of mice: healthy and tumor carrier mice. They reported that PEG coating enabled the formation of stealthy particles that were more uniformly distributed throughout the animal. It is important to note that PEGylation tends to increase the diameter of the NPs by about 10 nm, similar to other conjugation methods. However, an essential aspect of PEGylation is the characterization of NPs, and the dynamic light scattering (DLS) is a crucial technique that can provide three critical parameters: size; zeta potential that measures the surface charge of the NPs and determines their colloidal stability (values between −10 and +10 mV are neutral, while values greater than +30 mV or less than −30 mV are considered strongly cationic and strongly anionic, respectively), and size distribution [15]. The selection of ligands to bind the PEGylated-NPs may vary depending on the application. The purpose of having PEGylated-NPs with ligands is to target specific receptors on the surface of cancer cells and to allow for retention in the area due to the enhanced permeability and retention effect (EPR). A variety of ligands can be used, including molecules, peptides, proteins, antibodies, aptamers, among others [12, 16–19].

However, PEG may undergo degradation due to light, stress, or heat. Some authors have addressed this issue by combining PEG with copolymers such as PVP and poly(lactic-co-glycolic acid) or PLGA [20]. With these challenges, research with PEG continues to be relevant, as it has proven to be an important tool for achieving the biocompatibility of NPs.

2.2 Polyethyleneimine

Polyethyleneimine is a very versatile aliphatic polymer that contains primary, secondary, and tertiary amino groups, with a ratio of 1:2:1 [21]. It has found numerous applications in non-viral gene delivery and therapy for *in vitro* and *in vivo* models. In addition, PEI has been used for non-pharmaceutical applications, such as water purification and shampoo manufacturing. For instance, Ge and collaborators [22] developed near-infrared emitting nanoparticles coated with PEI and gold nanorods coated with dithiothreitol to detect arsenic (III), while Pan et al. [23] synthesized PEI-coated upconversion nanoparticles for use as an optical probe to determine the water content in organic solvents.

Polyethyleneimine-modified nanoparticles have also been explored for various biomedical applications. Mi et al. [24] developed luminescent NPs coated with PEI that can bind to antibodies through their amino groups, resulting in tunable colors. Xu et al. [25] functionalized NPs with folic acid and polycaprolactone/PEI for *in vivo* drug delivery in SKOV-3 cancer cells. Their results showed that their method was more effective in killing cancer cells than free doxorubicin. PEI-NPs have also been used for pulmonary gene delivery. Bivas-Benita et al. [26] developed a PLGA-PEI-NP that can deliver genes to the lung epithelium using Calu-3 cells. Huh et al. [27] used PEI-NPs composed with glycol chitosan and encapsulated with siRNA, which significantly inhibited red fluorescent protein (RFP) gene expression in B16-F10-bearing mice cells.

PEI nanoparticles represent an important tool especially for drug delivery of anticancer drugs and also gene therapy applications, among others.

2.3 Polyvinylpyrrolidone

PVP is commonly used as a coating for silver NPs and as a drug carrier [25, 28, 29]. However, several authors have also used PVP as a coating for UCNPs [25, 30–34].

PVP is a versatile coating because it can work as a NP dispersant or as a surface stabilizer, and it also has reducing properties. Its functional groups, which include C=O, C–N, and CH_2, enable it to control the growth of certain aspects by binding onto others, providing biocompatibility to the NPs [29, 35].

Johnson et al. [34] synthesized β-$NaYF_4$:Yb^{3+}/Er^{3+} UCNPs and used PVP to replace the oleate surface ligands. This modification makes the UCNPs water-dispersible, which is crucial for *in vivo* applications. Additionally, PVP is biocompatible, has a prolonged blood circulation time, and shows low accumulation in vital organs. Zou et al. [33] prepared UCNP $NaYF_4$:Yb^{3+}/Er^{3+} embedded into PVP nanotubes using the electrospinning method, resulting in an intense emission of the UCNPs compared to bare UCNPs. Due to their biocompatibility, these modified NPs may have important applications in biomedicine.

2.4 Polyacrylic acid

PAA is a hydrophilic and pH-responsive polymer that can replace hydrophobic ligands on the surface of NPs, making it an excellent candidate for *in vivo* and *in vitro* applications [36]. Its biocompatibility and other desirable qualities make it an attractive coating option for various types of NPs [37–41].

Hilderbrand et al. [42] synthesized UCNPs coated with PAA and linked amino-PEG to the carboxyl groups of the PAA. The resulting modified UCNPs were non-cytotoxic and displayed good NIR emission. Wang et al. [41] also prepared UCNPs YF_3:Yb^{3+}/Er^{3+} with NIR emission and coated with PAA, resulting in strong luminescence. In a study by Xiong et al. [40], PAA-coated UCNPs were shown to have excellent biodistribution and cellular uptake in mice, with no observed toxicity, suggesting that these NPs could be used for long-term therapy and bioimaging studies *in vivo*. Additionally, Jia et al. [36] investigated the effects of doxorubicin hydrochloride (DOX) and PAA-coated UCNPs (DOX@PAA-UCNPs) on HeLa cells and found that the UCNPs were biocompatible and effective as a drug carrier.

In summary, PAA is a very versatile polymer that can be used to coat on various types of NPs for a wide range of biomedical applications.

2.5 Silica

Silica (SiO_2) is a commonly used coating material for various types of NPs due to its favorable properties, including biocompatibility, thermodynamic stability, low toxicity, colloidal stability, ease synthesis, and scalability. Two main methods are generally used for producing the coating: sol-gel in a reverse micelle nanoreactor and the Stöber method [43, 44]. However, achieving a complete and homogeneous coating is a significant challenge, and Ureña-Horno et al. [45] developed a method for coating UCNPs with silica. By determining the optimal concentration of nanoparticles, they were able to achieve high yields of homogeneous functionalization and prevent agglomeration.

Hlaváček et al. [46] employed agarose gel electrophoresis for the purification of silica-coated UCNPs and for the separation of the protein-UCNPs from surplus reagents. This work represents a significant advancement in nanoparticle separation and measurement of their size and surface charge. In another study, also, Gnanasammandhan et al. [47] used silica-coated UCNPs for photoactivation in two specific applications: photodynamic therapy (PpDt) and photoactivated control of gene expression. The UCNPs were coated with PEG and functionalized with FA to target specific tumors, and their protocols for photoactivation therapy are valuable for future studies.

Overall, the efficient coating and functionalization of nanoparticles with silica are vital for their successful use in various applications, and these studies provide important insights and protocols for achieving these goals.

3. Toxicology of nanoparticles

The study of the toxicological effects elicited by NPs on cells and organisms is crucial in biomedical-nanotechnology applications. Thus, it is important to ensure that NPs are not cytotoxic or genotoxic. **Table 1** summarizes various approaches used by different authors for the toxicological assessment of nanoparticles.

3.1 Cytotoxicity assays

Assessing the cytotoxicity of new agents or nanomaterials is a crucial step in evaluating their potential biomedical applications. *In vitro* cell culture tests are preferred over *in vivo* animals test for ethical, speed, and cost reasons. However, cell cultures tend to be susceptible to various environmental factors, such as pH, nutrients, and temperature, which may interfere with the interpretation of the results. Therefore, it is important to ensure that the observed cell viability is observed solely due to the toxicity of the nanomaterials being tested, rather than environmental factors. Performing a range of tests with different concentrations of NPs and consistent experimental conditions enhances the validity of results [56, 57].

The MTT assay, based on the reduction of 3-(4,5-dimethyl-2-thiazolyl)-2,5-diphenyl-2H-tetrazolium bromide by dehydrogenase enzymes, is one of the most common methods to assess cell viability, as it measures mitochondrial activity in living cells [58–60]. This assay detects living cells, and the results are easily read using a multiwell scanning spectrophotometer (ELISA plate reader). Several authors have successfully used this assay, including those listed in **Table 1** [48, 50, 55].

Another variation of the MTT assay is the Cell Titer 96 Aqueous One Solution Cell Proliferation Assay, which uses MTS [3-(4,5-dimethylthiazol-2-yl)-5-(3-carboxymethoxyphenyl)-2-(4-sulfophenyl)-2H-tetrazolium], and phenazine ethosulfate, instead of MTT. Bahadar et al. [61] used both methods to evaluate the cytotoxicity of different metallic and non-metallic NPs on cells.

Other methods for measuring cell viability include the trypan blue and neutral red assays, which detect dead cells based on dye penetration into cell membrane. Ramírez-García et al. [62] used the trypan blue assay to measure the cell viability of zinc-gallium luminescent NPs; also, Zairov et al. [63] used gadolinium-based luminescent NPs with PC12 cells for obtaining low cytotoxicity, and the viability of the living cells was measured with a hemocytometer.

Live/dead viability assay, which measures the number of damaged cells, uses calcein acetoxymethyl (calcein AM) and ethidium homodimer. This method was mostly used to test the cytotoxicity exerted by gold nanoshells, silver, silica NPs, or fullerenes on cells [64]. The water-soluble tetrazolium (WST-1) assay is another method that measures mitochondrial activity by transforming the light-red tetrazolium salt into dark-red formazan salt due to the mitochondrial activity in living cells. Braun et al. [65] evaluated silica NPs with C2C12 cells using MTT and WST assays, and described that the MTT assay overestimated the low and medium cytotoxicity of the NPs, while the WST assay underestimates the particle concentrations studied.

NPs	Cytotoxicity assay	ROS quantification	Genotoxicity/gene expression	Stability or distribution	Cells/organism	References
Y_2O_3:Eu and Yb/Er	MTT assay	Comet assay	Comet assay	—	HeLa, MCF-7 cells	[48, 49]
$NaYF_4$:Yb/Er	MTT assay	—	—	—	RAW 264.7 cells	[50]
$NaYF_4$: Yb/Er	—	Oxidative stress assay	—	Stability assay/ biodistribution studied	Fetal bovine serum/ mice	[51, 52]
NPs-PEI-SiRNA	RFP	—	RFP expressing B16F10 cells	—	Murine melanoma/RFP/ B16-F10 cells	[27]
Yb_2O_3:Gd	A. cepa chromosomal aberration assay method	—	A. Cepa genotoxic studies	-	*E. coli* and *S. aureus*	[53]
Gd_2O_2: Yb/Tm	*In vitro* biodegradation assay	—	—	Biodistribution and toxicity in organs studied	Mice *in vivo*/blood *in vitro*	[54]
Y_2O_3	MTT assay	—	—	—	Human breast cancer	[55]

Table 1.
Toxicological assessment of NPs described by different authors.

There are alternative approaches to assess the cytotoxicity of NPs: For instance, Das et al. [66] carried out a study on the toxic effects of three types of functionalized UCNPs: oleate ligands-NPs, PEG-NPs, and bilayer PEG-oleate-NPs. They employed the calcein and propidium iodide viability assay and concluded that the bilayer NPs exhibited significant toxicity due to functionalization. In another study, Malvindi et al. [67] evaluated the cytotoxicity of silica-coated iron oxide NPs using the WST-8 ([2-(2-methoxy-4-nitrophenyl)-3-(4-nitrophenyl)-5-(2,4-disulfophenyl)-2H-tetrazolium, monosodium salt] assay and lactate dehydrogenase release (LDH assay) to analyze cell viability and cell membrane integrity. The NPs demonstrated good internalization in HeLa cells with no observed toxicity. Meindl et al. [68], on the other hand, assessed the cytotoxicity of NPs by measuring intracellular calcium levels, providing an example of an alternative approach for toxicity evaluation. It is important to select an appropriate method for assessing cytotoxicity, with suitable experimental parameters and consistent concentrations of NPs and exposition times across different studies. Otherwise, non-toxic NPs may yield misleading results due to factors such as cellular senescence.

3.2 Reactive oxygen species/reactive nitrogen species

The production of reactive nitrogen species (RNS), such as nitric oxide (NO), is closely associated with inflammatory responses and can react with oxygen to produce ROS. When NPs interact with cells, they may induce cell death by triggering the production of NO. The production of RNS is regulated by the enzyme nitric oxide synthase (NOS), while ROS production is regulated by NAD(P)H oxidase isoforms. Excessive ROS production can cause oxidative stress, leading to damage in the cell membrane, proteins, lipids, or DNA. However, low or moderate concentrations of ROS/RNS are beneficial, as they can help to defend against infections [69–71].

Several studies have demonstrated that metal and silica nanoparticles can induce oxidative stress and inflammation. The reactivity at the target sites and the surface area are two crucial factors affecting these outcomes. In a study conducted by Tran et al. [72], the effects of nanoparticles' surface area on lung health were investigated. They demonstrated that NPs with a higher surface area tend to be retained and accumulate in the lungs, reaching a saturation point where they become less susceptible to phagocytosis and exhibit reduced mobility. This overload effect stimulates macrophages, leading to the production of inflammatory responses, including tumor necrosis factor.

In a recent study, Wang [73] investigated the use of ROS probes to detect and visualize ROS production in living cells. The most commonly used ROS include H_2O_2, $^1O_2$, $O_2^{\bullet -}$, ClO−, ONOO−, and •OH. Luminescent NPs were found to be effective probes for detecting H_2O_2 and other ROS forms in living cell systems. The authors suggest that these nanoprobes may have promising therapeutic applications for sensing ROS.

3.3 Genotoxicity

When conducting deeper cytotoxicity studies, determining the genotoxic potential of NPs is often necessary. Various authors have employed different methods to ensure single- and double-stranded DNA breakage caused by NPs exposure. One of the most used methods is the flow cytometry that differentiates among various cell populations, between cell size, and complexity (granularity) through a laser beam [74]. Intercalating dyes such as propidium iodide can be used to measure DNA damage by counting apoptotic cells as the dye fluoresces in proportion to the increase in cell membrane permeability of damaged cells, and the presence of cell death-associated molecules in the cell membrane [56].

The comet assay, or alkaline single cell gel electrophoresis assay, is another widely used method, in which agarose gels are suspended and then lysed, electrophoresed, and stained with a fluorescent DNA-binding dye [75]. DNA damages can manifest in various forms, such as mutations, carcinogenesis, oxidative stress, or damage to the mitotic spindle and its components [76]. However, the assay has limitations in terms of processing multiple samples, as the electrophoresis can hold only 20 slides per run and generally each slide has one or two gels. Some researchers have attempted to increase throughput by increasing the size of the gels. For instance, Azqueta et al. [77] compared standard, medium- and high-throughput comet assays, by increasing throughput by an increase in the size of the gels to analyze more samples, however, scoring the results can be time-consuming.

There are still some issues with the analysis of the DNA damage that need to be addressed, such as the potential for NPs to induce additional DNA breaks during the assay, leading to false high damage results, or interfering with the scoring of the DNA-heads, resulting in less intensity.

3.4 Apoptosis/necrosis

Cell death, whether induced by apoptosis or necrosis, can be measured using flow cytometry. Upon injection of NPs into the bloodstream, the kidneys play a key role in clearing the NPs. However, depending on the timeg duration of therapy, NPs can potentially cause nephrotoxicity. The surface charge of NPs is an important factor in their physical stability. Positively charged NPs tend to interact more strongly with blood components and are easily cleared from the circulatory system. Conversely, NPs with a more negative surface charge exhibit lower interaction with plasma proteins [78].

In vivo studies have investigated the toxicity and DNA damage induced by various types of NPs, including metal, silver, and gold NPs. Some findings suggest that oxidative stress may be a mechanism of cytotoxicity and apoptosis induced by the NPs [78, 79]. Zhao et al. [80] conducted a study on nickel NPs in mouse epithelial (JB6) cells, revealing high cytotoxicity and apoptosis resulting from NPs interactions.

Moreover, Wang and Cho [81] discovered that the nuclear factor kappa B (NF-κB) plays a crucial role in regulating inflammatory responses that may induce DNA damage. In the process of evading apoptosis, reduced tissue capability to eliminate damaged cancer cells can occur. Therefore, NPs that cause DNA damage can lead to unwanted inflammatory responses.

One approach utilized in certain studies is the utilization of apoptosis as a targeted cancer treatment through customized NPs designed for this purpose. In some instances, these NPs have been observed to induce morphological changes and trigger autophagy through toxicity, such as cerium oxide NPs or iron oxide NPs [82, 83].

Necrosis, on the other hand, refers to the premature cell death caused by injury. Mohammadinejad et al. [84] conducted a study on the apoptotic, necrotic, and autophagic effects of several types of NPs. They concluded that during necroptosis, which is a regulated necrosis, the cell receptor apoptotic signaling pathway detects different stimuli, leading to a complex process. For instance, Schaeublin et al. [85] demonstrated that charged gold NPs induced apoptotic death in human keratinocyte cells (HaCaT), while non-charged NPs induced necrosis. Therefore, it is crucial to address the study of the effects of different types of NPs on cell death.

When evaluating cell death, it is important to assess the specific mode of death rather than solely relying on cell viability. This is because apoptosis, in some cases, may be followed by secondary necrosis, which can lead to inconclusive or erroneous results.

3.5 Hemocompatibility

The hemolysis test is employed to assess the acute hemolytic activity of nanomaterials intended for prolonged contact with soft tissue and cells [86]. Additionally, the hemocompatibility assay evaluates the effect on the blood components resulting from contact with certain materials, including NPs. *In vivo* assays are specifically designed to simulate clinical conditions, including flow dynamic and the geometry factors. The ISO 10993-4 is the standard that serves as the governing norm that establishes the parameters that must be fulfilled when conducting tests involving blood contact.

Liang et al. [87] developed $NaYF_4$:Yb, Er-FA UCNPs and tested their biocompatibility in L929 fibroblast cells using the hemolysis and coagulation tests. They found good compatibility and conducted additional tests using blood from rats stabilized with heparin sodium. Fresh plasma was obtained from the sample via centrifugation. T prothrombin time (PT) and activated partial thromboplastin time (APTT) were measured to evaluate the effects of the UCNPs on blood coagulation. The researchers detected almost no hemolysis effect due to the UCNPs interaction.

The response of organism in the bloodstream may vary depending on the exposure to NPs, but the liver and spleen are the most common sites of NPs accumulation. Some studies have suggested that quantum dots (QDs) tend to accumulate in the lymph nodes. The Bakalova [88] used polymersomes as carriers for lymph node mapping, employing QDs as a contrast agent. Other authors have investigated the effects of metal-NPs on blood, and their findings indicate primary accumulation in the liver [89, 90]. In a study conducted by Balasubramanian et al. [91], they examined the biodistribution of gold NPs in rats and observed a significant biodistribution after two months following a single intravenous injection, along with gene expression changes in organs such as the kidney, liver, and spleen.

Mehrizi studied the effects of polymeric, metallic, and nonmetallic NPs on red blood cells (RBCs), and multiple studies converge on the conclusion that PEGylated forms of NPs and negatively charged dendrimers exhibit the best hemocompatibility on RBCs [92]. Smaller NPs, higher concentrations, and longer exposure times tend to induce hemolysis. NPs can compromise RBCs integrity through hemolysis or hemagglutination, which refers to the agglutination of the RBCs. Colorimetric assays detecting the release of hemoglobin can be used to evaluate these issues, with the standard assay being conducted according to ASTM-F756 guidelines. According to Nemmar et al. [93], silica NPs exhibit a dose-dependent hemolytic behavior. Hemagglutination of RBCs can be measured using DLS. The Lima group employed this assay to evaluate chitosan-NPs in human erythrocytes [94]. Other authors have evaluated the hemagglutination in other types of NPs, such as gold, iron oxide, silver, or carbon, with results strongly dependent on the size, type, or the functionalization of a given nanoparticle [95–97].

Thrombogenicity is another assessment that can be conducted on certain types of NPs can trigger clotting in the blood. T platelet activation assays (flow cytometry), thrombin generation (computed tomography assay), platelet aggregation (light aggregometry), or clinical coagulation assays, such as thromboelastographic aPTT, can be employed to evaluate this property in accordance with ASTM standard ASTM F2382-18 [98].

Additionally, Saha et al. discussed NPs approved by the US Food and Drug Administration for cancer treatment, specifically protein-bound NPs, raising concerns about their hemocompatibility. These concerns include anemia, neutropenia, thrombocytopenia, bradycardia, hyper/hypotension, and the possibility to undergo severe cardiovascular events [98]. Nonetheless, further studies are necessary to understand the behavior of NPs in the bloodstream and their potential long-term consequences.

4. Conclusions

The impact of NPs on different cellular components, plasma, organelles, or membranes may elicit a variety of responses in both *in vitro* and *in vivo* studies. Therefore, nanotoxicological assessments must be conducted for each type of NP, cell, or organism, and at different doses, concentrations, and exposition times, to guarantee their safety. Some research groups have reported negligible or low toxicity at low doses of NPs, while others have found toxicity at high doses and long incubation times. Although these factors have varied depending on the type of NP used, as some nanomaterials tend to be more toxic, such as QDs or silver nanoparticles. Another crucial factor is that some NPs can induce the overproduction of ROS, also cellular uptake of NPs has been implicated in cellular toxicity, and significant changes in cellular responses including cell morphology and differentiation processes.

The function of the NPs relies on their matrix cell and the doping elements, which make them functional, biocompatibility is crucial to guarantee their safe use in cells and organisms, and most NPs require some coating to fulfill biocompatibility. Generally, the *in vitro* toxicity studies reported that NP exposure is toxic and causes cell death. Still, on some occasions, there may not be cell death, representing toxicity that is not correctly interpreted. Only a few groups have reported studies on cell signaling using genomic and proteomic array tests. The Coto-García et al. [99] investigated the bioanalysis for proteomics and genomics of several types of NPs and concluded that there is still a need for further discussion to approach the nanotoxicological effects of the different NPs used.

Nanoparticles can elicit various effects on cells, including cell proliferation, differentiation, cell cycle regulation, DNA damage, and cell death via apoptosis or necrosis. These effects are contingent upon factors such as nanoparticle type, size, surface charge, shape, and functionalization. Consequently, the assessment of NP toxicity encompasses a complex process that necessitates addressing the multitude of casual factors and interpreting the results. However, there are existing gaps in our understanding due to some inconsistencies among different studies, hindering conclusive determinations regarding the impact of NPs on living organisms. Furthermore, a standardized and systematic approach to testing has yet to be established.

Acknowledgements

The authors wish to acknowledge the support of CETYS University.

Conflict of interest

The authors declare no conflict of interest.

Author details

Dalia Chávez-García[1*] and Karla Juarez-Moreno[2]

1 Centro de Enseñanza Técnica y Superior (CETYS), Ensenada, Mexico

2 Centro de Física Aplicada y Tecnología Avanzada, Universidad Nacional Autónoma de México, México

*Address all correspondence to: dalia.chavez@cetys.mx

References

[1] Henri RB. In: Taylor & Francis Group L, editor. Biomedical application of nanoparticles. 1st ed. Taylor & Francis Group, LLC: Boca Raton, FL; 2018. p. 321

[2] Chatterjee K, Sarkar S, Jagajjanani Rao K, Paria S. Core/shell nanoparticles in biomedical applications. Advances in Colloid and Interface Science. 2014;**209**:8-39

[3] Su Y, Xue T, Liu Y, Qi J, Jin R, Lin Z. Luminescent metal nanoclusters for biomedical applications. Nano Research. 2019;**12**:1251-1265. DOI: 10.1007/s12274-019-2314-y

[4] Blasse G, Grabmaier BC, Blasse G, Grabmaier BC. Luminescent Materials. Berlin Heidelberg: Springer-Verlag; 1994. p. 230. DOI: 10.1021/ja965667t

[5] Manson J, Kumar D, Meenan BJ, Dixon D. Polyethylene glycol functionalized gold nanoparticles: the influence of capping density on stability in various media. Gold Bulletin. 2011;**44**(2):99-105

[6] Buchman YK, Lellouche E, Zigdon S, Bechor M, Michaeli S, Lellouche JP. Silica nanoparticles and polyethyleneimine (PEI)-mediated functionalization: a new method of pei covalent attachment for siRNA delivery applications. Bioconjugate Chemistry. 2013;**24**(12):2076-2087

[7] Gao L, Ge X, Chai Z, Xu G, Wang X, Wang C. Shape-controlled synthesis of octahedral α-NaYF4 and its rare earth doped submicrometer particles in acetic acid. Nano Research. 2009;**2**(7):565-574

[8] Liu C, Wang Z, Wang X, Li Z. Surface modification of hydrophobic NaYF4:Yb,Er upconversion nanophosphors and their applications for immunoassay. Science China. Chemistry. 2011;**54**(8):1292-1297

[9] Kang H, Mintri S, Menon AV, Lee HY, Choi HS, Kim J. Pharmacokinetics, pharmacodynamics and toxicology of theranostic nanoparticles graphical abstract HHS public access. Nanoscale. 2015;**7**(45):18848-18862

[10] Kumar V, Sharma N, Maitra SS. In vitro and in vivo toxicity assessment of nanoparticles. International Nano Letters. 2017;**7**(4):243-256

[11] Veronese FM, Pasut G. PEGylation, successful approach to drug delivery. Drug Discovery Today. 2005;**10**:1451-1458

[12] Abuchowski A, Van Es T, Palczuk NC, Davis FF. Alteration of immunological properties of bovine serum albumin by covalent attachment of polyethylene glycol. Journal of Biological Chemistry. 1977;**252**(11):3578-3581

[13] Zeng S, Tsang MK, Chan CF, Wong KL, Hao J. PEG modified BaGdF5:Yb/Er nanoprobes for multi-modal upconversion fluorescent, in vivo X-ray computed tomography and biomagnetic imaging. Biomaterials. 2012;**33**(36):9232-9238

[14] Maldiney T, Richard C, Seguin J, Wattier N, Bessodes M, Scherman D. Effect of core diameter, surface coating, and PEG chain length on the biodistribution of persistent luminescence nanoparticles in mice. ACS Nano. 2011;**5**(2):854-862

[15] Jokerst JV, Lobovkina T, Zare RN, Gambhir SS. Nanoparticle PEGylation for imaging and therapy. Nanomedicine. 2011;**7**:715-728

[16] Yang L, Peng XH, Wang YA, Wang X, Cao Z, Ni C, et al. Receptor-targeted nanoparticles for in vivo imaging of breast cancer. Clinical Cancer Research. 2009;**15**(14):4722-4732

[17] Smith BR, Cheng Z, De A, Koh AL, Sinclair R, Gambhir SS. Real-time intravital imaging of RGD-quantum dot binding to luminal endothelium in mouse tumor neovasculature. Nano Letters. 2008;**8**(9):2599-2606

[18] Lee ALZ, Wang Y, Ye WH, Yoon HS, Chan SY, Yang YY. Efficient intracellular delivery of functional proteins using cationic polymer core/shell nanoparticles. Biomaterials. 2008;**29**(9):1224-1232

[19] Zhang P, Liu S, Jiang S. Anti-PEG antibodies in the clinic: current issues and beyond PEGylation. Journal of Controlled Release. 2016;**244**:184-193. DOI: 10.1016/j.jconrel.2016.06.040

[20] Cheng J, Teply BA, Sherifi I, Sung J, Luther G, Gu FX, et al. Formulation of functionalized PLGA-PEG nanoparticles for in vivo targeted drug delivery. Biomaterials. 2007;**28**(5):869-876

[21] Vicennati P, Giuliano A, Ortaggi G, Masotti A. Polyethylenimine in medicinal chemistry. Current Medicinal Chemistry. 2008;**15**(27):2826-2834

[22] Ge K, Liu J, Wang P, Fang G, Zhang D, Wang S. Near-infrared-emitting persistent luminescent nanoparticles modified with gold nanorods as multifunctional probes for detection of arsenic(III). Microchimica Acta. 2019;**186**(3)

[23] Pan Z, Wen Y, Wang T, Wang K, Teng Y, Shao K. One-step synthesis of hollow PEI-NaBiF4:Yb3+/Er3+ upconversion nanoparticles for water-responsive luminescent probe. Journal of Rare Earths. 2020;**38**(4):362-368

[24] Mi CC, Tian ZH, Han BF, Bin MC, Xu SK. Microwave-assisted one-pot synthesis of water-soluble rare-earth doped fluoride luminescent nanoparticles with tunable colors. Journal of Alloys and Compounds. 2012;**525**:154-158

[25] Xu P, Van Kirk EA, Zhan Y, Murdoch WJ, Radosz M, Shen Y. Targeted charge-reversal nanoparticles for nuclear drug delivery. Angewandte Chemie. 2007;**119**(26):5087-5090

[26] Bivas-Benita M, Romeijn S, Junginger HE, Borchard G. PLGA-PEI nanoparticles for gene delivery to pulmonary epithelium. European Journal of Pharmaceutics and Biopharmaceutics. 2004;**58**(1):1-6

[27] Huh MS, Lee SY, Park S, Lee S, Chung H, Lee S, et al. Tumor-homing glycol chitosan/polyethylenimine nanoparticles for the systemic delivery of siRNA in tumor-bearing mice. Journal of Controlled Release. 2010;**144**(2):134-143

[28] Bharali DJ, Sahoo SK, Mozumdar S, Maitra A. Cross-linked polyvinylpyrrolidone nanoparticles: a potential carrier for hydrophilic drugs. Journal of Colloid and Interface Science. 2003;**258**(2):415-423

[29] Li J, Inukai K, Takahashi Y, Tsuruta A, Shin W. Effect of PVP on the synthesis of high-dispersion core–shell barium-titanate–polyvinylpyrrolidone nanoparticles. Journal of Asian Ceramic Societies. 2017;**5**(2):216-225

[30] Kormshchikov ID, Voronov VV, Burikov SA, Dolenko TA, Kuznetsov SV. Study of stability of luminescence intensity of β-nagdf4:Yb:Er nanoparticle colloids in aqueous solution. Nanosystems: Physics, Chemistry, Mathematics. 2021;**12**(2):218-223

[31] Zhang J, Mi C, Wu H, Huang H, Mao C, Xu S. Synthesis of NaYF4:Yb/Er/Gd up-conversion luminescent nanoparticles and luminescence resonance energy transfer-based protein detection. Analytical Biochemistry. 2012;**421**(2):673-679

[32] Chen H, Zhai X, Li D, Wang L, Zhao D, Qin W. Water-soluble Yb3+, Tm3+ codoped NaYF4 nanoparticles: aynthesis, characteristics and bioimaging. Journal of Alloys and Compounds. 2012;**511**(1):70-73

[33] Zou P, Hong X, Ding Y, Zhang Z, Chu X, Shaymurat T, et al. Up-conversion luminescence of NaYF 4:Yb 3+/Er 3+ nanoparticles embedded into PVP nanotubes with controllable diameters. Journal of Physical Chemistry C. 2012;**116**(9):5787-5791

[34] Johnson NJJ, Sangeetha NM, Boyer JC, Van Veggel FCJM. Facile ligand-exchange with polyvinylpyrrolidone and subsequent silica coating of hydrophobic upconverting β-NaYF4:Yb 3+/Er3+ nanoparticles. Nanoscale. 2010;**2**(5):771-777

[35] Koczkur KM, Mourdikoudis S, Polavarapu L, Skrabalak SE. The versatile role of PVP in nanoparticle synthesis polyvinylpyrrolidone (PVP) in nanoparticle synthesis. Dalton Transactions. 2015;**44**:17883-17905. DOI: 10.1039/C5DT02964C

[36] Jia X, Yin J, He D, He X, Wang K, Chen M, et al. Polyacrylic acid modified upconversion nanoparticles for simultaneous pH-triggered drug delivery and release imaging. Journal of Biomedical Nanotechnology. 2013;**9**(12):2063-2072

[37] Shen J, Zhao L, Han G. Lanthanide-doped upconverting luminescent nanoparticle platforms for optical imaging-guided drug delivery and therapy. Advanced Drug Delivery Reviews. 2013;**65**:744-755

[38] Lahtinen S, Lyytikäinen A, Sirkka N, Päkkilä H, Soukka T. Improving the sensitivity of immunoassays by reducing non-specific binding of poly(acrylic acid) coated upconverting nanoparticles by adding free poly(acrylic acid). Microchimica Acta. 2018;**185**(4):220

[39] Luo Y, Li H, Cai M, Liu Y, Chen L, Xu S, et al. Designing polyacrylic acid capped luminescent rare earth core–shell nanoparticles for simultaneous Cu(II) and temperature sensing. Materials and Design. 2022;**224**:111405

[40] Xiong L, Yang T, Yang Y, Xu C, Li F. Long-term in vivo biodistribution imaging and toxicity of polyacrylic acid-coated upconversion nanophosphors. Biomaterials. 2010;**31**(27):7078-7085

[41] Wang L, Zhang Y, Zhu Y. One-pot synthesis and strong near-infrared upconversion luminescence of poly(acrylic acid)-functionalized YF3:Yb3+/Er3+ nanocrystals. Nano Research. 2010;**3**(5):317-325

[42] Hilderbrand SA, Shao F, Salthouse C, Mahmood U, Weissleder R. Upconverting luminescent nanomaterials: application to in vivo bioimaging. Chemical Communications. 2009;**28**:4188-4190

[43] Liu JN, Bu WB, Shi JL. Silica coated upconversion nanoparticles: a versatile platform for the development of efficient theranostics. Accounts of Chemical Research. 2015;**48**(7):1797-1805

[44] Stöber W, Fink A, Bohn E. Controlled growth of monodisperse silica spheres in the micron size range. Journal of Colloid and Interface Science. 1968;**26**(1):62-69

[45] Ureña-Horno E, Kyriazi ME, Kanaras AG. A method for the growth of uniform silica shells on different size and morphology upconversion nanoparticles. Nanoscale Advances. 2021;**3**(12):3522-3529

[46] Hlaváček A, Sedlmeier A, Skládal P, Gorris HH. Electrophoretic characterization and purification of silica-coated photon-upconverting nanoparticles and their bioconjugates. ACS Applied Materials and Interfaces. 2014;**6**(9):6930-6935

[47] Gnanasammandhan MK, Idris NM, Bansal A, Huang K, Zhang Y. Near-IR photoactivation using mesoporous silica-coated NaYF4:Yb, Er/Tm upconversion nanoparticles. Nature Protocols. 2016;**11**(4):688-713

[48] Chávez-García D, Juárez-Moreno K, Campos CH, Alderete JB, Hirata GA. Upconversion rare earth nanoparticles functionalized with folic acid for bioimaging of MCF-7 breast cancer cells. Journal of Materials Research. 2018;**33**(2):191-200

[49] Chávez-García D, Juarez-Moreno K, Calderón-Osuna I, Navarro P, Hirata GA. Nanotoxicological study of downconversion Y2O3:Eu3+luminescent nanoparticles functionalized with folic acid for cancer cells bioimaging. Journal of Biomedical Materials Research. Part B, Applied Biomaterials 2020;(December 2019):1-11

[50] Kembuan C, Oliveira H, Graf C. Effect of different silica coatings on the toxicity of upconversion nanoparticles on RAW 264.7 macrophage cells. Beilstein. Journal of Nanotechnology. 2021;**12**:35-48

[51] Zhou M, Ge X, Ke DM, Tang H, Zhang JZ, Calvaresi M, et al. The bioavailability, biodistribution, and toxic effects of silica-coated upconversion nanoparticles in vivo. Frontiers in Chemistry. 2019:7. DOI: 10.3389/fchem.2019.00218

[52] Kokubo T, Takadama H. How useful is SBF in predicting in vivo bone bioactivity? Biomaterials [Internet]. 2006;**27**(15):2907-2915

[53] Chaudhary S, Kumar S, Chaudhary GR. Tuning of structural, optical and toxicological properties of Gd3+ doped Yb2O3 nanoparticles. Ceramics International. 2019;**45**(15):19307-19315

[54] Santelli J, Lechevallier S, Calise D, Marsal D, Siegfried A, Vincent M, et al. Multimodal gadolinium oxysulfide nanoparticles for bioimaging: a comprehensive biodistribution, elimination and toxicological study. Acta Biomaterialia. 2020;**108**:261-272

[55] Isacfranklin M, Ameen F, Ravi G, Yuvakkumar R, Hong SI, Velauthapillai D, et al. Y2O3 nanorods for cytotoxicity evaluation. Ceramics International. 2020;**46**(12):20553-20557

[56] Lewinski N, Colvin V, Drezek R. Cytotoxicity of nanopartides. Small. 2008;**4**:26-49

[57] Weyermann J, Lochmann D, Zimmer A. A practical note on the use of cytotoxicity assays. International Journal of Pharmaceutics. 2005;**288**(2):369-376

[58] Mosmann T. Rapid colorimetric assay for cellular growth and survival: application to proliferation and cytotoxicity assays. Journal of Immunological Methods. 1983;**65**(1-2):55-63

[59] Kumar P, Nagarajan A, Uchil PD. Analysis of cell viability by the MTT

assay. Cold Spring Harbor Protocols. 2018;**2018**(6):469-471

[60] Ciapetti G, Cenni E, Pratelli L, Pizzoferrato A. In vitro evaluation of cell/biomaterial interaction by MTT assay. Biomaterials. 1993;**14**(5):359-364. DOI: 10.1016/0142-9612(93)90055-7

[61] Bahadar H, Maqbool F, Niaz K, Abdollahi M. Toxicity of nanoparticles and an overview of current experimental models. Iran Biomed J [Internet]. 2016;**20**(1):1-11

[62] Ramírez-García G, Martínez-Alfaro M, d'Orlyé F, Bedioui F, Mignet N, Varenne A, et al. Photo-stimulation of persistent luminescence nanoparticles enhances cancer cells death. International Journal of Pharmaceutics. 2017;**532**(2):696-703

[63] Zairov R, Mustafina A, Shamsutdinova N, Nizameev I, Moreira B, Sudakova S, et al. High performance magneto-fluorescent nanoparticles assembled from terbium and gadolinium 1,3-diketones. Scientific Reports. 2017;**16**:7

[64] Sayes CM, Fortner JD, Guo W, Lyon D, Boyd AM, Ausman KD, et al. The differential cytotoxicity of water-soluble fullerenes. Nano Letters. 2004;**4**(10):1881-1887

[65] Braun K, Stürzel CM, Biskupek J, Kaiser U, Kirchhoff F, Lindén M. Comparison of different cytotoxicity assays for in vitro evaluation of mesoporous silica nanoparticles. Toxicology in Vitro. 2018;**52**:214-221

[66] Das GK, Stark DT, Kennedy IM. Potential toxicity of up-converting nanoparticles encapsulated with a bilayer formed by ligand attraction. Langmuir. 2014;**30**(27):8167-8176

[67] Malvindi MA, De Matteis V, Galeone A, Brunetti V, Anyfantis GC, Athanassiou A, et al. Toxicity assessment of silica coated iron oxide nanoparticles and biocompatibility improvement by surface engineering. PLoS One. 2014;**9**(1):e85835

[68] Meindl C, Kueznik T, Bösch M, Roblegg E, Fröhlich E. Intracellular calcium levels as screening tool for nanoparticle toxicity. Journal of Applied Toxicology. 2015;**35**(10):1150-1159

[69] Hsiao IL, Hsieh YK, Wang CF, Chen IC, Huang YJ. Trojan-horse mechanism in the cellular uptake of silver nanoparticles verified by direct intra- and extracellular silver speciation analysis. Environmental Science & Technology. 2015;**49**(6):3813-3821

[70] Liang H, Nacharaju P, Friedman A, Friedman JM. Nitric oxide generating/releasing materials. Future Science OA. 2015;**1**(1):FSO54

[71] Wink DA, Hines HB, Cheng RYS, Switzer CH, Flores-Santana W, Vitek MP, et al. Nitric oxide and redox mechanisms in the immune response. Journal of Leukocyte Biology. 2011;**89**(6):873-891

[72] Tran CL, Buchanan D, Cullen RT, Searl A, Jones AD, Donaldson K. Inhalation of poorly soluble particles. Influence of particle surface area on inflammation and clearance. Inhalation Toxicology. 2008;**12**(12):1113-1126. DOI: 10.1080/08958370050166796

[73] Wang HS. Development of fluorescent and luminescent probes for reactive oxygen species. TrAC, Trends in Analytical Chemistry. 2016;**85**:181-202

[74] Kumar A, Pandey AK, Singh SS, Shanker R, Dhawan A. A flow cytometric method to assess nanoparticle

uptake in bacteria. Cytometry. Part A. 2011;**79**(9):707-712

[75] Speit G, Hartmann A. The Comet Assay. In: Henderson DS. (eds) DNA Repair Protocols. Methods in Molecular Biology. Humana Press. 2016;**314**. DOI: 10.1385/1-59259-973-7:275

[76] Karlsson HL, Di Bucchianico S, Collins AR, Dusinska M. Can the comet assay be used reliably to detect nanoparticle-induced genotoxicity?. Environ. Mol. Mutagen. 2015;**56**:82-96. DOI: 10.1002/em.21933

[77] Azqueta A, Gutzkow KB, Priestley CC, Meier S, Walker JS, Brunborg G, et al. A comparative performance test of standard, medium- and high-throughput comet assays. Toxicology in Vitro. 2013;**27**(2):768-773

[78] Wu W, Wu Z, Yu T, Jiang C, Kim WS. Recent progress on magnetic iron oxide nanoparticles: synthesis, surface functional strategies and biomedical applications. Science and Technology of Advanced Materials. 2015;**16**(2):023501

[79] Magaye R, Zhao J, Bowman L, Ding M. Genotoxicity and carcinogenicity of cobalt-, nickel- and copper-based nanoparticles (review). Experimental and Therapeutic Medicine. 2012;**4**(4):551-561

[80] Zhao J, Bowman L, Zhang X, Shi X, Jiang B, Castranova V, et al. Metallic nickel nano- and fine particles induce JB6 cell apoptosis through a caspase-8/AIF mediated cytochrome c-independent pathway. J Nanobiotechnology. 2009;7:2

[81] Wang H, Cho CH. Effect of NF-B Signaling on Apoptosis in Chronic Inflammation-Associated Carcinogenesis. 2010

[82] Wu QH, Jin RR, Feng T, Liu L, Yang L, Tao YH, et al. Iron oxide nanoparticles and induced autophagy in human monocytes. International Journal of Nanomedicine. 2017;**12**:3993-4005

[83] Hussain S, Al-Nsour F, Rice AB, Marshburn J, Yingling B, Ji Z, et al. Cerium dioxide nanoparticles induce apoptosis and autophagy in human peripheral blood monocytes. ACS Nano. 2012;**6**(7):5820-5829

[84] Mohammadinejad R, Moosavi MA, Tavakol S, Vardar DÖ, Hosseini A, Rahmati M, et al. Necrotic, apoptotic and autophagic cell fates triggered by nanoparticles. Autophagy. 2019;**15**:4-33

[85] Schaeublin NM, Braydich-Stolle LK, Schrand AM, Miller JM, Hutchison J, Schlager JJ, et al. Surface charge of gold nanoparticles mediates mechanism of toxicity. Nanoscale. 2011;**3**(2):410-420

[86] Niom H-P. Comparison of the methods available for assessing cytotoxicity. International Endodontic Journal. 1988;**21**:89-99

[87] Liang X, Fan J, Wang Y, Zhao Y, Jin R, Sun T, et al. Synthesis of hollow and mesoporous structured NaYF4:Yb,Er upconversion luminescent nanoparticles for targeted drug delivery. Journal of Rare Earths. 2017;**35**(5):419-429

[88] Bakalova R. Lymph node mapping using quantum dot-labeled polymersomes designing new generation anticancer agents by theranostic functionalization view project multifunctional Magneto-DDS for single drug-based anticancer effect view project. General Physiology and Biophysics. 2015;**34**(4):393-398. DOI: 10.4149/gpb_2015007

[89] Ballou B, Andreko SK, Osuna-Highley E, McRaven M, Catalone T, Bruchez MP, et al. Nanoparticle transport from mouse vagina to adjacent lymph nodes. PLoS One [Internet]. 2012;7(12):e51995. DOI: 10.1371/journal.pone.0051995

[90] Paek HJ, Lee YJ, Chung HE, Yoo NH, Lee JA, Kim MK, et al. Modulation of the pharmacokinetics of zinc oxide nanoparticles and their fates in vivo. Nanoscale. 2013;**5**(23):11416-11427. DOI: 10.1039/C3NR02140H

[91] Balasubramanian SK, Jittiwat J, Manikandan J, Ong CN, Yu LE, Ong WY. Biodistribution of gold nanoparticles and gene expression changes in the liver and spleen after intravenous administration in rats. Biomaterials. 2010;**31**(8):2034-2042

[92] Mehrizi TZ. Hemocompatibility and hemolytic effects of functionalized nanoparticles on red blood cells: a recent review study. Nano [Internet]. 2021;**16**(08):2130007. DOI: 10.1142/S1793292021300073

[93] Nemmar A, Beegam S, Yuvaraju P, Yasin J, Shahin A, Ali BH. Interaction of amorphous silica nanoparticles with erythrocytes in vitro: role of oxidative stress. Cellular Physiology and Biochemistry. 2014;**34**(2):255-265

[94] Lima JM De, Sarmento RR, Souza JR De, Brayner FA, Feitosa APS, Padilha R, et al. Evaluation of hemagglutination activity of chitosan nanoparticles using human erythrocytes. BioMed Research International 2015;**2015**:247965-247971

[95] Choi J, Reipa V, Hitchins VM, Goering PL, Malinauskas RA. Physicochemical Characterization and in vitro hemolysis evaluation of silver nanoparticles. Toxicological Sciences. 2011;**123**(1):133-143

[96] Love SA, Thompson JW, Haynes CL. Development of screening assays for nanoparticle toxicity assessment in human blood: preliminary studies with charged Au nanoparticles. Nanomedicine [Internet]. 2012;**7**(9):1355-1364. DOI: 10.2217/nnm.12.17

[97] Aseichev AV, Azizova OA, Beckman EM, Skotnikova OI, Dudnik LB, Shcheglovitova ON, et al. Effects of gold nanoparticles on erythrocyte hemolysis. Bulletin of Experimental Biology and Medicine. 2014;**156**(4):495-498. DOI: 10.1007/s10517-014-2383-6

[98] Saha AK, Zhen MYS, Erogbogbo F, Ramasubramanian AK. Design considerations and assays for hemocompatibility of FDA-approved nanoparticles. Seminars in Thrombosis and Hemostasis. 2020;**46**(5):637-652

[99] Coto-García AM, Sotelo-González E, Fernández-Argüelles MT, Pereiro R, Costa-Fernández JM, Sanz-Medel A. Nanoparticles as fluorescent labels for optical imaging and sensing in genomics and proteomics. Analytical and Bioanalytical Chemistry. 2011;**399**:29-42

Chapter 3

Toxicity of 2D Materials and Their Future Prospect

Subash Adhikari

Abstract

Miniaturization of the devices in terms of size and the necessity of high speed device performance have created opportunities as well as challenges in the material research community. Nanomaterials like 0D and 2D materials are one of such material choices that can help realize the nanosize and ultrafast devices. However, the growth process of these materials, especially emerging 2D materials, needs to be reviewed in terms of human, animal and environmental toxicity along with the economic cost for synthesizing material. Moreover, the green and sustainable alternatives for minimizing or eliminating the toxicity should also be considered for the commercial scale nanomaterials synthesis and device fabrication. This topic will thus highlight the currently developed 2D materials, their growth process, application prospective, toxicity effect and their possible sustainable alternatives.

Keywords: nanostructures, 2D materials, growth toxicity, nanomaterial toxicity, green synthesis

1. Introduction

The need of efficient and convenient human lifestyle has induced new and innovative research in fields of food, agriculture, technology, construction, transportation, environment and health [1]. These modern global needs are possible with the use of new and more efficient materials and technology. In such, wide range of nanomaterials from zero dimensional (0D) quantum dots to one dimensional (1D) nanorods/nanowire and two dimensional (2D) sheet having large surface to volume ratio; ability to bind with various other materials; ballistic transport of charges; tunable optical, electrical and magnetic properties; low volume of material consumption; and superior overall performance than its bulk counterpart have shown various new possibilities in the field of health, technology developments, energy devices, transport, communication, computation and agricultural productivity [2]. Moreover, the rise of these nano-dimensional materials and nano-based technology have replaced many of the traditional industries ranging from electronic, optics, energy storage/production, pharmaceuticals, transport, cosmetics, agriculture, food production to material processing [3, 4]. At present the global nanotechnology market is worth 75.8 billion USD which mainly comprises industries related to electronic, sporting goods, automotive, energy storage, aerospace, defense, food and pharmaceutical [5, 6]. However, having superior physical and chemical properties than pre-existing bulk materials, the nanotechnology based products should have been a global need with high economic prospective. The major factors restraining the development of global nanotechnology market from its rise in early 2000 are toxicity of nanomaterial on human health and environment and stringent requirements of the government bodies on adopting the nanomaterials and nanotechnology in commercial products [3, 7, 8]. Nanomaterials have adverse toxicological effect on both animal and plant cells [9]. Nanomaterials upon interaction with body cell of an organism can have inflammatory response, dysfunction of organs, tissue damage, tumors and upon interaction with nervous system can accumulate in the brain leading to neurological diseases. Similarly nanomaterials upon interaction with plants can prevent plants protein, chlorophyll, carotenoids and biomass content as well as extend the plant harvest period [10]. However these effects are mostly observed in plants if high concentration of nanomaterials are used and more importantly, the use of synthetic nanomaterials which are non-biodegradable are primarily harmful for plants and humans.

Various global acts and forums like Pollution Prevention Act of 1990, regulations of European Union Observatory for Nanomaterials, Food and Drug Administration in United States (US), US Environmental Protection Agency, Intergovernmental Forum on Chemical Safety including many others have initiated governmental and global alliance to minimize the rising issue of chemical toxicity and hazard from nanostructure materials and its synthesis techniques on human, animal and environmental health [11]. The additional initiative of these acts and agencies are to prevent chemical waste, synthesize safer and less hazardeous chemicals and chemical products, use of natural resources against high toxicity materials and procedures, design of energy efficient procedures for material production, reduce derivative products and promote biodegradable and sustainable materials [12]. Currently, based on these policies and acts, the industrial application of various nanomaterials, nanocompounds and nanocomposites either as a final product or as an additive supplement have been approved after extensive research on toxicity, stability and the wide scale applicability [13].

These nanomaterial are synthesized mainly from physical, chemical and biological synthesis procedures. Among these, physical vapor deposition, chemical vapor deposition, sol-gel and colloidal methods are currently commercialized for large scale synthesis [14]. Moreover, the high cost required in the material synthesis, toxicity of the synthetic material and the synthesis procedure along with the environmental issues arising due to toxicity have led to the development of biological synthesis method which utilizes green chemistry approach for synthesizing nanomaterials [15]. Hence as an alternative, bio-nanomaterials produced from biological sources using green synthesis techniques are being extensively explored. Also green synthesis employs clean, cost effective, safe and environment friendly process of constructing nanomaterials using natural substrates like bacteria, yeast, fungi, algae and plants or agricultural resources [16]. Especially, carbon materials derived from plant and agricultural byproducts along with the organic chemicals derived from natural products can be used for synthesizing nanomaterial and nanocompounds. Similarly, thermal and hydrothermal synthesis techniques like biosorption, pyrolysis and hydrolysis can also be used for biologically synthesizing 2D nanostructures. Hence both these synthesis techniques produces bio-degradable nanomaterials without the use of synthetic chemicals. Moreover, biomass derived nanomaterials are non-toxic bio-nanomaterials that can be used safely in medicines, fertilizers, devices and cosmetic products.

In this review, we will mainly focus on the toxicity arising from the material synthesis procedure as well as the toxicity of the materials itself on human, animal and environment health mainly focusing on the 2D material family. We will also highlight the current research trend in utilizing green and sustainable procedure for synthesizing 2D materials.

2. Growth mechanism of 2D materials and their toxic effects

Among the class of nanomaterials, 2D materials are recently developed materials that exists in various materials forms. Present in layered form and bonded by the van der Waals (vdW) interaction, these materials in various form from metallic like graphene to semiconducting like Transition metal Di-chalcogenides (TMDs), insulating like hBN, superconducting like NbSe2, thermoelectric like PbTe and topological insulator like HgTe quantum wells [17, 18]. Also depending on the layer number and doping the optical, electrical and magnetic properties of these materials can be tuned. Because of these wide range of physical, chemical and structural properties that does not exist in bulk form and the additional advantage of flexibility, high strength and tunable bandgap, these materials are being considered for electronics, optics, opto-electronics, computing, transport, energy storage/production, health, drug design and agriculture industries [19–21].

The development of 2D materials started with the exfoliation process wherein bulk form of crystals arranged in layered structure through van der Waals (vdWs) interaction are exfoliated using a simple scotch tape. These exfoliated crystals were used for successful demonstration of various interesting low dimensional electrica, optical, optoelectronic phenomenon in room temperature like integer quantum hall effect, extremely high intrinsic mobility, ballistic transport, tunable bandgap, excitonic features, high optical transmittance and high quantum yield [22–27]. However, the limitation of material size within few micrometers in the exfoliated crystal and the manual operation process to obtain the 2D flakes are one of the major bottleneck for large scale device integration. Hence various other alternative synthesis techniques were developed using top-down or bottom-up synthesis approach. Moreover with the competitive research on developing large area and high quality 2D materials, various new and emerging synthesis techniques are developed. These techniques however, comes with various bottlenecks including quality of the 2D materials, size of 2D materials produced and more importantly the cost of the system and the human, animal and environmental toxicity that is associated with the synthesis techniques. The chapters below summarize the two main 2D material synthesis techniques based on their toxicity and cost.

2.1 Toxicity from top-down synthesis

Top-down synthesis based on solution is one of the cost effective, fast and scalable methods to obtain commercial scale 2D crystals [28, 29]. Here 2D crystals like graphene, various TMD materials like MoS2, WS2, Black Phosphorous and MXenes are obtained either using chemical based intercalation or liquid based exfoliation (LPE) [30–34]. Intercalation method mainly usages ion or molecule that can penetrate into the layers of bulk 2D crystal by weakening the van der Waals interaction either through driving force of reaction, attractive force of van der Waals and repulsive force of polarization between ions of same charge to disintegrate the 2D layers into individual 2D layers [35, 36]. The intercalation method can be carried out through electrochemical process, solution based and vapor based. In the electrochemical process ions like Li^+, Na^+, Co^{2+}, cetyl-trimethylammonium (CTA^+), tetra-heptyl

(THA^+), K^+, tetrabutylammonium (TBA^+) obtained from source electrolytes like $LiC/LiO_2/LiPF_6/LiOxNy/LiOx$, NaxOy, $CoSO_4$, cetyl-trimethylammonium bromide (CTAB), tetra-heptyl-ammonium bromide/N-methyl-2-pyrrolidone (THAB)/(NMP), KPF_6, tetra-butyl-ammonium bromide/dimethylglyoxime (TBAB)/(DMG) are used [37–40]. Among these Li ions the major ionic source that is commonly used for chemical exfoliation of 2D crystals. Similarly, in the solution based process metallic ions or molecules like Cu, Ag, Au, Sn, Pt, Mo, Fe, N2H4, $Co(Cp)_2$, etc. are used from the metal salts precursors. Moreover, in the chemical based intercalation process disintegration of ions occurs in presence of solvents like water, hexane, DMF, DEC, acetonitrile, toluene, acetone, isopropanol, etc. and the 2D layers obtained after exfoliation have to undergo treatment process like multiple washing in DI water, ultrasonication, AND centrifugation in organic solvents to obtain clean and isolated 2D crystals. Hence even though chemical based intercalation is much convenient process than mechanical exfoliation for large scale 2D materials synthesis, the use of toxic chemicals limits its wide scale application prospective. It is mainly because, commonly used intercalating metals like Li and Co has an increasing demand in battery manufacturing, aerospace industry, missile systems, radar and sensor industries [41]. Along with this, metallic Lithium is also considered health, physiochemical and/or ecotoxicological hazard according to the National Occupational Health and Safety Commission (NOHSC) [42]. Beside the metals that are used for intercalations, the organic solvents that are used for disintegration of metal ions as well as for cleaning the exfoliated 2D materials are also considered hazardous for human health and environment. Commonly used organic solvents like hexane, benzene, acetonitrile, DMF, toluene, acetone, isopropanol (IPA), chloroform and ethyl alcohol have shown acute oral toxicity in rats [43] as well as some of these commonly used organic solvents like benzene and chloroform are considered carcinogen by United States and European Union with sever toxicity to humans and environment [44]. Similarly alcohols like ethanol, methanol, IPA are considered toxic alcohols as it can damage human organs like retina, liver, kidney and brain with excess consumption and it can also increase risk of certain types of cancers [44–46].

Liquid phase exfoliation can be a substitute to chemical intercalation method for 2D material exfoliation as it is simple physical method that usages external forces like high intensity ultrasound mediums like high power sonic probes, sonic baths and tip ultrasonicators [47–49]. This has been used in exfoliation of graphene and other 2D materials like black phosphorous, boron nitride including various TMDs and topological insulators [50–53]. However, due to the high external vibrations, the exfoliated 2D layers are usually non-uniform in size and thickness and there is a high possibility for the phase transition of 2D materials during exfoliation process [54]. Hence the process requires stabilization and sorting after exfoliation [55]. These process are carried out using various organic solvents including polymers, co-polymers and alcohols like DMF, odichlorobenzene, N-methyl-2-pyrrolidone, octylbenzene, poly(styrene co-butadiene), polystyrene, poly(vinyl acetate), polycarbonate, ethanol and IPA for exfoliation of various 2D materials including graphene, TMDs and topological insulators [44, 56–61]. Moreover, all these organic polymers, co-polymers and alcohols have high toxicity [43, 44, 62] and many of these organic solvents even have high boiling points thus producing aggregated and unstable 2D layers [55]. Beside these, other top-down method that can be used for exfoliating 2D materials are laser based exfoliation and ball milling assisted exfoliation. However, ball milling produces non-uniform flake size of 2D crystals [63, 64] and laser based exfoliation is known to create various defects during the exfoliation process mainly from the heat generated in the exfoliation process [65, 66]. Hence even though there is no toxicological effect in these 2D materials synthesis process, the quality and size of flake is the major issue that has inhibited in using these techniques.

2.2 Toxicity from bottom-up synthesis

Various limiting factors like grain size, defects, uniformity and mainly the toxicological effect of top-down synthesis techniques has inhibited its commercialization prospective. However, with the development of chemical vapor based 2D materials synthesis technique, wafer scale and layer controlled grown of various kind of 2D materials like graphene, MoS_2, $MoTe_2$, WS_2, WSe_2 and hBN have been possible [67, 68]. Additionally, high purity and commercial scale 2D materials with controlled morphology, crystallinity and defect engineering is possible with the CVD. Here 2D materials are synthesized at high temperature through the chemical reaction of gaseous substances which react in the gas phase to produce 2D materials on various metallic and insulating substrates. For example CVD graphene having 3–4 layers was first grown in nickel surface [69] and later high quality, centimeter scale monolayer graphene using methane gas as a carbon precursor were grown on copper foil [70]. Moreover, with the use of high quality copper film with large grain boundaries and using polished copper surface to enhance the graphene grain size are currently practiced to produce few tens of centimeter scale polycrystalline graphene flakes [71]. However, these large scale graphene are grown in high temperature and high pressure systems which are expensive, requires complex preparation process and more importantly usages gaseous materials like methane which are toxic in nature [72, 73]. Listed as a greenhouse gas, methane is highly flammable and can ignite even a relatively low pressure and low concentration levels and possesses various health hazards including coma and death due to deprived oxygen level when inhaled [74–76]. Hence either high level of safety has to be ensured in CVD based graphene growth system which required additional expenses in producing graphene or alternate synthesis routes has to be considered.

Besides graphene, CVD is also one of the primary tools for the synthesis of TMD materials. Millimeter scale poly-crystalline monolayer MoS_2 on SiO_2/Si were first reported by Lee et al. [77] using MoO_3 and S powders as a precursor material synthesized at 650°C in a nitrogen environment. After the successful demonstration of large area monolayer MoS_2, the same method were replicated for the synthesis of other TMD materials like $MoSe_2$, WS_2, WSe_2 and $MoTe_2$ simply by tuning the metal precursor and chalcogenides precursors. However, melting temperature of pure chalcogenides are lower than 500°C while transition metals have melting temperature >2000°C [78]. Hence high quality and uniform crystals in large area are difficult to achieve using a single zone furnace in TMDs synthesis. For this, instead of using a single heating zone for both transition metal and chalcogenide, two or three temperature zone CVD furnace for heating the transition metal, chalcogenide precursor and/or the substrate separately are used for growing wafer scale high quality TMDs [79]. Thus controlling various parameters like temperature, pressure, gas flow rate and precursor concentrations are key to obtaining high quality and large scale monolayer TMDs. Moreover, the complexity of the CVD system and the gas precursors required for growth like Ar, N and Hydrogen as activating agent, surface cleaning, impurities/defects reduction, transport agent and surface states/precursor molecule reduction makes TMD synthesis an expensive process [80, 81].

In addition to this, the toxicity of some transition metals and chalcogens are another important issue that needs to be considered. Especially molybdenum, selenium and tellurium are the major elements that are known to be toxic. A toxicology study conducted by Franke and Moxon [82] on Albino rats fed with salts of molybdenum, tellurium and selenium at levels of 25 and 50 parts per million of the elements in their diets showed that selenium salts had the highest toxicity causing distinct disturbance of the hematopoietic systems while tellurium and then molybdenum showed lower toxicity effects. Additionally, another study conducted by Larner [83] in rats showed that tellurium can damage mitochondria and produce defects in mitochondrial energy metabolism that can eventually produce cognitive impairment and cerebral lipofuscinosis. The high concentration of Selenium is known to cause blindness and premature deaths [84] while acute Tellurium can cause glad malfunctioning and paralysis of nervous systems [85].

Also elemental tellurium has been known to cause significant neurotoxicological effects in animals even though they are poorly absorbed by gastrointestinal tract [86]. It was also observed from the study that fumes of tellurium dioxide and volatile tellurium esters can be easily absorbed through the lungs and skin. Beside elemental toxicity, hydrogen selenide is also known to be toxic for human health which when absorbed through lungs can cause pallor, nervousness, depression, languor, dermatitis and gastrointestinal disturbances [87]. Moreover, CVD based growth of TMDs relies mostly on compound like tellurium and selenium which undergoes vapor based reaction under nitrogen, argon and hydrogen environment. Hence the possibility of fume production as well as production of toxic ternary compounds of calcogenide including hydrogen selenide and hydrogen telluride are possible [88]. Moreover, there are other bottom-up synthesis techniques which can produce high quality 2D materials like Molecular Beam Epitaxy, Magnetron Sputtering and Pulsed Laser deposition. But these techniques can produce 2D materials within few tens of micrometer and the operation of these systems requires high vacuum well as the system are complex and expensive to operate [89–91].

3. Toxicity of 2D materials on human, animal and environment

The interesting optical, electrical and magnetic phenomenon in 2D materials at an atomic dimension have developed new research avenues in industrial sectors of electronic, opto-electronic, energy harvesting/production and health. More importantly, in the field of biology for drug discovery/design, drug delivery, nano biomedical equipment development and cellular level research, nanomaterials is considered an important tool. It is mainly because, nanomaterials due to their atomic dimension; high surface area; tunable electrical, optical and magnetic properties; ballistic transport of carriers and high optical gain can enhance the optical/electrical response in disease detection, increase the drug loading efficiency, ensure targeted drug delivery within specific region of cell and organs for disease cure, promote the development of nano-chips and nano-kits for disease diagnostic and more importantly minimize the cost of disease detection and prevent organ damages in animals [21, 92–94]. However, 2D materials are in its early stage of development. Even though various new materials are being developed in material research communities around the globe, many of these materials have not been tested for its toxic effect in human and animals [95]. Besides the toxicity effect of 2D materials on animal, environmental toxicity is another important aspect that needs to be considered with the rapid development of new and emerging 2D materials. Since 2D materials have high strength, high melting temperature and are confined in atomic dimensions, they can neither be filtered, incinerated or collected for disposal with normal equipment and techniques. Hence, once these 2D materials are released or dumped in environment as a waste, they will pollute the soil, air and aquatic environment. Moreover, the toxicity induced by 2D materials can further enhance the soil, air and aquatic health in addition to the material based toxicity. In such, specific and in-depth toxicological studies are being carried out in order to find the material toxicity on both human and environment. These finding are summarized below.

Primary research on the toxicity effect of graphene started in 2010, soon after the discovery of graphene in 2004. A study by Zhang et al., in 2010 studied the in vitro toxicity of graphene using neuronal PC12 cells [96]. Reactive oxygen species (ROS) which are usually generated in human cell during mitochondrial oxidative metabolism as well as during cellular response to xenobiotics, cytokines and bacterial invasion [97], were generated even at a low concentration of 0.1 μg/ml concentration. ROS were also generated with graphene studied on human primary umbilical vein endothelial cells [98]. The study also concluded that besides ROS, few layer graphene can exert cellular toxicity employing oxidative stress, in HUVEC cells altering critical cell parameters like cytoskeletal dysfunction, reduction in metabolic activity, compromised plasma, membrane integrity, lipid peroxidation, ionized calcium and deposition of mitochondrial membrane potential. Moreover, ROS generation is known one of the major cyctotoxicity from graphene based 2D materials. Especially functionalized graphene or reduced graphene oxide (rGO) are one of the primary 2D materials that are known to generate ROS activities in human and animal. Graphene oxide based study on human using HEK 293T cells showed activated ROS generation with increase DNA damage [99], in vitro and in vivo studies on human corneal epithelium cells and human conjunctiva epithelium cells showed increased intracellular ROS [100], graphene oxide incubated with human plasma having different diseases exhibited ROS production together with lipid production and increased nitrogen oxide levels [101], and in vitro study in GLC-82 pulmonary adenocarcinoma cells showed ROS production and apoptosis with dysregulation of cell cycles [102]. Similarly, studies of graphene oxide on animal using chorion of zebrafish embryos also showed high generation of ROS including DNA damage and apoptosis [103, 104]. W1118 flies studied with low concentration of graphene oxide showed excessive accumulation of ROS with rapid weight loss, developmental delay thus reducing lifespan [105] and industrial organism like Pichia Pastoris under graphene oxide showed accumulation of ROS with cell membrane damage [106]. Other forms of carbon like graphene nanosheets also showed increased ROS activities in studies carried out on embryonic stem cells derived cells [107].

Apart from graphene based 2D materials, cytotoxicity studies using TMDs also have shown cellular toxicities in animal and environment. However, few of the 2D TMDs like MoS_2, hBN, WS_2 and WSe_2 have been studied for toxicity effects. MoS_2 dispersed in Pluronic F87 in liver cells have shown to induces dose-dependent cytotoxicity [108], human macrophages with MoS_2 can trigger cell stress and inflammatory response [109], and analysis of pulmonary hazard with aggregated MoS_2 induces strong proinflammatory and profibreogenic responses [110]. Similarly, studies with MoS_2 on Zebrafish embryos showed high toxicity affecting amino acid and protein biosynthesis and energy metabolism [111], eggs of Gallus gallus domesticus with MoS_2 showed growth defects and deaths [112], and mice exposed to MoS_2 via food showed Mo accumulation in mouse organs changing the intenstinal microbiota [113].

In study with chemically exfoliated WS_2 nanosheets in Algae showed oxidative stress, lipid peroxidation, membrane damage and photosynthesis inihibition [114], WSe_2 nanosheets on A549 cells showed reduced cell viability [115], hBN induced oxidative stress in rats by modulating thiol/disulfide homeostasis [116].

Similarly, the toxicological effects of 2D materials on plant and species have also shown various toxic effects on plant cells. For example, graphene can affect root and shot growth, biomass shape, cell death and influence ROS of cabbage, tomato, red spinach, and lettuce inhibiting plant growth [117], germination and growth of rice seeds can be affected by graphene by inhibiting stem length [118], grassland soil exposed to graphene can alter the bacterial communities [119], graphene oxide exposure to leaves of cabbage, spinach and tomato decreased size and number due to oxidative stress-mediated cell death [117], absorption of graphene oxide by roots of *Vicia faba* showed increased oxidative stress [120], wheat germination was inhibited with graphene oxide exposure by accumulating in the root and inducing oxidative stress [121], effect of MoS_2 on *Chlorella vulgaris* showed cell distortion and deformation [122], and unstable 2D materials like 1T MoS_2 in the environment can easily oxidize releasing the Mo ions which have high bioactivity in plants [123].

Besides these 2D materials there are several other 2D materials like topological insulators, perovskites, Metal-Organic Frameworks and 2D Oxides that are identified in research community. However, those materials are still in their early phase of research and still far from the commercialization prospective. Moreover, toxicity is one of the major issue that needs to be addressed before the materials are commercialized for practical applications. Hence simultaneous research on the application prospective as well as their toxicity study on human, animal and environment for specific materials that have high potential to be used by industries needs to be initiated.

4. Sustainable alternatives for synthesis of 2D materials

As mentioned in the previous section, toxicity is one of the key aspects of these emerging and highly potent 2D materials family which might inhibit its future practical applications. In such, either minimizing the hazard in synthesizing the materials, using less toxic chemical during material synthesis, completely replacing the material sources and synthesis process into sustainable alternatives or using other alternative materials having similar properties but with less hazard and toxicity are the only prospective for commercializing the materials and their functional devices. Various acts on environment protection have already directed the industrial communities to adhere to the principle of minimizing health and environmental hazard to human, animal and environment. Low material toxicity will also eventually help realize the biological application of 2D materials in drug synthesis and target drug delivery which is expected to provide better health treatment to living organisms [124, 125]. To achieve 2D materials with low toxicity, there are primarily two different resources: (i) using natural product resources like forest and agricultural waste as the source materials to derive 2D materials and (ii) using green synthesis techniques like plants based phytochemicals or microbes to synthesize 2D materials. Especially for synthesizing carbon materials, abundantly available biomass resources like forest and agricultural waste which contains around 55 wt% of carbon can be utilized [126]. Similarly, various phytochemicals derived from forest and agricultural resources and microbes like bacteria, fungi and yeast can be used in reducing various metal salts and compounds into nanostructure forms [127, 128]. Both these synthesis mediums utilizes green and sustainable resources hence producing toxic free and green nanomaterials.

Among the 2D materials family, carbon based 2D materials like graphene, graphene oxide and reduced graphene oxide have been successfully synthesized using the green synthesis technique. Using biomass as a source of carbon precursor, carbon gas can be generated which acts a natural carbon source to graphene synthesis. Mostly, low temperature carbonization of biomass to obtain pure carbon followed by high temperature graphitization yields layered graphene using the green synthesis technique [129, 130]. The successful demonstration of large area monolayers graphene synthesis on copper foil using forest resources like Camphor leaves was carried out by Kalita et al. using pyrolysis of camphor leaves in CVD at 1020°C [131]. Using similar approach, various biomass resources like rice husk, sugarcane bagasse, wheat straw, palm oil waste, fruit waste, soyabeans, newspaper, populous woods and various other lignin based biomass have been used to synthesize monolayer and few layer graphene. This has been reviewed in depth by Safian et al. and [132] and Saha and Dutta [133].

Apart from carbon based materials which are abundant natural resources, there are not much report on the green synthesis of TMDs and other 2D materials. This can be primarily because there are no natural resources for TMDs that are abundant and easily available in nature, as like carbon. Beside this, the reaction kinetics of transition metal with chalcogenide occurs with specific molecular structure which might be difficult to achieve with nature based resources. However, nanomaterial form of TMDs like CdS, CdSe, CdTe and its other ternary phases have been reported using green synthesis technique [134–136]. Here, greener synthesis are achieved primarily by using either non-toxic synthetic chemicals like hexylphosphonic acid (HPA) or tetradecylphosphonic acid (TDPA) or by aqueous or hydrothermal synthesis techniques using non-toxic chemical solvents like thiogycerol and mercaptopropionic acid (MPA) with metal salt precursors. This successful demonstration of nanomaterial form of TMDs further suggests that using less toxic synthetic chemicals and environmental conditions, synthesis of 2D materials are possible.

High temperature synthesis is another important factor that can induce reaction of precursor gases with oxidizing agents and environment to produce toxic gases and toxic chemicals. However, in the case of green synthesis the low temperature required for synthesis as well non-toxic precursor gases and solvents assures that there is no induced toxicity during material synthesis. This is thus an important prospective of selecting green synthesis techniques that is safe, non-toxic and economic to human and environment.

5. Conclusion

Tunable and quantum level electronic, optical and magnetic phenomenon in monolayer and few layer 2D materials makes it an unique and important material choice of twenty-first century. It is believed that the next generation of technology will be driven by low cost, low material volume, high speed, compact, flexible and tunable material properties and 2D materials are the only material choice that can incorporate all these interesting features. Moreover, growing concern of human and animal health is becoming a global issue. New and effective drug that can penetrate the specific region of human bodies; smart, effective and non-invasive biomedical and medical equipment; low cost and effective health care and management can only be achieved by using nanoscopic, high speed and highly efficient technologies using nanomaterials and nanodevices. Another important aspect of human development is quality of food and life. Global scale infertility of soil and chemically modified food using pesticides and insecticides have polluted soil, water and environment. Considering all these issues and choosing a right solution can be effective use of nanomaterials and nanotechnology. Moreover, this can only be possible if the material does not add any additional toxicity to the human, animal and environment health. Hence along with the development of nanomaterials and nanodevices using 0D, 1D or 2D materials, consideration should be taken in understanding the toxicity aspect of the material itself. The fast and growing research trend in 2D and 0D materials should be consistent with the toxicity study and long term toxicological effect of these materials on various global aspect. In doing so, high prospective materials that have instrumental effect on health of human, animal and environment can be industrially developed together with the pace of human needs and requirements.

Acknowledgements

The author would like to acknowledge Sudarshan GC, Media Lab Nepal: REDX Nepal, Prof. Sanjay Nath Khanal, Madan Bhandari University of Science and Technology and Prof. Nabeen Kumar Shrestha, Dongguk University, South Korea and Usha Adhikari, Unnati Data Mapping and Mining for comments and suggestions on the chapter.

Author details

Subash Adhikari[1,2]

1 Madan Bhandari University of Science and Technology, Lalitpur, Nepal

2 Unnati Data Mapping and Mining, Hetauda, Nepal

*Address all correspondence to: subashadhikari@hotmail.com

References

[1] Köhler J et al. An agenda for sustainability transitions research: State of the art and future directions. Environmental Innovation and Societal Transitions. 2019;**31**:1-32

[2] Navya PN, Daima HK. Rational engineering of physicochemical properties of nanomaterials for biomedical applications with nanotoxicological perspectives. Nano Convergence. 2016;**3**(1):1-1

[3] Inshakova E, Inshakov O. World market for nanomaterials: Structure and trends. MATEC web of conferences. EDP Sciences. 2017;**129**:02013

[4] Mongillo JF. Nanotechnology 101. Greenwood Press; 2007. ISSN 1931-3950

[5] Markets Ra. Global Nanotechnology Market Outlook 2024. 2020. Available from: https://www.researchandmarkets.com/reports/4991720/global-nanotechnology-market-outlook-2024

[6] Divyanshi Tewari SB. Nanotechnology Market by Type (Nanodevices and Nanosensors) and Application (Electronics, Energy, Chemical Manufacturing, Aerospace & Defense, Healthcare, and Others): Global Opportunity Analysis and Industry Forecast, 2018-2025. 2019. Available from: https://www.alliedmarketresearch.com/nanotechnology-market

[7] Foulkes R et al. The regulation of nanomaterials and nanomedicines for clinical application: Current and future perspectives. Biomaterials Science. 2020;**8**(17):4653-4664

[8] Pietroiusti A et al. Nanomaterial exposure, toxicity, and impact on human health. Wiley Interdisciplinary Reviews: Nanomedicine and Nanobiotechnology. 2018;**10**(5):e1513

[9] Najahi-Missaoui W, Arnold RD, Cummings BS. Safe nanoparticles: Are we there yet? International Journal of Molecular Sciences. 2020;**22**(1):385

[10] Zhu Y et al. Nanomaterials and plants: Positive effects, toxicity and the remediation of metal and metalloid pollution in soil. Science of the Total Environment. 2019;**662**:414-421

[11] Picecchi D. Tiny things with a huge impact: The international regulation of nanomaterials. Michigan Journal of Environmental and Administrative Lae. 2017;7:447

[12] Pinto AH et al. Green Chemistry Applied to Transition Metal Chalcogenides through Synthesis. Design of Experiments, Life Cycle Assessment, and Machine Learning. In: Kumar B, Debut A, editors. Green Chemistry – New Perspectives. Intechopen: 2022. DOI: 10.5772/intechopen.104432

[13] Amenta V et al. Regulatory aspects of nanotechnology in the agri/feed/food sector in EU and non-EU countries. Regulatory Toxicology and Pharmacology. 2015;**73**(1):463-476

[14] Goutam SP et al. Green synthesis of nanoparticles and their applications in water and wastewater treatment. In: Saxena G, Bharagava R, editors. Bioremediation of Industrial Waste for Environmental Safety. Springer; 2020:349-379. DOI: 10.1007/978-981-13-1891-7_16

[15] Eckelman MJ, Zimmerman JB, Anastas PT. Toward green nano: E-factor analysis of several nanomaterial

syntheses. Journal of Industrial Ecology. 2008;**12**(3):316-328

[16] Devatha CP, Thalla AK. Green synthesis of nanomaterials. In: Bhagyaraj S M et al. Synthesis of Inorganic Nanomaterials. Advances and Key Technologies. Elsevier; 2018:169-184. DOI: 10.1016/C2016-0-01718-7

[17] Gupta A, Sakthivel T, Seal S. Recent development in 2D materials beyond graphene. Progress in Materials Science. 2015;**73**:44-126

[18] Wang Y et al. Surface modification of monolayer MoS2 by baking for biomedical applications. Frontiers in Chemistry. 2020;**8**:741

[19] Song J-H et al. Multi-functionalization strategies using nanomaterials: A review and case study in sensing applications. International Journal of Precision Engineering and Manufacturing-Green Technology. 2022;**9**(1):323-347

[20] Glavin NR et al. Emerging applications of elemental 2D materials. Advanced Materials. 2020;**32**(7):1904302

[21] Das S et al. Synthesis, properties, and applications of 2-D materials: A comprehensive review. Critical Reviews in Solid State and Materials Sciences. 2014;**39**(4):231-252

[22] Novoselov KS et al. Unconventional quantum Hall effect and Berry's phase of 2π in bilayer graphene. Nature Physics. 2006;**2**(3):177-180

[23] Nair RR et al. Fine structure constant defines visual transparency of graphene. Science. 2008;**320**(5881):1308-1308

[24] Shytov A et al. Atomic collapse, Lorentz boosts, Klein scattering, and other quantum-relativistic phenomena in graphene. Solid State Communications. 2009;**149**(27-28):1087-1093

[25] Chaves A et al. Bandgap engineering of two-dimensional semiconductor materials. Materials and Applications. 2020;**4**(1):29

[26] Amani M et al. Near-unity photoluminescence quantum yield in MoS2. Science. 2015;**350**(6264):1065-1068

[27] Salehzadeh O et al. Exciton kinetics, quantum efficiency, and efficiency droop of monolayer MoS_2 light-emitting devices. Nano Letters. 2014;**14**(7):4125-4130

[28] Das S et al. A self-limiting electro-ablation technique for the top-down synthesis of large-area monolayer flakes of 2D materials. Scientific Reports. 2016;**6**(1):28195

[29] Cai X et al. Preparation of 2D material dispersions and their applications. Chemical Society Reviews. 2018;**47**(16):6224-6266

[30] Stankovich S et al. Synthesis of graphene-based nanosheets via chemical reduction of exfoliated graphite oxide. Carbon. 2007;**45**(7):1558-1565

[31] Coleman JN. Liquid exfoliation of defect-free graphene. Accounts of Chemical Research. 2013;**46**(1):14-22

[32] Niu L et al. Production of two-dimensional nanomaterials via liquid-based direct exfoliation. Small. 2016;**12**(3):272-293

[33] Zhang Q et al. Intercalation and exfoliation chemistries of transition metal dichalcogenides. Journal of Materials Chemistry A. 2020;**8**(31):15417-15444

[34] Du F et al. Environmental friendly scalable production of colloidal 2D titanium carbonitride MXene with minimized nanosheets restacking for

excellent cycle life lithium-ion batteries. Electrochimica Acta. 2017;**235**:690-699

[35] Allart D, Montaru M, Gualous H. Model of lithium intercalation into graphite by potentiometric analysis with equilibrium and entropy change curves of graphite electrode. Journal of the Electrochemical Society. 2018;**165**(2):A380

[36] Rajapakse M et al. Intercalation as a versatile tool for fabrication, property tuning, and phase transitions in 2D materials. Materials and Applications. 2021;**5**(1):30

[37] Zhang J et al. Reversible and selective ion intercalation through the top surface of few-layer MoS2. Nature Communications. 2018;**9**(1):5289

[38] Kim S et al. Orientation-dependent intercalation channels for lithium and sodium in black phosphorus. Advanced Materials. 2019;**31**(46):1904623

[39] Wang C et al. Monolayer atomic crystal molecular superlattices. Nature. 2018;**555**(7695):231-236

[40] He Q et al. In situ probing molecular intercalation in two-dimensional layered semiconductors. Nano Letters. 2019;**19**(10):6819-6826

[41] Liu B et al. Geochemistry of carboniferous coals from the Laoyaogou mine, Ningwu coalfield, Shanxi Province, northern China: Emphasis on the enrichment of valuable elements. Fuel. 2020;**279**:118414

[42] NOHSC. Available from: https://www.ioha.net/national-governmental/

[43] Kimura ET, Ebert DM, Dodge PW. Acute toxicity and limits of solvent residue for sixteen organic solvents. Toxicology and Applied Pharmacology. 1971;**19**(4):699-704

[44] Joshi DR, Adhikari N. An overview on common organic solvents and their toxicity. Journal of Pharmaceutical Research International. 2019;**28**(3):1-18

[45] Brooks PJ. DNA damage, DNA repair, and alcohol toxicity—A review. Alcoholism: Clinical and Experimental Research. 1997;**21**(6):1073-1082

[46] Rusyn I, Bataller R. Alcohol and toxicity. Journal of Hepatology. 2013;**59**(2):387-388

[47] Green AA, Hersam MC. Solution phase production of graphene with controlled thickness via density differentiation. Nano Letters. 2009;**9**(12):4031-4036

[48] Kang J et al. Solvent exfoliation of electronic-grade, two-dimensional black phosphorus. ACS Nano. 2015;**9**(4):3596-3604

[49] Nicolosi V et al. Liquid exfoliation of layered materials. Science. 2013;**340**(6139):1226419

[50] Hernandez Y et al. High-yield production of graphene by liquid-phase exfoliation of graphite. Nature Nanotechnology. 2008;**3**(9):563-568

[51] Song H et al. Mode-locked ytterbium-doped all-fiber lasers based on few-layer black phosphorus saturable absorbers. Optics Communications. 2017;**394**:157-160

[52] Coleman JN et al. Two-dimensional nanosheets produced by liquid exfoliation of layered materials. Science. 2011;**331**(6017):568-571

[53] Sun L et al. Preparation of few-layer bismuth selenide by

liquid-phase-exfoliation and its optical absorption properties. Scientific Reports. 2014;**4**(1):4794

[54] Bang GS et al. Effective liquid-phase exfoliation and sodium ion battery application of MoS2 nanosheets. ACS Applied Materials & Interfaces. 2014;**6**(10):7084-7089

[55] Zhang A et al. Recent Progress of two-dimensional materials for ultrafast photonics. Nanomaterials. 2021;**11**(7):1778

[56] Ghanbari H, Shafikhani MA, Daryalaal M. Graphene nanosheets production using liquid-phase exfoliation of pre-milled graphite in dimethylformamide and structural defects evaluation. Ceramics International. 2019;**45**(16):20051-20057

[57] Ogilvie SP et al. Considerations for spectroscopy of liquid-exfoliated 2D materials: Emerging photoluminescence of N-methyl-2-pyrrolidone. Scientific Reports. 2017;7(1):16706

[58] Haar S et al. Enhancing the liquid-phase exfoliation of graphene in organic solvents upon addition of n-octylbenzene. Scientific Reports. 2015;**5**(1):1-9

[59] Li J et al. Black phosphorus: A two-dimension saturable absorption material for mid-infrared Q-switched and mode-locked fiber lasers. Scientific Reports. 2016;**6**(1):30361

[60] Zhou KG et al. A mixed-solvent strategy for efficient exfoliation of inorganic graphene analogues. Angewandte Chemie International Edition. 2011;**50**(46):10839-10842

[61] May P et al. Role of solubility parameters in understanding the steric stabilization of exfoliated two-dimensional nanosheets by adsorbed polymers. The Journal of Physical Chemistry C. 2012;**116**(20):11393-11400

[62] Heyndrickx A. Toxicology of new synthetic organic polymers used in containers for food, pharmaceutical compounds, and drinkable water. 1974

[63] Vasudevan N et al. Effect of Ni addition on the densification of TiC: A comparative study of conventional and microwave sintering. International Journal of Refractory Metals and Hard Materials. 2020;**87**:105165

[64] Sadeghi B, Cavaliere P. Progress of flake powder metallurgy research. Metals. 2021;**11**(6):931

[65] Lin Z et al. 2D materials advances: From large scale synthesis and controlled heterostructures to improved characterization techniques, defects and applications. 2D Materials. 2016;**3**(4):042001

[66] Wang G et al. Direct laser irradiation and modification of 2D Te for development of volatile memristor. Materials. 2023;**16**(2):738

[67] Shen P-C et al. CVD technology for 2-D materials. IEEE Transactions on Electron Devices. 2018;**65**(10):4040-4052

[68] Novoselov KS et al. 2D materials and van der Waals heterostructures. Science. 2016;**353**(6298):aac9439

[69] Yu Q et al. Graphene segregated on Ni surfaces and transferred to insulators. Applied Physics Letters. 2008;**93**(11)

[70] Li X et al. Large-area synthesis of high-quality and uniform graphene films on copper foils. Science. 2009;**324**(5932):1312-1314

[71] Nguyen VL et al. Seamless stitching of graphene domains on polished

copper (111) foil. Advanced Materials. 2015;**27**(8):1376-1382

[72] Chen H et al. Transition-metal dichalcogenides heterostructure saturable absorbers for ultrafast photonics. Optics Letters. 2017;**42**(21):4279-4282

[73] Blessy Rebecca P et al. Biomass-derived graphene-based nanocomposites: A futuristic material for biomedical applications. ChemistrySelect. 2022;7(5):e202104013

[74] Jo JY et al. Acute respiratory distress due to methane inhalation. Tuberculosis and Respiratory Diseases. 2013;**74**(3):120-123

[75] Duncan IJ. Does methane pose significant health and public safety hazards?—A review. AAPG/DEG Environmental Geosciences. 2015;**22**(3):85-96

[76] CCOHS. Available from: https://www.ccohs.ca/oshanswers/chemicals/chem_profiles/methane.html#section-4-hdr

[77] Lee Y-H et al. Synthesis of large-area MoS2 atomic layers with chemical vapor deposition. Advanced Materials. 2012;**24**(17):2320-2325

[78] Li H et al. Epitaxial growth of two-dimensional layered transition-metal dichalcogenides: Growth mechanism, controllability, and scalability. Chemical Reviews. 2017;**118**(13):6134-6150

[79] Zheng B, Chen Y. Controllable growth of monolayer MoS2 and MoSe2 crystals using three-temperature-zone furnace. IOP Conference Series: Materials Science and Engineering. 2017;**274**(1):012085

[80] Ruiz AR et al. . In: Graphene I, Mujtaba MA, Aneeqa B, editors. Chemical Vapor Deposition Synthesis of Graphene on Copper Foils. London, UK, Rijeka: IntechOpen; 2022

[81] Kang K et al. Synthesis of transition metal dichalcogenides (TMDs). In: Ünlü H, Horing NJM, editors. Progress in Nanoscale and Low-Dimensional Materials and Devices: Properties, Synthesis, Characterization, Modelling and Applications. Cham: Springer International Publishing; 2022. pp. 155-179

[82] Franke KW, Moxon AL. The toxicity of orally ingested arsenic, selenium, tellurium, vanadium and molybdenum. Journal of Pharmacology and Experimental Therapeutics. 1937;**61**(1):89-102

[83] Larner AJ. Alzheimer's disease, kuf's disease, tellurium and selenium. Medical Hypotheses. 1996;**47**(2):73-75

[84] Saha U et al. Selenium in animal nutrition: Deficiencies in soils and forages, requirements, supplementation and toxicity. International Journal of Applied Agricultural Science. 2016;**2**(6):112-125

[85] Nurunnabi M, McCarthy J. Biomedical Applications of Graphene and 2D Nanomaterials. In: Nurunnabi Md, McCarthy J R, editors. Micro and Nano Technologies. Elsevier; 2019:1-383. DOI: 10.1016/C2017-0-02937-3

[86] Larner A. Biological effects of tellurium: A review. Trace Elements and Electrolytes. 1995;**12**(1):26-31

[87] Cerwenka EA Jr, Cooper WC. Toxicology of selenium and tellurium and their compounds. Archives of Environmental Health: An International Journal. 1961;**3**(2):189-200

[88] Nath NCD et al. In vitro Toxicity of 2D Materials. In: Nurunnabi Md,

McCarthy J R, editors. Micro and Nano Technologies. Elsevier; 2019:165-186. DOI: 10.1016/C2017-0-02937-3

[89] Duta L, Mihailescu IN. Advances and challenges in pulsed laser deposition for complex material applications. MDPI. 2023;**2023**:393

[90] Gautam S et al. Structural, electronic and thermoelectric properties of Bi2Se3 thin films deposited by RF magnetron sputtering. Journal of Electronic Materials. 2022;**51**(5):2500-2509

[91] Somdock N et al. Simultaneous stoichiometric composition and highly (00l) orientation of flexible Bi2Te3thin films via optimising the DC magnetron sputter-deposition process. Journal of Alloys and Compounds. 2019;**773**:78-85

[92] Akinwande D et al. Graphene and two-dimensional materials for silicon technology. Nature. 2019;**573**(7775):507-518

[93] Zhang P et al. Two-dimensional materials for miniaturized energy storage devices: From individual devices to smart integrated systems. Chemical Society Reviews. 2018;**47**(19):7426-7451

[94] Fiori G et al. Electronics based on two-dimensional materials. Nature Nanotechnology. 2014;**9**(10):768-779

[95] EUON. Assessment of the potential impact of graphene, graphene oxide and other 2D materials on health, and the environment. 2022. Available from: https://euon.echa.europa.eu/documents/2435000/3268573/echa_2021_286_graphene_study.pdf/16cc871b-ca06-caf4-fdb0-3c14113b3e2a?t=1670235623023

[96] Zhang Y et al. Cytotoxicity effects of graphene and single-wall carbon nanotubes in neural phaeochromocytoma-derived PC12 cells. ACS Nano. 2010;**4**(6):3181-3186

[97] Ray PD, Huang B-W, Tsuji Y. Reactive oxygen species (ROS) homeostasis and redox regulation in cellular signaling. Cellular Signalling. 2012;**24**(5):981-990

[98] Sasidharan A et al. Cellular and molecular mechanistic insight into the DNA-damaging potential of few-layer graphene in human primary endothelial cells. Nanomedicine: Nanotechnology, Biology and Medicine. 2016;**12**(5):1347-1355

[99] Deakin S, Markesinis B. Markesinis and Deakin's Tort Law. USA: Oxford University Press; 2019

[100] Wu N et al. Investigating the influence of MoS2 nanosheets on E. coli from metabolomics level. PLoS One. 2016;**11**(12):e0167245

[101] Hajipour MJ et al. Personalized disease-specific protein corona influences the therapeutic impact of graphene oxide. Nanoscale. 2015;**7**(19):8978-8994

[102] Li Y et al. Response of microRNAs to in vitro treatment with graphene oxide. ACS Nano. 2014;**8**(3):2100-2110

[103] Chen Y et al. Specific nanotoxicity of graphene oxide during zebrafish embryogenesis. Nanotoxicology. 2016;**10**(1):42-52

[104] Zhang P et al. Bioaccumulation and effects of sediment-associated gold- and graphene oxide nanoparticles on Tubifex tubifex. Journal of Environmental Sciences. 2017;**51**:138-145

[105] Guo Q et al. Graphene oxide toxicity in W1118 flies. Science of the Total Environment. 2022;**805**:150302

[106] Zhang M et al. Graphene oxide induces plasma membrane damage,

reactive oxygen species accumulation and fatty acid profiles change in Pichia pastoris. Ecotoxicology and Environmental Safety. 2016;**132**:372-378

[107] Hu L et al. Evaluating the cytotoxicity of graphene oxide using embryonic stem cells-derived cells. Journal of Biomedical Materials Research Part A. 2020;**108**(6):1321-1328

[108] Li J et al. Lateral size of graphene oxide determines differential cellular uptake and cell death pathways in Kupffer cells, LSECs, and hepatocytes. Nano Today. 2021;**37**:101061

[109] Lin H et al. Comparative effects of graphene and molybdenum disulfide on human macrophage toxicity. Small. 2020;**16**(35):2002194

[110] Wang X et al. Differences in the toxicological potential of 2D versus aggregated molybdenum disulfide in the lung. Small. 2015;**11**(38):5079-5087

[111] Li K et al. Surface atomic arrangement of nanomaterials affects nanotoxicity. Nanotoxicology. 2021;**15**(1):114-130

[112] Scalisi EM et al. Toxicity assessment of two-dimensional nanomaterials molybdenum disulfide in Gallus gallus domesticus. Ecotoxicology and Environmental Safety. 2020;**200**:110772

[113] Wu B et al. Combined effects of graphene oxide and zinc oxide nanoparticle on human A549 cells: Bioavailability, toxicity and mechanisms. Environmental Science: Nano. 2019;**6**(2):635-645

[114] Yuan P, Zhou Q, Hu X. The phases of WS2 nanosheets influence uptake, oxidative stress, lipid peroxidation, membrane damage, and metabolism in algae. Environmental Science & Technology. 2018;**52**(22):13543-13552

[115] Teo WZ et al. Cytotoxicity of exfoliated transition-metal dichalcogenides (MoS2, WS2, and WSe2) is lower than that of graphene and its analogues. Chemistry–A European Journal. 2014;**20**(31):9627-9632

[116] Kar F et al. Hexagonal boron nitride nanoparticles trigger oxidative stress by modulating thiol/disulfide homeostasis. Human & Experimental Toxicology. 2021;**40**(9):1572-1583

[117] Begum P, Ikhtiari R, Fugetsu B. Graphene phytotoxicity in the seedling stage of cabbage, tomato, red spinach, and lettuce. Carbon. 2011;**49**(12):3907-3919

[118] Liu S et al. Effects of graphene on germination and seedling morphology in rice. Journal of Nanoscience and Nanotechnology. 2015;**15**(4):2695-2701

[119] Ge Y et al. Long-term effects of multiwalled carbon nanotubes and graphene on microbial communities in dry soil. Environmental Science & Technology. 2016;**50**(7):3965-3974

[120] Anjum NA et al. Single-bilayer graphene oxide sheet impacts and underlying potential mechanism assessment in germinating faba bean (*Vicia faba* L.). Science of the Total Environment. 2014;**472**:834-841

[121] Chen L et al. Bioaccumulation and toxicity of 13C-skeleton labeled graphene oxide in wheat. Environmental Science & Technology. 2017;**51**(17):10146-10153

[122] Zeng H et al. Nanocolloids, but not humic acids, augment the phytotoxicity of single-layer molybdenum disulfide nanosheets. Environmental Science & Technology. 2021;**55**(2):1122-1133

[123] Zhao L et al. Environmental implications of MoS2 nanosheets on

rice and associated soil microbial communities. Chemosphere. 2022;**291**:133004

[124] Zhang H et al. Recent advances of two-dimensional materials in smart drug delivery nano-systems. Bioactive Materials. 2020;**5**(4):1071-1086

[125] Wang Y et al. Emerging 2D material-based nanocarrier for cancer therapy beyond graphene. Coordination Chemistry Reviews. 2019;**400**:213041

[126] Xie X, Goodell B. Thermal degradation and conversion of plant biomass into high value carbon products. In: Deterioration and Protection of Sustainable Biomaterials. American Chemical Society; 2014;**8**:147-158. DOI: 10.1021/bk-2014-1158.ch008

[127] Jeevanandam J, Chan YS, Danquah MK. Biosynthesis of metal and metal oxide nanoparticles. ChemBioEng Reviews. 2016;**3**(2):55-67

[128] Singh J et al. 'Green'synthesis of metals and their oxide nanoparticles: Applications for environmental remediation. Journal of Nanobiotechnology. 2018;**16**(1):1-24

[129] Widiatmoko P et al. Increasing yield of graphene synthesis from oil palm empty fruit bunch via two-stages pyrolysis. In: IOP Conference Series: Materials Science and Engineering; 4-6 October 2018; Indonesia: IOP; 2019;**543**:012032

[130] Akhavan O, Bijanzad K, Mirsepah A. Synthesis of graphene from natural and industrial carbonaceous wastes. RSC Advances. 2014;**4**(39):20441-20448

[131] Kalita G, Wakita K, Umeno M. Monolayer graphene from a green solid precursor. Physica E: Low-dimensional Systems and Nanostructures. 2011;**43**(8):1490-1493

[132] Safian MT, Haron US, Ibrahim MM. A review on bio-based graphene derived from biomass wastes. BioResources. 2020;**15**(4):9756

[133] Saha JK, Dutta A. A review of graphene: Material synthesis from biomass sources. Waste and Biomass Valorization. 2022;**13**(3):1385-1429

[134] Peng ZA, Peng X. Formation of high-quality CdTe, CdSe, and CdS nanocrystals using CdO as precursor. Journal of the American Chemical Society. 2001;**123**(1):183-184

[135] Unni C et al. Aqueous synthesis and characterization of CdS, CdS: Zn2+ and CdS: Cu2+ quantum dots. Spectrochimica Acta Part A: Molecular and Biomolecular Spectroscopy. 2009;**72**(4):827-832

[136] Xia Y, Zhu C. Aqueous synthesis of type-II core/shell CdTe/CdSe quantum dots for near-infrared fluorescent sensing of copper (II). Analyst. 2008;**133**(7):928-932

Chapter 4

Potential Toxicity of Nanoparticles for the Oral Delivery of Therapeutics

Iman M. Alfagih

Abstract

Nanoparticles (NPs) offer a promising solution for orally delivering therapeutic substances due to their capability to surpass traditional drug delivery system (DDS) limitations like low solubility, bioavailability, and stability. However, the possible toxic effects of using NPs for oral therapeutic delivery raise significant concerns, as they might interact with biological systems unexpectedly. This chapter aims to comprehensively understand the potential toxicity of NPs employed in oral therapeutic delivery. Factors such as size, surface area, surface charge, and surface chemistry of NPs can impact their toxicity levels. Both in vitro and in vivo models have been utilised to evaluate NPs toxicity, with in vivo models being more suitable for anticipating human toxicity. The possible toxic consequences of different NPs varieties, including polymer, lipid, and metal NPs, have been documented. Ultimately, grasping the potential toxicity of NPs in oral therapeutic delivery is essential for creating safe and effective DDS.

Keywords: nanoparticles, oral drug delivery, cytotoxicity, inflammation, intestinal microbiota

1. Introduction

The use of nanoparticles (NPs) for oral drug delivery is a ground-breaking and rapidly growing area of research [1]. NPs are solid colloidal drug delivery systems (DDS) with sizes ranging from 1 nm to 1000 nm. They comprise polymers, lipids, carbon, silica, or metal encapsulating the drug moiety [2–5]. Encapsulating drugs within NPs can increase stability, solubility, and bioavailability, ultimately boosting their therapeutic effectiveness [6]. Recent advances in biomedical research have led to the successful enhancement of therapeutic agents for treating various diseases. However, a significant challenge remains in efficiently delivering these agents to the target site [7]. NPs can be engineered with specific characteristics to optimise the DDS [4]. For example, biocompatible polymer NPs are coated to enhance their presence in the bloodstream or be equipped with unique ligands or antibodies that focus on specific target cells or tissues [2]. By directing drug delivery to specific locations within the body, NPs can reduce negative side effects and increase the effectiveness of treatments [8].

Oral drug administration is the most prevalent method and the favoured option for patients, as it is non-invasive, user-friendly, and well-accepted [9]. However, traditional forms like tablets and capsules may have a rapid and inadequately regulated drug release, potentially leading to drug degradation and alteration due to the gastrointestinal tract (GIT) environment (like, changes in pH, digestive enzymes, and microbiota) [10]. NPs can address these issues by safeguarding drugs from degradation and promoting their absorption through the intestinal barrier. Moreover, oral drugs can be directed to specific areas within the GIT to provide localised treatment for stomach and colorectal cancers, infections, inflammation, bowel disorders, gastro-duodenal ulcers, and gastroesophageal reflux issues [11]. GIT fluid consists of enzymes, acids, and other compounds that quickly decompose drugs, resulting in inadequate absorption and decreased effectiveness. Furthermore, a mucus layer in the GI tract can hinder drug absorption by acting as a barrier, restricting their access to the intestinal epithelium [12, 13]. Thus, using NPs as an oral DDS holds great promise due to their unique properties (which include good therapeutic properties, self-assembly, enhanced biocompatibility, serving as vehicles for antimicrobial agents, controlled drug release, reducing off-target effects, etc.) [2, 4, 14].

Although NPs-based DDS offer promising benefits, their safety must be thoroughly evaluated. The small size of NPs can result in interactions with biological systems that are not yet fully understood, possibly causing toxic consequences. Furthermore, NPs can accumulate in specific tissues or organs, leading to long-lasting damage. A recent study by Cabellos et al. [15] showed that mice exposed to silica NPs suffered inflammation and epithelial damage in the intestines and showed the expression of autophagy proteins [4, 16]. Various mechanisms can contribute to the potential toxicity of NPs. For example, they can cause oxidative stress by generating reactive oxygen species (ROS) that harm cell components [17]. NPs can also trigger inflammatory reactions by activating immune cells, causing tissue injury and malfunction [18]. Moreover, certain NPs might exhibit genotoxic effects by damaging DNA, potentially resulting in mutations and cancer [19]. However, it is crucial to meticulously assess the possible toxicity of NPs to ensure their safety and effectiveness [20, 21]. By refining the development and analysis of NPs, one can reduce toxicity and increase therapeutic advantages by modifying functional groups, coating with cell membranes, or using nanorobots [22]. For example, DDS can be developed using stable intestinal cell-derived exosomes that target colonic cells. It is even known that these exosomes can be found in faces. Lipid NPs derived from plants can also be used as an alternative strategy, thus providing a safe NPs delivery [23, 24]. Thereby, conducting comprehensive pre-clinical and clinical research is vital to evaluate the safety of NPs in the context of orally administered treatments. Such investigations should examine the biodistribution, pharmacokinetics, and toxicity of NPs in both animal models and human subjects [21, 25]. Specifically, close observation of NPs accumulation in different organs and tissues is necessary, along with monitoring any negative impacts on cellular and molecular functions [26]. Therefore, the present chapter delves into the diverse toxicological issues of orally administered NPs (from polymer, lipid, protein, carbon, silica, and metal-based NPs) and summarises the in vitro and *in vivo* assessments of the toxicity of oral NPs.

2. Structure of GIT and the interaction of nanoparticles

The GIT is a complex system responsible for nutrient absorption, immune function, and waste elimination, divided into several regions, including the mouth, oesophagus, stomach, small intestine, and large intestine. Each region of GIT presents unique anatomical characteristics and physiological conditions that can influence the interaction of NPs. The GIT represents a selective mucosal barrier, with an estimated surface area of 200 m^2 in adult humans, that can potentially interact with ingested NPs [27]. The anatomical and physiological characteristics of each part of the GIT can affect NPs absorption and elimination [28]. For example, the stomach presents a tough barrier to drug absorption, with a strong acid environment (pH range of 1.0–2.5) that can degrade food, acid-labile drugs, and pathogenic microorganisms. Furthermore, the stomach has extrinsic epithelial cells and a mucin-bicarbonate barrier, which, combined with tight junctions beneath the intrinsic barrier, limits drug absorption. Furthermore, stomach pepsins can inactivate protein drugs [13]. The small intestine has a huge surface area due to the villi and microvilli in the intestinal lumen [29]. The small intestine is considered a major site for oral drug delivery due to its enormous surface and various transport routes [13]. The intestinal mucosa can recognise and transfer ectogenic antigens to the immune system. However, some challenges to drug delivery in the small intestine still arise from its unique physiology [30]. DDS that can increase their retention time in villi and microvilli, improve lipid solubility, and interact with a specific receptor or carrier can increase their overall bioavailability [13, 31]. Therefore, the unique anatomical and physiological features of each part of the GIT play a crucial role in the interaction and potential toxicity of NPs in the GIT.

The interaction of NPs with GIT depends on their physicochemical properties, including size, shape, surface charge, and surface chemistry (**Table 1**) [58]. Following oral administration, NPs encounter the mucous layer covering the GIT, which

Nanoparticles composition	Toxic effects	In vitro models	In vivo models	Reference
Polymers such as PLGA, PEG, chitosan, and others	Cytotoxicity, genotoxicity, oxidative stress, inflammation	Cell lines such as HEK293, A549, and others	Mice, rats, and Zebrafish	[32–35]
Phospholipids, cholesterol, and other lipids	Cytotoxicity, genotoxicity, oxidative stress, inflammation	Cell lines such as HEK293, A549, and others	Mice and rats	[36–39]
Silicon dioxide or silicon-based compounds	Cytotoxicity, genotoxicity, oxidative stress, inflammation	Cell lines such as HepG2, BEAS-2B, and others	Mice and rats; *Caenorhabditis elegans*	[40–45]
Gold	Cytotoxicity, genotoxicity, oxidative stress, inflammation	Cell lines such as MCF-7, HEK293, and others	Mice and rats	[46–51]
Silver	Cytotoxicity, genotoxicity, oxidative stress, inflammation	Cell lines such as A549, RAW264.7, and others	Mice and rats	[52–55]
Iron oxide	Cytotoxicity, genotoxicity, oxidative stress, inflammation	Cell lines such as RAW264.7, MCF-7, and others	Mice and rats	[56, 57]

Table 1.
Some studies on the toxic effects of various nanoparticles on different models.

protects against harmful agents, such as pathogens and toxins [59]. However, NPs can interact with the mucous layer and penetrate the underlying tissue, resulting in potential toxicity [60].

2.1 Influence of NPs size on GIT toxicity

Smaller NPs possess a higher surface area-to-volume ratio, promoting better engagement with the mucous layer and increased absorption into the tissue [61]. Conversely, larger NPs may interact less with the mucous layer and could be eliminated from the GIT before reaching the tissue. Zhao et al. [62] noted that a 100 nm nano-vaccine showed superior pharmacokinetic effectiveness compared to a 500 nm nano-vaccine when an alum adjuvant was administered. The toxicities of NPs are inversely proportional to their size, and they are generally more toxic than large particles of the same chemical substance [63].

2.2 Influence of NPs shape and aspect ratio on GIT toxicity

It was reported that the shape of NPs can impact their association with the GIT as well. NPs with round shapes have been discovered to exhibit a higher level of interaction with the mucous layer compared to alternative forms, such as cylindrical or cubic shapes [61]. This occurs because round NPs possess a larger surface area-to-volume ratio, which can increase their connection with the mucous layer [4]. In addition to size, the configuration and proportion of NPs are essential factors in determining their cytotoxic effects *in vivo* [64]. Vedhanayagam et al. [65] investigated the impact of various zinc oxide NPs, such as spheres, needle, rod, hexagonal, star, flower, doughnut, circular discs, and cube, on the healing process of wounds. The researchers found that the spherical structure of zinc dioxide (ZnO) within a cross-linked collagen framework leads to improved re-epithelization and more rapid collagen accumulation. A higher aspect ratio of NPs is believed to be linked to increased cytotoxicity due to decreased clearance and enhanced bioavailability of these particles [66]. NPs with higher aspect ratios often exhibit cytotoxicity patterns that resemble those of asbestos. These particles can cause macrophage cell death during phagocytosis and, similar to asbestos fibres, can contribute to cancer formation [67].

2.3 Influence of NPs charge on GIT toxicity

NPs with positive charges have been shown to have a stronger connection with the negatively charged mucous layer compared to those with a negative charge or no charge at all. Researchers have explored using charged polymers to temporarily open tight junctions and enhance drug delivery across the intestinal epithelial barrier. For example, chitosan (a cationic polymer) has been shown to promote the paracellular transport of NPs [68]. NPs' bioavailability and absorption have been studied only a few times after oral administration with regard to surface charge, hydrophobicity, and shape. NPs with a positive charge demonstrated increased absorption and movement by enterocytes compared to those with a negative or neutral charge, as well as a notable increase in oral bioavailability *in vivo* [69]. Furthermore, rod-shaped gold NPs (AuNPs) and DNA struggle to permeate or enter cells because of their charge. To enhance uptake, both AuNPs and DNA have undergone surface modification by adding lipid layers, while DNA has also been electrostatically attached to cationic liposomes, facilitating transport into cells [21].

2.4 Influence of NPs surface chemistry on GIT toxicity

Numerous NPs undergo modifications to alter their surface chemistry for specific objectives [70]. For instance, applying a polyethylene glycol (PEG) coating on NPs surfaces to establish water-attracting surface chemistry lessened the intense interaction with mucus components and increased particle movement across the mucus and mucosa [69]. Furthermore, Hasse and colleagues reported that peptide-coated silver NPs (20 nm) were more cytotoxic than citrate-coated silver NPs of the same size. They evaluated cytotoxicity in a human leukaemia cell line using the WST-1 assay [71]. It was perceived that the toxic effects of AgHECp (Silver hexagonal close placed), Ag pristine, and AgPVP (silver polyvinylpyrrolidone) were examined in A431 cell lines (human epidermoid carcinoma) and HaCaT (human keratinocytes) cell lines [71]. It has been revealed that smaller particles with negative surface charge and increased ionic content, such as AgPVP, were highly toxic. Similarly, toxic effects were observed for the larger particles, which were pure AgNP with a negative surface charge and ion content comparable to AgPVP [71]. Additionally, NPs featuring a hydrophobic surface exhibit a more robust interaction with the mucous layer than their hydrophilic counterparts. A cationic polysaccharide with mucoadhesive properties, chitosan has been extensively investigated. Chitosan's mucoadhesive potency is limited due to its low water solubility and low mucoadhesive strength, so it must be chemically modified to optimise its mucoadhesive properties [72].

3. Toxicological concern with oral nanoparticles

3.1 Polymer-based nanoparticles

Polymer-based NPs (PNPs) can interact with GIT cells and tissues, resulting in inflammation, oxidative stress, and damage to the intestinal barrier. Various poly(lactide-co-glycolic acid) NPs (PLGA) were examined for toxicity in THP-1 macrophages similar to human cells. When used as an NPs stabiliser, chitosan polymer conferred significant cytotoxicity to PLGA NPs, despite being slightly cytotoxic [73]. Research has shown that NPs exhibit varying levels of impact on intestinal cells, intestinal flora, and the intestinal barrier, confirming that NPs cause damage to the digestive system [45, 74]. Polystyrene NPs administration has been found to trigger the TOS-MAPK/NF-kB signalling pathway in macrophage RAW 264.7, leading to inflammation with pro-inflammatory and cytotoxic potential activity [35]. Furthermore, exposure to polystyrene NPs have been shown to disrupt the balance of cell populations within the intestinal cells of zebrafish [75]. Existing data on the toxic effects of ingested zinc oxide (ZnO) NPs on intestinal models need to be more consistent. Several studies have found negative biological effects, such as increased intestinal inflammation, reduced cell viability, and mitochondrial membrane depolarization, resulting from the treatment with ZnO NPs. However, modified ZnO NPs do not cause significant cell damage [76, 77].

3.2 Lipid-based nanoparticles

Lipid-based nanoparticles (LNPs) have gained attention as potential DDS due to their ability to encapsulate hydrophobic medications and shield them from degradation [78]. Cationic lipids, for instance, show significant potential as carriers for delivering delicate substances such as nucleic acids, but certain cationic lipids can lead to cytotoxicity [36]. The impact of hydrophobic chains on lipid toxicity has yet to be thoroughly investigated, impeding the development of less harmful lipids [79]. Solid lipid nanoparticles (SLNs) are among the most prevalent types of LNPs employed in drug delivery. SLNs have been discovered to cause toxicity in the GIT, resulting in inflammation, oxidative stress, and damage to the intestinal barrier [80]. Numerous studies have indicated that the interaction between LNPs and GIT can trigger an immune response, modify the gut microbiota, and induce toxicity [79, 81]. A recent study discovered that oral administration of altered RNA-LNPs led to the expression of pro-inflammatory cytokines (such as interleukin (IL) -6 and macrophage inflammation protein 2 (MIP-2)) and chemokines in the intestines of unexposed mice [39]. Ball et al. [82] documented that LNPs can compromise the stability of the intestinal epithelial barrier and result in intestinal inflammation.

3.3 Silica-based nanoparticles

Silicon-based nanoparticles (SiNPs) can interact with GIT cells and tissues, causing inflammation, oxidative stress, and damage to the intestinal barrier [83]. Various factors influence the toxicity of SiNPs in GIT, including their dimensions, surface area, surface charge, and chemical makeup [84]. Cellular absorption of silica dioxide NPs (SiO2NPs) depends on the size of the particles, particularly in the range of 30-50 nm. SiO_2 NPs within the 30-50 nm size range can be carriers in multiple applications [40, 41]. An *in vivo* study revealed that ingesting SiO_2NPs led to GIT inflammation and increased permeability, causing intestinal contents to leak into the bloodstream and disrupting the microbiota-gut-brain axis [85]. Guo et al. [83] discovered that exposure to SiO_2NPs affected nutrient transportation, ROS production, barrier function, gene expression, and microvilli structure. To assess the potential toxicity of SiNPs, examining their interactions with gut cells, uptake, and impact on GIT function and microbiota is important. The nanoscale SiO_2 found in the E551 food additive could uniquely impact the absorption and distribution of SiO_2 within the human body [86]. Numerous in vitro studies have shown that SiO_2NPs can produce cytotoxic effects in cultured human cell lines, such as glioblastoma cells, depending on their size, shape, and dose [87]. Furthermore, SiO_2NPs, inhaled or ingested, can infiltrate cells and engage with cellular membranes or organelles, leading to mammalian cell death through oxidative stress, endoplasmic reticulum stress, and apoptosis [42, 43]. A different instance of SiNPs that exhibit toxicity in the GIT involves mesoporous silica NPs (M-SiNPs) [44]. M-SiNPs have been identified to induce toxicity in the GIT, with inflammation, oxidative stress, and harm to the intestinal barrier. SiNPs administration has been found to cause intestinal inflammation by interfering with the hydrolysis and metabolism of nutrient peptides in *Caenorhabditis elegans* (an invertebrate nematode) [45]. Furthermore, Ogawa et al. [88] revealed that administering 10-nm SiNPs provoked intestinal inflammation by activating an apoptosis-associated speck-like protein with a CARD (caspase activation and recruitment domain) inflammasome.

3.4 Metallic-based nanoparticles

Numerous scientific investigations have been conducted on the therapeutic applications of gold nanoparticles (AuNPs), silver nanoparticles (AgNPs) and superparamagnetic iron oxide nanoparticles (SPIONPs) [55, 89]. They possess several characteristics that make them attractive as DDS. Specifically, they can be easily synthesised in various sizes and shapes, their surfaces can be functionalized with different elements such as polymers, peptides, targeting ligands, imaging probes, and more, and they are generally considered safe for certain *in vivo* biological applications [55].

AgNPs exhibit distinct properties related to toxicity, surface plasmon resonance, and electrical resistance [90]. In a study by Vandebriel et al. [91], repeated exposure to AgNPs resulted in cytotoxic effects on various rat cells. Intending to create new anticancer treatments, Azizi et al. [92] developed albumin-coated AgNPs, which were found to be specifically taken up by cancerous cells and trigger apoptosis. Furthermore, through numerous pathways, AgNPs are highly effective against gram-positive and gram-negative bacteria. They help address the problem of drug resistance often seen with traditional antibiotics due to their unique mode of action [54]. Clear evidence of systemic toxicity from AgNPs ingested orally or through intravenous routes has been demonstrated, and the toxic effects are related to the amount of Ag^+ released during exposure [93]. Moreover, the significant surface area of AgNPs that release Ag^+ ions is a critical aspect that contributes to the cytotoxic behaviour. As widely recognised, smaller AgNPs exhibit a faster pace of silver ion (Ag^+) dissolution in the nearby microenvironment, owing to their greater surface area-to-volume ratio. This results in increased bioavailability, improved distribution, and increased toxicity compared to larger AgNPs [40, 94]. Research examining the anti-inflammatory properties of AgNPs has shown a considerable decrease in wound inflammation, adjustment of fibrogenic cytokines, a reduction of pro-inflammatory cytokines, and cell death in inflammatory cells [52, 53]. NPs with diameters below 100 nm have been documented to be mainly taken up by endocytosis in epithelial cells. Within these cells, AgNPs can induce oxidative stress, DNA damage, and inflammation [95]. Jeong et al. [96] observed an increase in goblet cells in the intestines and a significant release of mucus granules in mice treated with oral AgNPs (60 nm) at a 30 mg/kg body weight per day for 28 days. Furthermore, AgNPs administered orally (5-20 nm) for 21 days in mice (20 mg/kg body weight) disrupted the microvilli of epithelial cells and affected the intestinal glands [93].

Various *in vivo* studies have been conducted to assess the possible toxic effects of AuNPs, but the findings still need to be definitive [51]. Factors such as size, shape, surface properties, stabilising coatings, and administration aspects (dosage, duration, and method of administration) can lead to the varying toxic effects of AuNP *in vivo* [46]. Research has shown that smaller AuNPs (5–15 nm) exhibit a more extensive organ distribution in rodents compared to larger AuNPs (50–100 nm), suggesting a higher risk of toxicity *in vivo* for smaller AuNPs [47]. Research conducted by Goodman and colleagues [97] indicated that cationic AuNPs interact with the negatively charged cellular membrane, causing damage to the intestinal membrane. Furthermore, citrate-capped AuNP (13 nm in diameter) were harmful to human lung carcinoma cells while not affecting human liver carcinoma cells at the same dose [98]. As additional information is collected, it has been recommended that proper consideration be given to surface chemistry and dosages of Au and AgNPs to utilise them in biomedical applications efficiently [49, 50].

SPIONPs have become increasingly popular in numerous biomedical applications, such as magnetic resonance imaging, targeted drug or gene delivery, and hyperthermia [57]. However, SPIONPs can potentially cause cytotoxicity, negatively impacting vital cellular components such as mitochondria, the nucleus, and DNA [99]. Research examining the influence of various surface coatings on cell behaviour and structure revealed that dextran-magnetite (Fe_3O_4) NPs lead to cell death and decreased proliferation, similar to the effects of uncoated iron oxide particles [56]. Contact with SPIONPs has been linked to considerable harmful consequences, including inflammation, developing apoptotic structures, compromised mitochondrial functioning (MTT), membrane leakage, ROS production, extensive chromosomal aberration, and condensation [57].

4. In vitro tests to evaluate the toxicity of nanoparticles

The in vitro evaluation of the toxicity of NPs serves as a standard approach for identifying acute hazards associated with potentially harmful NPs during this screening process. However, these methods can only partially replace *in vivo* evaluations. At the same time, in vitro analyses report the immediate harmful effects of NPs in specific cellular settings, *in vivo* animal models monitor biodistribution and bioaccumulation pathways, which are not accessible through in vitro observations. A comprehensive understanding of NPs toxicity and potential risk requires using both methods [100]. In vitro, NPs toxicity assessment is a critical technique that offers benefits such as reduced cost, quicker results, and minimal ethical issues [101].

4.1 Size and surface charge evaluation

Various analytical approaches exist to examine the toxicological properties of NPs, with two key techniques frequently employed to provide crucial quantitative data: dynamic light scattering (DLS) and zeta potential (ZP) analysis. The ZP and, subsequently, the surface charge of a particle play an important role in inferring potential toxicity, promoting combined DLS-ZP systems as vital tools for preliminary biocompatibility assessments. Other in vitro tests utilised to determine the size and surface charge of NPs encompass scanning electron microscopy (SEM), transmission electron microscopy (TEM), atomic force spectroscopy (AFM), Fourier transform infrared spectroscopy (FTIR), X-ray diffraction (XRD), surface-enhanced Raman spectroscopy (SERS) and solid-state nuclear magnetic resonance spectroscopy (SSNMR) for compositional analysis, as well as fluorometry for photonic characteristics [102].

4.2 Proliferation assay

Proliferation analysis is conducted by evaluating cell metabolic function to determine cell metabolism using 3-(4-dimethylthiazol-2-yl)-2,5-diphenyltetrazolium bromide (MTT), the most commonly utilised tetrazolium bromide for evaluating NPs toxicity [103]. This test uses tetrazolium salt to gauge cellular metabolism. The intricate chemical preparation of the MTT assay has led to its substitution for the Alamar blue assay, which is simpler to prepare and measures the cellular redox potential [104]. However, its success has been hindered due to an unclear operating mechanism. Other evaluations include the [3H] thymidine incorporation method to assess cell proliferation, which is typically avoided due to its toxicity [105]. The cologenic assay is another method that counts proliferating cells by visually inspecting them after exposure to NPs [106].

4.3 Apoptosis assay

One primary indicator of nanoparticles (NPs)-induced cell toxicity is apoptosis. This process, triggered by NPs, can lead to cellular renewal and accelerated ageing in mammalian hosts [100]. Commonly used markers for apoptosis include phosphatidylserine (PS), which moves to the cell membrane's outer layer, and caspase activation. Research has shown that silver NP can induce apoptosis in cells such as mouse embryonic stem cells [107] and stimulate the activation of Caspase 3 and 9 when applied to Drosophila melanogaster larvae [108]. Once host signalling pathways are initiated, active caspases provide various assays, such as cleavable substrates with fluorogenic and chromogenic labels, immunoblotting, immunofluorescence, and affinity assays with connected reporter units [109].

Numerous in vitro tests evaluate NPs toxicity, such as the Annexin V assay, which binds to PS and is a barrier against coagulation cascades. There are also various other techniques for examining apoptosis in cells or tissues exposed to NPs, including the comet assay, the TUNEL (terminal deoxynucleotidyl transferase (TdT) Nick-End labelling) staining technique, and analysing morphological alterations [110–112]. A decrease in cell size and DNA fragmentation are signs of apoptosis in NPs-treated cells. DNA gel electrophoresis is the most straightforward test for identifying cellular abnormalities. DNA fragmentation with uneven DNA sizes in agarose gel signifies necrosis-mediated cell death, while ladder-like electrophoretic DNA patterns indicate apoptosis-induced cell death in NPs-treated cells [113, 114]. Such changes were observed in human HepG2 hepatoma cells exposed to silica NPs [115]. Furthermore, the single-cell gel electrophoresis assay (SCGE), also known as the comet assay, is used to detect the mutagenic potential of NPs-treated cells by identifying DNA breaks in *Drosophila tertiolecta* exposed to SiO and TiO NPs [70, 116].

4.4 In vitro assessment of oxidative stress

As a result of the high surface area-to-volume ratio imposed by NPs, enhanced reactivity promotes intracellular damage due to oxidative stress. Detection of oxidative events through ROS measurement, protein carbonyl content, genotoxicity, and inflammatory markers reveal the potential of NPs to produce harmful toxicants in the host [100]. Beyond the natural levels, ROS levels that are produced in the cytoplasm can be detected using ROS-sensitive dyes such as nonionic, non-polar, membrane permeable fluorophore 2′,7′-dichlorofluorescein diacetate (DCFH-DA) [117]. After cellular entry of DCFH-DA, the fluorophore is hydrolysed enzymatically by cytosolic esterase into non-fluorescent polar analogue dichlorofluorescein (DCFH). It is oxidised by cellular ROS into highly fluorescent dichlorofluorescein (DCF) that can be monitored [118]. The use of other agents, such as 2,2,6,6-tetramethylpiperidine (TEMP), detect ROS by reacting with the stable O_2 radical, which can be detected using X-band electron paramagnetic resonance (EPR) [119]. However, the use of DCFH-DA is advantageous over that of TEMP because of its high cost. Other available assays to detect ROS induced due to NPs can be detected using lipid hydroperoxide, Amplex Red assay, measurement of antioxidant depletion by 5,5′-dithiobis-(2-nitrobenzoic acid) (DTNB) and superoxide dismutase (SOD) activity by Nitro blue tetrazolium assay [120].

4.5 Necrosis assay

Necrosis disrupts the integrity of the membrane and is commonly used to determine the viability of cells. Necrosis is measured using two dyes Neutral Red and trypan Blue. Neutral red (2-amino-3 methyl-7-dimethyl-amino-phenazonium-chloride) is a weakly cationic supravital dye that, at slightly acid pH, yields a deep red colour [121]. The mechanism of action of neutral red includes diffusion through the plasma membrane and concentrates on the binding of lysosomes through electrostatic, hydrophobic bonds [101]. Any disruption in the cell membrane brought about by NPs results in reduced uptake of neutral red, differentiating live and necrotic cells. Studies have used neutral red to detect necrotic cells induced when exposed to silver NPs [122]. The other dye used for detecting necrotic cells induced through NPs is trypan blue which enters dead cells while being excluded by live cells and thus is used to detect the stability of the cell membrane. Trypan blue has been used to detect the cytotoxicity of cells exerted by Zn NPs [123].

4.6 Viability assays

Lactate dehydrogenase (LDH) is an enzyme produced by living cells that regulates pyruvate and lactate levels through the oxidation of nicotinamide adenine dinucleotide (NAD). When cells are lysed due to the toxic effects of NPs and other agents, LDH is released into the surroundings and maintains the enzyme activity. The usual conversion of pyruvate to lactate in the presence of LDH follows the NADH + pyruvate NAD + + lactate. Spectroscopic methods can monitor the NADH complementary reaction with tetrazolium salts. This time-dependent change in spectroscopic absorbance through enzyme-linked immunosorbent assay (ELISA) tests infers the extent of cellular trauma [100]. Common tetrazolium salts include iodonitrotetrazolium (INT), 3-(4,5-dimethylthiazol-2-yl)-2,5-diphenyltetrazolium bromide (MTT), and 3-(4,5-dimethylthiazol-2-yl)-5-(3-carboxymethoxy-phenyl)-2-(4-sulfophenyl)-2H-tetrazolium (MTS) which rival the throughput of the 51Cr assay and exhibit a threefold increase in sensitivity over ultraviolet NADH assays [124]. In addition to these, luminogenic glycylphenylalanyl-aminofluorocoumarin (GF-AFC) and firefly luciferase ATP assays provide non-toxic and highly sensitive options for assessing cell viability [125].

5. *In vivo* tests to evaluate the toxicity of nanoparticles

In vivo toxicity assessment methods provide more reliable data than in vitro models, and often yield conflicting results [126]. Some examples of *in vivo* toxicity assessment methods for NPs include biodistribution, clearance, haematology, serum chemistry, and histopathology [101]. For example, researchers have studied the toxicity of aluminium oxide NPs. These NPs have been analysed for their cell toxicity, immunotoxicity, and genotoxicity using *in vivo* models [126]. Tetrazolium-based assays such as MTT, MTS, and WST-1 have been used to evaluate the cytotoxicity of these NPs [126]. Manganese oxide nanomaterials have also been examined for their potentially harmful effects [127]. One study evaluated the toxicity of a newly synthesised nanomaterial called GNA35 and its precursor Mn_3O_4 using in vitro models representing the respiratory, GIT, and skin systems of the human body [127]. The study provided information on the potential health risks associated with these nanomaterials, highlighting the need for further evaluation using *in vivo* models [127].

In vivo toxicity assessment of various NPs is crucial for assessing their safety in biomedical and industrial applications. Here are some examples of *in vivo* toxicity assessment of various NPs:

Polymer NPs, often used for drug delivery systems, have been evaluated for toxicity using *in vivo* models. A study utilised zebrafish embryos as a model to investigate the toxicity of polystyrene NPs. The study found that NPs demonstrated dose-dependent toxicity, leading to developmental and behavioural abnormalities in zebrafish embryos [46].

Lipid-based NPs are commonly used in drug delivery systems because of their biocompatibility and low toxicity. In a study by Lama et al., researchers evaluated the *in vivo* toxicity of lipid NPs in rats by measuring their biodistribution, clearance, and histopathology [128]. They found lipid NPs were well tolerated, with no significant adverse effects observed [101].

Protein NPs, such as albumin-based NPs, have been developed for targeted drug delivery. *In vivo* studies in mice have shown that these NPs can accumulate in tumour tissues and effectively deliver drugs without causing significant systemic toxicity. However, further evaluation of protein NPs using *in vivo* models is necessary to ensure the safety of these drug delivery systems [129].

Silica NPs have many applications, from drug delivery to cosmetics. In a study by Yu et al., researchers assessed the *in vivo* toxicity of silica NPs in mice by assessing their biodistribution, clearance, and inflammatory response. They found that silica NPs were distributed primarily in the liver, spleen, and kidneys and induced a transient inflammatory response, suggesting potential risks associated with their use [130].

AuNPs have been studied for their potential applications in various fields, including drug delivery and diagnostics. *In vivo* toxicity evaluation of gold NPs has been carried out using different animal models. For example, a study investigated the biodistribution and toxicity of AuNPs in mice using various doses and sizes. The results showed that, depending on size and dose, AuNPs could accumulate in organs such as the liver, spleen, lungs, and kidneys, leading to potential toxicity problems [46, 129].

AgNPs have also been investigated for potential applications in medicine and industry. An example of *in vivo* toxicity assessment of AgNPs is a study that evaluated silver NPs stabilised with gum Arabic protein (AgNPs-GP) in Daphnia, a small freshwater crustacean model. The results indicated that AgNPs-GP exhibited dose-dependent toxicity, causing adverse effects on Daphnia's survival and reproduction [131].

Iron oxide NPs (SPIONPs) are widely used in biomedical applications, such as drug delivery and magnetic resonance imaging. In a study by Mahmoudi et al., the researchers evaluated the *in vivo* toxic effects of SPIONPs in mice by observing their biodistribution, retention, and clearance [132]. They found that SPIONPs did not exhibit significant toxicity, but NPs tended to accumulate in the liver and spleen, which can cause long-term adverse effects [57].

6. Conclusion

Throughout the years, the remarkable properties of NPs have led to their recognition as marvels of contemporary science because of their extensive applications. The employment of NPs in the biomedical field has contributed significantly to overcoming numerous diseases and improving the quality of human life. Unlike other delivery methods, orally administering NPs exposes them to various biological and chemical conditions, such as enzyme and microbiota-related digestion and fluctuations in ionic strength and pH levels. Nanotoxicity generally stems from the instability of the NPs and distinct physicochemical properties. Numerous factors, such as the type of NPs, particle and pore size, the extent of modification, frequency and amount of dosage, and administration timing, influence them. Other crucial factors include cell type, cell condition, organ distribution, animal condition, and delivery duration, presenting individual differences in cellular and *in vivo* contexts. NPs toxicity mechanisms include oxidative stress, inflammation, and apoptosis; however, studies of the implicated signal pathways are still relatively scarce. Due to the numerous variables involved, there currently needs to be a standardised approach to toxicity research. Investigations of NPs toxicity should consider their actual applications and refine safety evaluation indicators for both in vitro and in vivo testing. As observed, a limited amount of data is available for specific NPs types used in DDS. Consequently, before introducing NPs into the therapeutic market, it is essential to evaluate the risk-benefit ratio.

Author details

Iman M. Alfagih
Department of Pharmaceutics, College of Pharmacy, King Saud University, Riyadh, Saudi Arabia

*Address all correspondence to: fagih@ksu.edu.sa

References

[1] Yetisgin AA, Cetinel S, Zuvin M, Kosar A, Kutlu O. Therapeutic nanoparticles and their targeted delivery applications. Molecules. 2020;**25**(9):2193. DOI: 10.3390/molecules25092193

[2] Patra JK et al. Nano based drug delivery systems: Recent developments and future prospects. Journal of Nanobiotechnology. 2018;**16**(1):71. DOI: 10.1186/s12951-018-0392-8

[3] Nasrollahzadeh M, Sajadi SM, Sajjadi M, Issaabadi Z. An Introduction to Nanotechnology, Interface Science and Technology. Amsterdam, The Netherlands: Elsevier; 2019. pp. 1-27. DOI: 10.1016/b978-0-12-813586-0.00001-8

[4] Mitchell MJ, Billingsley MM, Haley RM, Wechsler ME, Peppas NA, Langer R. Engineering precision nanoparticles for drug delivery. Nature Reviews. Drug Discovery. 2021;**20**(2):101-124. DOI: 10.1038/s41573-020-0090-8

[5] Carissimi G, Montalbán MG, Fuster MG, Víllora G. Silk Fibroin Nanoparticles: Synthesis and Applications as Drug Nanocarriers, 21st Century Nanostructured Materials - Physics, Chemistry, Classification, and Emerging Applications in Industry, Biomedicine, and Agriculture. London: IntechOpen; 2022. DOI: 10.5772/intechopen.100386

[6] Palmerston Mendes L, Pan J, Torchilin VP. Dendrimers as Nanocarriers for nucleic acid and drug delivery in cancer therapy. Molecules. 2017;**22**(9):1401. DOI: 10.3390/molecules22091401

[7] Kadam RS, Bourne DWA, Kompella UB. Nano-advantage in enhanced drug delivery with biodegradable nanoparticles: Contribution of reduced clearance. Drug Metabolism and Disposition. 2012;**40**(7):1380-1388. DOI: 10.1124/dmd.112.044925

[8] Obeid MA, Al Qaraghuli MM, Alsaadi M, Alzahrani AR, Niwasabutra K, Ferro VA. Delivering natural products and biotherapeutics to improve drug efficacy. Therapeutic Delivery. 2017;**8**(11):947-956. DOI: 10.4155/tde-2017-0060

[9] Zhan S, Paik A, Onyeabor F, Ding B, Prabhu S, Wang J. Oral bioavailability evaluation of Celastrol-encapsulated silk fibroin nanoparticles using an optimized LC-MS/MS method. Molecules. 2020;**25**(15):3422. DOI: 10.3390/molecules25153422

[10] Alqahtani MS, Kazi M, Alsenaidy MA, Ahmad MZ. Advances in Oral drug delivery. Frontiers in Pharmacology. 2021;**12**:618411. DOI: 10.3389/fphar.2021.618411

[11] Mudie DM, Amidon GL, Amidon GE. Physiological parameters for oral delivery and in vitro testing. Molecular Pharmaceutics. 2010;7(5):1388-1405. DOI: 10.1021/mp100149j

[12] Majumder J, Taratula O, Minko T. Nanocarrier-based systems for targeted and site specific therapeutic delivery. Advanced Drug Delivery Reviews. 2019;**144**:57-77. DOI: 10.1016/j.addr.2019.07.010

[13] Lou J et al. Advances in Oral drug delivery systems: Challenges and opportunities. Pharmaceutics. 2023;**15**(2):484. DOI: 10.3390/pharmaceutics15020484

[14] Joseph TM et al. Nanoparticles: Taking a unique position in medicine. Nanomaterials (Basel). 2023;**13**(3):574. DOI: 10.3390/nano13030574

[15] Cabellos J et al. Short-term oral administration of non-porous and mesoporous silica did not induce local or systemic toxicity in mice. Nanotoxicology. 2020;**14**(10):1324-1341. DOI: 10.1080/17435390.2020.1818325

[16] Sharma S, Parveen R, Chatterji BP. Toxicology of nanoparticles in drug delivery. Current Pathobiology Reports. 2021;**9**(4):133-144. DOI: 10.1007/s40139-021-00227-z

[17] Zhang Q et al. A superoxide dismutase/catalase mimetic nanomedicine for targeted therapy of inflammatory bowel disease. Biomaterials. 2016;**105**:206-221. DOI: 10.1016/j.biomaterials.2016.08.010

[18] Vong LB, Mo J, Abrahamsson B, Nagasaki Y. Specific accumulation of orally administered redox nanotherapeutics in the inflamed colon reducing inflammation with dose–response efficacy. Journal of Controlled Release. 2015;**210**:19-25. DOI: 10.1016/j.jconrel.2015.05.275

[19] Geiser M et al. Ultrafine particles cross cellular membranes by nonphagocytic mechanisms in lungs and in cultured cells. Environmental Health Perspectives. 2005;**113**(11):1555-1560. DOI: 10.1289/ehp.8006

[20] Fu PP, Xia Q, Hwang H-M, Ray PC, Yu H. Mechanisms of nanotoxicity: Generation of reactive oxygen species. Journal of Food and Drug Analysis. 2014;**22**(1):64-75. DOI: 10.1016/j.jfda.2014.01.005

[21] Yusuf A, Almotairy ARZ, Henidi H, Alshehri OY, Aldughaim MS. Nanoparticles as drug delivery systems: A review of the implication of nanoparticles' physicochemical properties on responses in biological systems. Polymers (Basel). 2023;**15**(7):1596. DOI: 10.3390/polym15071596

[22] Lu W, Yao J, Zhu X, Qi Y. Nanomedicines: Redefining traditional medicine. Biomedicine & Pharmacotherapy. 2021;**134**:111103. DOI: 10.1016/j.biopha.2020.111103

[23] Hosseini A et al. Exosome-inspired targeting of cancer cells with enhanced affinity. Journal of Materials Chemistry B. 2016;**4**(4):768-778. DOI: 10.1039/c5tb01741f

[24] Teng Y et al. Plant-derived Exosomal MicroRNAs shape the gut microbiota. Cell Host & Microbe. 2018;**24**(5):637-652.e8. DOI: 10.1016/j.chom.2018.10.001

[25] Xia W et al. Targeted delivery of drugs and genes using polymer Nanocarriers for cancer therapy. International Journal of Molecular Sciences. 2021;**22**(17):9118. DOI: 10.3390/ijms22179118

[26] Quinson J et al. Investigating particle size effects in catalysis by applying a size-controlled and surfactant-free synthesis of colloidal nanoparticles in alkaline ethylene glycol: Case study of the oxygen reduction reaction on Pt. ACS Catalysis. 2018;**8**(7):6627-6635. DOI: 10.1021/acscatal.8b00694

[27] Bergin IL, Witzmann FA. Nanoparticle toxicity by the gastrointestinal route: Evidence and knowledge gaps. International Journal of Biomedical Nanoscience and Nanotechnology. 2013;**3**(1-2):163-210. DOI: 10.1504/IJBNN.2013.054515

[28] Greenwood-Van Meerveld B, Johnson AC, Grundy D. Gastrointestinal Physiology and Function,

Gastrointestinal Pharmacology. New York, USA: Springer International Publishing; 2017. pp. 1-16. DOI: 10.1007/164_2016_118

[29] Lim YF et al. An exploration of the microrheological environment around the distal ileal villi and proximal colonic mucosa of the possum (Trichosurus vulpecula). Journal of the Royal Society, Interface. 2013;**10**(81):20121008. DOI: 10.1098/rsif.2012.1008

[30] Azman M, Sabri AH, Anjani QK, Mustaffa MF, Hamid KA. Intestinal absorption study: Challenges and absorption enhancement strategies in improving Oral drug delivery. Pharmaceuticals (Basel). 2022;**15**(8):975. DOI: 10.3390/ph15080975

[31] Mowat AM, Agace WW. Regional specialization within the intestinal immune system. Nature Reviews. Immunology. 2014;**14**(10):667-685. DOI: 10.1038/nri3738

[32] Brannon-Peppas L, Blanchette JO. Nanoparticle and targeted systems for cancer therapy. Advanced Drug Delivery Reviews. 2012;**64**:206-212. DOI: 10.1016/j.addr.2012.09.033

[33] Padhi S, Behera A. Cellular Internalization and Toxicity of Polymeric Nanoparticles, Environmental Chemistry for a Sustainable World. New York, USA: Springer International Publishing; 2022. pp. 473-488. DOI: 10.1007/978-3-031-14848-4_17

[34] Yu J, Chen L, Gu W, Liu S, Wu B. Heterogeneity effects of nanoplastics and lead on zebrafish intestinal cells identified by single-cell sequencing. Chemosphere. 2022;**289**:133133. DOI: 10.1016/j.chemosphere.2021.133133

[35] Chen J et al. Surface functionalization-dependent inflammatory potential of polystyrene nanoplastics through the activation of MAPK/NF-κB signaling pathways in macrophage raw 264.7. Ecotoxicology and Environmental Safety. 2023;**251**:114520. DOI: 10.1016/j.ecoenv.2023.114520

[36] Dass CR. Lipoplex-mediated delivery of nucleic acids: Factors affecting in vivo transfection. Journal of Molecular Medicine. 2004;**82**(9):579-591. DOI: 10.1007/s00109-004-0558-8

[37] El Moukhtari SH, Rodríguez-Nogales C, Blanco-Prieto MJ. Oral lipid nanomedicines: Current status and future perspectives in cancer treatment. Advanced Drug Delivery Reviews. 2021;**173**:238-251. DOI: 10.1016/j.addr.2021.03.004

[38] Syama K, Jakubek ZJ, Chen S, Zaifman J, Tam YYC, Zou S. Development of lipid nanoparticles and liposomes reference materials (II): Cytotoxic profiles. Scientific Reports. 2022;**12**(1):18071. DOI: 10.1038/s41598-022-23013-2

[39] Parhiz H et al. Added to pre-existing inflammation, mRNA-lipid nanoparticles induce inflammation exacerbation (IE). Journal of Controlled Release. 2022;**344**:50-61. DOI: 10.1016/j.jconrel.2021.12.027

[40] Shang L, Nienhaus K, Nienhaus GU. Engineered nanoparticles interacting with cells: Size matters. Journal of Nanobiotechnology. 2014;**12**:5. DOI: 10.1186/1477-3155-12-5

[41] McClements DJ, DeLoid G, Pyrgiotakis G, Shatkin JA, Xiao H, Demokritou P. The role of the food matrix and gastrointestinal tract in the assessment of biological properties of ingested engineered nanomaterials (iENMs): State of the science and knowledge gaps. NanoImpact. 2019;**3-4**:47-57. DOI: 10.1016/j.impact.2016.10.002

[42] Hu H et al. Silicon dioxide nanoparticles induce insulin resistance through endoplasmic reticulum stress and generation of reactive oxygen species. Particle and Fibre Toxicology. 2019;**16**(1):41. DOI: 10.1186/s12989-019-0327-z

[43] Liu J et al. Silica nanoparticles induce spermatogenesis disorders via L3MBTL2-DNA damage-p53 apoptosis and RNF8-ubH2A/ubH2B pathway in mice. Environmental Pollution. 2020;**265**:114974. DOI: 10.1016/j.envpol.2020.114974

[44] Huang Y et al. Silica nanoparticles: Biomedical applications and toxicity. Biomedicine and Pharmacotherapy. 2022;**151**:113053. DOI: 10.1016/j.biopha.2022.113053

[45] Piechulek A, Berwanger LC, von Mikecz A. Silica nanoparticles disrupt OPT-2/PEP-2-dependent trafficking of nutrient peptides in the intestinal epithelium. Nanotoxicology. 2019;**13**(8):1133-1148. DOI: 10.1080/17435390.2019.1643048

[46] Adewale OB, Davids H, Cairncross L, Roux S. Toxicological behavior of gold nanoparticles on various models: Influence of physicochemical properties and other factors. International Journal of Toxicology. 2019;**38**(5):357-384. DOI: 10.1177/1091581819863130

[47] De Jong WH, Hagens WI, Krystek P, Burger MC, Sips AJAM, Geertsma RE. Particle size-dependent organ distribution of gold nanoparticles after intravenous administration. Biomaterials. 2008;**29**(12):1912-1919. DOI: 10.1016/j.biomaterials.2007.12.037

[48] Kim JH, Kim JH, Kim K-W, Kim MH, Yu YS. Intravenously administered gold nanoparticles pass through the blood–retinal barrier depending on the particle size, and induce no retinal toxicity. Nanotechnology. 2009;**20**(50):505101. DOI: 10.1088/0957-4484/20/50/505101

[49] Alkilany AM, Murphy CJ. Toxicity and cellular uptake of gold nanoparticles: What we have learned so far? Journal of Nanoparticle Research. 2010;**12**(7):2313-2333. DOI: 10.1007/s11051-010-9911-8

[50] Ferdous Z, Nemmar A. Health impact of silver nanoparticles: A review of the biodistribution and toxicity following various routes of exposure. International Journal of Molecular Sciences. 2020;**21**(7):23750. DOI: 10.3390/ijms21072375

[51] Sun Q et al. Cytotoxicity and cellular responses of gold Nanorods to smooth muscle cells dependent on surface chemistry coupled action. Small. 2018;**14**(52):1803715. DOI: 10.1002/smll.201803715

[52] David L et al. Green synthesis, characterization and anti-inflammatory activity of silver nanoparticles using European black elderberry fruits extract. Colloids and Surfaces. B, Biointerfaces. 2014;**122**:767-777. DOI: 10.1016/j.colsurfb.2014.08.018

[53] Hebeish A, El-Rafie MH, EL-Sheikh MA, Seleem AA, El-Naggar ME. Antimicrobial wound dressing and anti-inflammatory efficacy of silver nanoparticles. International Journal of Biological Macromolecules. 2014;**65**:509-515. DOI: 10.1016/j.ijbiomac.2014.01.071

[54] Le Ouay B, Stellacci F. Antibacterial activity of silver nanoparticles: A surface science insight. Nano Today. 2015;**10**(3):339-354. DOI: 10.1016/j.nantod.2015.04.002

[55] Pasparakis G. Recent developments in the use of gold and silver nanoparticles in biomedicine.

Wiley Interdisciplinary Reviews. Nanomedicine and Nanobiotechnology. 2022;**14**(5):e1817-e1817. DOI: 10.1002/wnan.1817

[56] Berry CC, Wells S, Charles S, Curtis ASG. Dextran and albumin derivatised iron oxide nanoparticles: Influence on fibroblasts in vitro. Biomaterials. 2003;**24**(25):4551-4557. DOI: 10.1016/s0142-9612(03)00237-0

[57] Singh N, Jenkins GJS, Asadi R, Doak SH. Potential toxicity of superparamagnetic iron oxide nanoparticles (SPION). Nano Reviews. 2010;**1**:1-15. DOI: 10.3402/nano.v1i0.5358

[58] Sabourian P et al. Effect of physico-chemical properties of nanoparticles on their intracellular uptake. International Journal of Molecular Sciences. 2020;**21**(21):8019. DOI: 10.3390/ijms21218019

[59] Date AA, Hanes J, Ensign LM. Nanoparticles for oral delivery: Design, evaluation and state-of-the-art. Journal of Controlled Release. 2016;**240**:504-526. DOI: 10.1016/j.jconrel.2016.06.016

[60] Lai SK, Wang Y-Y, Hanes J. Mucus-penetrating nanoparticles for drug and gene delivery to mucosal tissues. Advanced Drug Delivery Reviews. 2009;**61**(2):158-171. DOI: 10.1016/j.addr.2008.11.002

[61] Wang Y, Pi C, Feng X, Hou Y, Zhao L, Wei Y. The influence of nanoparticle properties on Oral bioavailability of drugs. International Journal of Nanomedicine. 2020;**15**:6295-6310. DOI: 10.2147/ijn.s257269

[62] Zhao Z et al. A nanoparticle-based nicotine vaccine and the influence of particle size on its immunogenicity and efficacy. Nanomedicine. 2017;**13**(2):443-454. DOI: 10.1016/j.nano.2016.07.015

[63] Mostafalou S, Mohammadi H, Ramazani A, Abdollahi M. Different biokinetics of nanomedicines linking to their toxicity; an overview. Daru. 2013;**21**(1):14. DOI: 10.1186/2008-2231-21-14

[64] Abbasi R, Shineh G, Mobaraki M, Doughty S, Tayebi L. Structural parameters of nanoparticles affecting their toxicity for biomedical applications: A review. Journal of Nanoparticle Research. 2023;**25**(3):43. DOI: 10.1007/s11051-023-05690-w

[65] Vedhanayagam M, Unni Nair B, Sreeram KJ. Collagen-ZnO scaffolds for wound healing applications: Role of dendrimer functionalization and nanoparticle morphology. ACS Applied Bio Materials. 2018;**1**(6):1942-1958. DOI: 10.1021/acsabm.8b00491

[66] Shukla S et al. The impact of aspect ratio on the biodistribution and tumor homing of rigid soft-matter Nanorods. Advanced Healthcare Materials. 2015;**4**(6):874-882. DOI: 10.1002/adhm.201400641

[67] Fubini B, Fenoglio I, Tomatis M, Turci F. Effect of chemical composition and state of the surface on the toxic response to high aspect ratio nanomaterials. Nanomedicine. 2011;**6**(5):899-920. DOI: 10.2217/nnm.11.80

[68] Liu M et al. Efficient mucus permeation and tight junction opening by dissociable 'mucus-inert' agent coated trimethyl chitosan nanoparticles for oral insulin delivery. Journal of Controlled Release. 2016;**222**:67-77. DOI: 10.1016/j.jconrel.2015.12.008

[69] Du X-J et al. The effect of surface charge on oral absorption of polymeric nanoparticles. Biomaterials Science. 2018;**6**(3):642-650. DOI: 10.1039/c7bm01096f

[70] González-García LE et al. Nanoparticles surface chemistry influence on protein Corona composition and inflammatory responses. Nanomaterials (Basel). 2022;**12**(4): 682. DOI: 10.3390/nano12040682

[71] Marassi V et al. Silver nanoparticles as a medical device in healthcare settings: A five-step approach for candidate screening of coating agents. Royal Society Open Science. 2018;**5**(1):171113. DOI: 10.1098/rsos.171113

[72] Prasher P et al. Targeting mucus barrier in respiratory diseases by chemically modified advanced delivery systems. Chemico-Biological Interactions. 2022;**365**:110048. DOI: 10.1016/j.cbi.2022.110048

[73] Grabowski N et al. Surface coating mediates the toxicity of polymeric nanoparticles towards human-like macrophages. International Journal of Pharmaceutics. 2015;**482**(1-2):75-83. DOI: 10.1016/j.ijpharm.2014.11.042

[74] Henson TE, Navratilova J, Tennant AH, Bradham KD, Rogers KR, Hughes MF. In vitro intestinal toxicity of copper oxide nanoparticles in rat and human cell models. Nanotoxicology. 2019;**13**(6):795-811. DOI: 10.1080/17435390.2019.1578428

[75] Yu Z et al. Corrigendum: Yu et al. PPy@Fe(3)O(4) nanoparticles inhibit the proliferation and metastasis of CRC via suppressing the NF-κB signaling pathway and promoting ferroptosis. Frontiers in Bioengineering and Biotechnology. 2022;**10**:1094064. DOI: 10.3389/fbioe.2022.1094064

[76] De Berardis B et al. Exposure to ZnO nanoparticles induces oxidative stress and cytotoxicity in human colon carcinoma cells. Toxicology and Applied Pharmacology. 2010;**246**(3):116-127. DOI: 10.1016/j.taap.2010.04.012

[77] Mittag A, Singer A, Hoera C, Westermann M, Kämpfe A, Glei M. Impact of in vitro digested zinc oxide nanoparticles on intestinal model systems. Particle and Fibre Toxicology. 2022;**19**(1):39. DOI: 10.1186/s12989-022-00479-6

[78] Seo Y et al. Recent Progress of lipid nanoparticles-based lipophilic drug delivery: Focus on surface modifications. Pharmaceutics. 2023;**15**(3):772. DOI: 10.3390/pharmaceutics15030772

[79] Tenchov R, Bird R, Curtze AE, Zhou Q. Lipid nanoparticles–from liposomes to mRNA vaccine delivery, a landscape of research diversity and advancement. ACS Nano. 2021;**15**(11):16982-17015. DOI: 10.1021/acsnano.1c04996

[80] Satapathy MK et al. Solid lipid nanoparticles (SLNs): An advanced drug delivery system targeting brain through BBB. Pharmaceutics. 2021;**13**(8):1183. DOI: 10.3390/pharmaceutics13081183

[81] Lv H, Zhang S, Wang B, Cui S, Yan J. Toxicity of cationic lipids and cationic polymers in gene delivery. Journal of Controlled Release. 2006;**114**(1):100-109. DOI: 10.1016/j.jconrel.2006.04.014

[82] Ball RL, Bajaj P, Whitehead KA. Oral delivery of siRNA lipid nanoparticles: Fate in the GI tract. Scientific Reports. 2018;**8**(1):1-12. DOI: 10.1038/s41598-018-20632-6

[83] Guo Z, Martucci NJ, Liu Y, Yoo E, Tako E, Mahler GJ. Silicon dioxide nanoparticle exposure affects small intestine function in an in vitro model. Nanotoxicology. 2018;**12**(5):485-508. DOI: 10.1080/17435390.2018.1463407

[84] Sakai-Kato K, Hidaka M, Un K, Kawanishi T, Okuda H. Physicochemical properties and in vitro intestinal

permeability properties and intestinal cell toxicity of silica particles, performed in simulated gastrointestinal fluids. Biochimica et Biophysica Acta (BBA) - General Subjects. 2014;**1840**(3):1171-1180. DOI: 10.1016/j.bbagen.2013.12.014

[85] Diao J et al. Silicon dioxide nanoparticles induced neurobehavioral impairments by disrupting microbiota-gut-brain axis. Journal of Nanobiotechnology. 2021;**19**(1):174. DOI: 10.1186/s12951-021-00916-2

[86] Fruijtier-Pölloth C. The safety of nanostructured synthetic amorphous silica (SAS) as a food additive (E 551). Archives of Toxicology. 2016;**90**(12):2885-2916. DOI: 10.1007/s00204-016-1850-4

[87] Kusaczuk M, Krętowski R, Naumowicz M, Stypułkowska A, Cechowska-Pasko M. Silica nanoparticle-induced oxidative stress and mitochondrial damage is followed by activation of intrinsic apoptosis pathway in glioblastoma cells. International Journal of Nanomedicine. 2018;**13**: 2279-2294. DOI: 10.2147/IJN.S158393

[88] Ogawa T et al. Oral intake of silica nanoparticles exacerbates intestinal inflammation. Biochemical and Biophysical Research Communications. 2021;**534**:540-546. DOI: 10.1016/j.bbrc.2020.11.047

[89] Abedin MR et al. Polymer coated gold-ferric oxide superparamagnetic nanoparticles for theranostic applications. Journal of Nanobiotechnology. 2018;**16**(1):80. DOI: 10.1186/s12951-018-0405-7

[90] Syafiuddin A, Salmiati, Salim MR, Kueh ABH, Hadibarata T, Nur H. A review of silver nanoparticles: Research trends, global consumption, synthesis, properties, and future challenges. Journal of the Chinese Chemical Society. 2017;**64**(7):732-756. DOI: 10.1002/jccs.201700067

[91] Vandebriel RJ et al. Immunotoxicity of silver nanoparticles in an intravenous 28-day repeated-dose toxicity study in rats. Particle and Fibre Toxicology. 2014;**11**:21. DOI: 10.1186/1743-8977-11-21

[92] Azizi M, Ghourchian H, Yazdian F, Bagherifam S, Bekhradnia S, Nyström B. Anti-cancerous effect of albumin coated silver nanoparticles on MDA-MB 231 human breast cancer cell line. Scientific Reports. 2017;**7**(1):5178. DOI: 10.1038/s41598-017-05461-3

[93] van der Zande M et al. Distribution, elimination, and toxicity of silver nanoparticles and silver ions in rats after 28-day Oral exposure. ACS Nano. 2012;**6**(8):7427-7442. DOI: 10.1021/nn302649p

[94] Recordati C et al. Tissue distribution and acute toxicity of silver after single intravenous administration in mice: Nano-specific and size-dependent effects. Particle and Fibre Toxicology. 2016;**13**:12. DOI: 10.1186/s12989-016-0124-x

[95] De Matteis V. Exposure to inorganic nanoparticles: Routes of entry, immune response, biodistribution and In vitro/In vivo toxicity evaluation. Toxics. 2017;**5**(4):29. DOI: 10.3390/toxics5040029

[96] Jeong GN, Jo UB, Ryu HY, Kim YS, Song KS, Yu IJ. Histochemical study of intestinal mucins after administration of silver nanoparticles in Sprague–Dawley rats. Archives of Toxicology. 2009;**84**(1):63-69. DOI: 10.1007/s00204-009-0469-0

[97] Goodman CM, McCusker CD, Yilmaz T, Rotello VM. Toxicity of gold

nanoparticles functionalized with cationic and anionic side chains. Bioconjugate Chemistry. 2004;**15**(4):897-900. DOI: 10.1021/bc049951i

[98] Patra HK, Banerjee S, Chaudhuri U, Lahiri P, Kr A. Dasgupta, cell selective response to gold nanoparticles. Nanomedicine. 2007;**3**(2):111-119. DOI: 10.1016/j.nano.2007.03.005

[99] Oberdörster G, Stone V, Donaldson K. Toxicology of nanoparticles: A historical perspective. Nanotoxicology. 2007;**1**(1):2-25. DOI: 10.1080/17435390701314761

[100] Savage DT, Hilt JZ, Dziubla TD. In vitro methods for assessing nanoparticle toxicity. Methods in Molecular Biology. 2019;**1894**:1-29. DOI: 10.1007/978-1-4939-8916-4_1

[101] Kumar V, Sharma N, Maitra SS. In vitro and in vivo toxicity assessment of nanoparticles. International Nano Letters. 2017;**7**(4):243-256. DOI: 10.1007/s40089-017-0221-3

[102] Lin P-C, Lin S, Wang PC, Sridhar R. Techniques for physicochemical characterization of nanomaterials. Biotechnology Advances. 2014;**32**(4):711-726. DOI: 10.1016/j.biotechadv.2013.11.006

[103] Sayes CM, Reed KL, Warheit DB. Assessing toxicity of fine and nanoparticles: Comparing In vitro measurements to In vivo pulmonary toxicity profiles. Toxicological Sciences. 2007;**97**(1):163-180. DOI: 10.1093/toxsci/kfm018

[104] Punshon G, Vara D, Sales K, Kidane A, Salacinski H, Seifalian A. Interactions between endothelial cells and a poly(carbonate-silsesquioxane-bridge-urea)urethane. Biomaterials. 2005;**26**(32):6271-6279. DOI: 10.1016/j.biomaterials.2005.03.034

[105] Hussain SM. Cellular toxicity of hydrazine in primary rat hepatocytes. Toxicological Sciences. 2002;**69**(2):424-432. DOI: 10.1093/toxsci/69.2.424

[106] Casey A, Herzog E, Davoren M, Lyng FM, Byrne HJ, Chambers G. Spectroscopic analysis confirms the interactions between single walled carbon nanotubes and various dyes commonly used to assess cytotoxicity. Carbon N Y. 2007;**45**(7):1425-1432. DOI: 10.1016/j.carbon.2007.03.033

[107] Ahamed M et al. DNA damage response to different surface chemistry of silver nanoparticles in mammalian cells. Toxicology and Applied Pharmacology. 2008;**233**(3):404-410. DOI: 10.1016/j.taap.2008.09.015

[108] Ahamed M, Posgai R, Gorey TJ, Nielsen M, Hussain SM, Rowe JJ. Silver nanoparticles induced heat shock protein 70, oxidative stress and apoptosis in drosophila melanogaster. Toxicology and Applied Pharmacology. 2010;**242**(3):263-269. DOI: 10.1016/j.taap.2009.10.016

[109] Kaufmann SH et al. Apoptosis-associated caspase activation assays. Methods. 2008;**44**(3):262-272. DOI: 10.1016/j.ymeth.2007.11.005

[110] Jin Y, Kannan S, Wu M, Zhao JX. Toxicity of luminescent silica nanoparticles to living cells. Chemical Research in Toxicology. 2007;**20**(8):1126-1133. DOI: 10.1021/tx7001959

[111] Mo Y, Lim L-Y. Paclitaxel-loaded PLGA nanoparticles: Potentiation of anticancer activity by surface conjugation with wheat germ agglutinin. Journal of Controlled Release. 2005;**108**(2-3):244-262. DOI: 10.1016/j.jconrel.2005.08.013

[112] Pan Y et al. Size-dependent cytotoxicity of gold nanoparticles. Small.

2007;**3**(11):1941-1949. DOI: 10.1002/smll.200700378

[113] Suman S, Pandey A, Chandna S. An improved non-enzymatic 'DNA ladder assay' for more sensitive and early detection of apoptosis. Cytotechnology. 2012;**64**(1):9-14. DOI: 10.1007/s10616-011-9395-0

[114] Abdel-Khalek AA. Comparative evaluation of genotoxic effects induced by CuO bulk and Nano-particles in Nile tilapia, Oreochromis niloticus. Water, Air, & Pollution. 2015;**227**(1). DOI: 10.1007/s11270-015-2737-3

[115] Yuan H et al. Sulfur nanoparticles improved plant growth and reduced mercury toxicity via mitigating the oxidative stress in Brassica napus L. Journal of Cleaner Production. 2021;**318**:128589. DOI: 10.1016/j.jclepro.2021.128589

[116] Handy RD et al. Practical considerations for conducting ecotoxicity test methods with manufactured nanomaterials: What have we learnt so far? Ecotoxicology. 2012;**21**(4):933-972. DOI: 10.1007/s10646-012-0862-y

[117] Keston AS, Brandt R. The fluorometric analysis of ultramicro quantities of hydrogen peroxide. Analytical Biochemistry. 1965;**11**(1):1-5. DOI: 10.1016/0003-2697(65)90034-5

[118] Bass DA, Parce JW, Dechatelet LR, Szejda P, Seeds MC, Thomas M. Flow cytometric studies of oxidative product formation by neutrophils: A graded response to membrane stimulation. The Journal of Immunology. 1983;**130**(4):1910-1917. DOI: 10.4049/jimmunol.130.4.1910

[119] Li M, Yin J-J, Wamer WG, Lo YM. Mechanistic characterization of titanium dioxide nanoparticle-induced toxicity using electron spin resonance. Journal of Food and Drug Analysis. 2014;**22**(1):76-85. DOI: 10.1016/j.jfda.2014.01.006

[120] Hussain SM, Javorina AK, Schrand AM, Duhart HM, Ali SF, Schlager JJ. The interaction of manganese nanoparticles with PC-12 cells induces dopamine depletion. Toxicological Sciences. 2006;**92**(2):456-463. DOI: 10.1093/toxsci/kfl020

[121] Monteiro-Riviere NA, Nemanich RJ, Inman AO, Wang YY, Riviere JE. Multi-walled carbon nanotube interactions with human epidermal keratinocytes. Toxicology Letters. 2005;**155**(3):377-384. DOI: 10.1016/j.toxlet.2004.11.004

[122] Miranda RR et al. Toxicological interactions of silver nanoparticles and non-essential metals in human hepatocarcinoma cell line. Toxicology In Vitro. 2017;**40**:134-143. DOI: 10.1016/j.tiv.2017.01.003

[123] Kononenko V et al. Comparative in vitro genotoxicity study of ZnO nanoparticles, ZnO macroparticles and ZnCl2 to MDCK kidney cells: Size matters. Toxicology In Vitro. 2017;**40**:256-263. DOI: 10.1016/j.tiv.2017.01.015

[124] Babson AL, Babson SR. Kinetic colorimetric measurement of serum lactate dehydrogenase activity. Clinical Chemistry. 1973;**19**(7):766-769. DOI: 10.1093/clinchem/19.7.766

[125] Niles AL, Moravec RA, Worzella TJ, Evans NJ, Riss TL. High-Throughput Screening Assays for the Assessment of Cytotoxicity, High-Throughput Screening Methods in Toxicity Testing. New Jersey, USA: John Wiley & Sons, Inc.; 2013. pp. 107-127. DOI: 10.1002/9781118538203.ch5

[126] Bahadar H, Abdollahi M, Maqbool F, Baeeri M, Niaz K. Mechanistic overview

of immune modulatory effects of environmental toxicants. Inflammation & Allergy-Drug Targets. 2015; **13**(6):382-386. DOI: 10.2174/1871528114666150529103003

[127] Fernández-Pampín N et al. Toxicology assessment of manganese oxide nanomaterials with enhanced electrochemical properties using human in vitro models representing different exposure routes. Scientific Reports. 2022;**12**(1):20991. DOI: 10.1038/s41598-022-25483-w

[128] Lama S, Merlin-Zhang O, Yang C. In vitro and In vivo models for evaluating the oral toxicity of nanomedicines. Nanomaterials (Basel). 2020;**10**(11):2177. DOI: 10.3390/nano10112177

[129] Carnovale C, Bryant G, Shukla R, Bansal V. Identifying trends in gold nanoparticle toxicity and uptake: Size. ACS Omega. 2019;**4**(1):242-256. DOI: 10.1021/acsomega.8b03227

[130] Adam A, Mertz D. Iron oxide@ mesoporous silica Core-Shell nanoparticles as multimodal platforms for magnetic resonance imaging, magnetic hyperthermia, near-infrared light Photothermia, and drug delivery. Nanomaterials (Basel). 2023;**13**(8):1342. DOI: 10.3390/nano13081342

[131] Maziero JS et al. Species-specific in vitro and in vivo evaluation of toxicity of silver nanoparticles stabilized with gum Arabic protein. International Journal of Nanomedicine. 2020;**15**:7359-7376. DOI: 10.2147/IJN.S250467

[132] Mahmoudi M, Simchi A, Imani M, Milani AS, Stroeve P. An in vitro study of bare and poly(ethylene glycol)-co-fumarate-coated superparamagnetic iron oxide nanoparticles: A new toxicity identification procedure. Nanotechnology. 2009;**20**(22):225104. DOI: 10.1088/0957-4484/20/22/225104

Section 2

Chemical State-of-the-Art

Chapter 5

Phyto-Metallic Nanoparticles: Biosynthesis, Mechanism, Therapeutics, and Cytotoxicity

Akeem Omolaja Akinfenwa and Ahmed Abdelfattah Hussein

Abstract

Nanoparticles synthesized from noble metals have wide applications in therapeutics and medicine due to their excellent properties. Properties such as surface plasmon resonance, low toxicity, biocompatibility, and ease of surface modification account for the recent surge in nanoscience and technology. Noble metals such as gold, silver, copper, iron, and platinum with nano size are well-known metallic nanoparticles with increasing applications in nanomedicine. Biomedical applications of these particles especially for diagnosis and targeted drug delivery in living systems require considering the toxicity level. Because of their surface electrons, metal ions in solution affect cellular activities via cell division, apoptosis, DNA replication, homeostasis etc. They influence cell viability through metabolic outputs in both desired and undesired paths which may result in chemical and or biological toxicity to cells. Phyto-metallic nanoparticles biosynthesised from plant extracts are reported with improved functionalities for biomedical applications over those from chemical and physical methods. The synergies from the natural organic properties of the plant and the metallic properties elicit minimal cytotoxicity paving way for their different biomedical applications. This chapter is intended to provide an overview of recent advances and new perspectives in phyto-metallic nanoparticles, their biosynthesis and mechanism, therapeutics, and cytotoxicity to biomedical industries, research centres, and academia.

Keywords: biosynthesis, cytotoxicity, phytochemicals, nanoparticles, metal precursors, therapeutics

1. Introduction

Medicinal application of plants and the development of green nanotechnology stimulated the use of plant phytochemicals in the green synthesis of metallic nanoparticles (MNPs). Metals such as gold, silver, zinc, tin, platinum, lead, copper, palladium and so forth are easily bio-reduced to phyto-metallic nanoparticles (PM-NPs) in the range 1 nm and 100 nm by phytochemicals from plants through biosynthesis and green nanotechnology methods [1]. These methods are reported to eliminate the use of toxic solvents and chemicals, minimize cost, and the loss of atom economy associated with chemical and physical methods [1, 2]. In the biosynthesis of nanoparticles, bio-nanomaterials are synthesized using natural products of plant and plant-derived compounds (phytochemicals), microbes (bacterial, yeast, fungi, viruses, and algae), and animals. Plant phytochemicals are prevalently used due to their bioavailability and relatively low experimental costs. The phytochemicals possess functional groups such as hydroxyl (–OH), amide ($-NH_2C=O$), carbonyl (C=O), carboxylic (–COOH) etc. with reducing/stabilizing abilities through π-π dative bonding, hydrogen bonding, and electrostatic interactions [3]. This stabilization is reported to enhance the activities of a metal-based drug through size reduction, increase surface area, drug durability and biodistribution through tissue/cell binding [4]. During biosynthesis with phytochemicals, crude extract, or compounds isolated act as reducing/capping agents for single or bimetallic precursors to form their respective single PM-NPs or phyto-bimetallic nanoparticles (PBM-NPs) [1, 5]. These nanoparticles can interface with biological systems as drugs, diagnostic tools or as implants due to their physical, chemical, and biological properties.

For therapeutics and nanomedicine, metallic nanoparticles (MNPs) capped with plant extracts rich in phenolics, alkaloids, and terpenoids contents through biological methods have shown increased acceptance over chemical and physical methods [6]. The plant-based synthesis method of metallic nanoparticles as biomaterials is less expensive and eco-friendly. The method involves bio-reduction, nucleation and growth, capping and stabilization of metals and transition metals, notably gold (Au), silver (Ag), platinum (Pt), palladium (Pd), copper (Cu), iron (Fe), zinc (Zn) precursors using phytochemical constituents of plants. A unique property of noble metals is the surface plasmon resonance (SPR) phenomenon, due to the availability of free electrons oscillating at the metal surface near the visible frequency range (260–800 nm). Absorption of UV light in this region is responsible for the color changes observed from their aqueous solutions [7–9]. Interactions of surface electrons with reducing agents from phytochemicals cause the free electrons to collectively oscillate at a unique frequency specific to each metal. With the interaction between phytochemicals (capping agents) and metal precursors, the surface chemistry of the metals after synthesis is enhanced making PBM-NPs as good drug carrier for targeted drug delivery and improving biodistribution and clearance in the body [10, 11]. Biomedical applications of these materials in living systems require consideration of the toxicity level of the nanoparticles. The synergies provided by phytochemicals in the synthesized metallic nanoparticles are believed to reduce the inherent toxicity of metal nanoparticles with living cells and elicit selective cytotoxicity on bacterial and cancer cells [12]. Based on recent findings, schematic studies of PM-NPs in biomedical applications covers their activities as bactericidal, anticancer, antioxidants, antifungal, cytotoxicity, formation, and mechanism of action [13].

2. Biosynthesis of metallic nanoparticles

Synthesis of nanoparticles generally involves two approaches: the top-down and bottom-up approaches. The top-down approach requires physical methods to disintegrate bulk materials into smaller units in the nano range. Typical methods such as electron beam lithography, laser ablation, spray pyrolysis, arc discharge, milling and so forth are physical methods used in a top-down approach. On the other hand, the bottom-up is simply a build-up of nanomaterials from smaller atoms/molecules through chemical or biological methods. In the chemical method, chemical substances with reducing properties are used to convert the metallic salt solution to a nanoscale [14]. Commonly used chemical reducing agents are sodium citrate, Cetyltrimethylammonium bromide (CTAB), sodium borohydride ($NaBH_4$), sodium tetrachloroaurate dihydrate salt ($NaAuCl_4.2H_2O$) etc. The physical and chemical methods are highly expensive and non-green methods resulting in increased environmental pollution and the release of toxic chemicals. Over the past few decades, there has been a surge in green chemistry through plant-based metallic nanoparticles research-based. The concept of green chemistry seeks to ensure the sustainable safety of humans, and the environment and process efficiency with the use of biodegradable and eco-friendly materials [15]. Thus, the biosynthetic route to PM-NPs helps to achieve this purpose with wide application in nanomedicine.

The biosynthesis of metal nanoparticles is a paradigm shift from the conventional physical and chemical methods due to their drawbacks. It is a biological method that applies green technology in the production of chemical substances using natural products of plants, microorganisms, and biomolecules from animals. The method has attracted huge attention as it displays efficient and effective utilization of the principle of green chemistry. Synthesis using plant extract and microorganisms is well documented with the advantage of being eco-friendly and does not involve the use of hazardous chemicals [2, 16]. However, the higher experimental cost and slow synthesis time due to microbial culture and growth, and the risk of infection with microorganisms are major concerns with the use of microorganisms. Biosynthesis through plant phytochemicals is relatively fast, safe, and suitable for large-scale synthesis [17]. Another biological method reported involves the use of natural products of animal origin. Although rarely used, this method involves biomolecules such as peptides or blood serum extracted from animals in nanoparticle synthesis [18, 19]. Animal blood serum consists of fibrinogen, globulin, albumin blood proteins and polypeptides that are biocompatible for use in the synthesis of NPs. These biomolecules are potentially reducing agents for metal ion reduction due to their involvement in oxidation/reduction reactions in animals. The redox properties, availability from slaughterhouses and their biocompatibility are reasons researchers used the blood serum as an alternative to the conventional synthesis of metallic nanoparticles [17].

In PM-NPs, extracts, and isolated bioactive compounds from plants serve as both reducing and stabilizing agents in place of chemical-reducing agents during biosynthesis. The effectiveness of biosynthesis with phytochemicals over non-biogenic synthesis was evidenced by Buono et al. [20] in the biogenic zinc oxide nanoparticles (ZnO-NPs) synthesis from the extract of *Lemna minor* (duckweed). The bioactive compounds of plants are mainly secondary metabolites (alkaloids, terpenoids, flavonoids, phenolics, steroids) that are not involved in the vegetative growth of plants but have an important function in the survival of plants as defense agents against herbivores, metal transporting agents, antibiotic agents, enzymes inhibitors etc. [3]. Activities of these secondary metabolites in green nanotechnology are reported to elicit bioreduction of metal ions to stable oxidation states through electron donations to metal ions to form stable atoms [14]. Because of this role of phytochemicals in green nanotechnology, much attention has been given to their extraction and isolation of the pure compounds.

2.1 Extraction and isolation of phytochemicals

Plant phytochemicals used for the synthesis of PM-NPs are mostly obtained from the leaves, stem, or tuber of the plant through decoction, infusion in water or aqueous ethanol [21]. The plant materials collected are cleaned of dirt and dried under shade for a few days until completely dried before use. The shade-dried material is ground to particulate matter to achieve complete extraction and a high percentage yield. The extract obtained is then sieved and filtered using Whatman No.1 filter paper. Further purification by centrifugation (600–800 rpm for 20 minutes), washing and filtration with an appropriate syringe filter give a pure extract of the plant. The freshly prepared aqueous extract can be made to powder by heating at 70°C or through lyophilization [5]. The above method is considered simple and conventional but with the limitation of involving the use of excess solvent and longer extraction time. The non-conventional methods for extraction of plant phytochemicals involve the application of improved technology like microwave, ultrasound, pressurized liquid, and enzymes assisted extraction [21, 22]. Both methods yield crude extract which serves as a bio-reducing/capping agent for the synthesis of PM-NPs. To identify specific phytocompounds present in the crude extract and their roles in biosynthesis, the crude extract is subjected to further extractions/isolation and characterization procedures such as HPLC, LC-MS, Automated Flash chromatography, NMR etc. [3, 23]. The identification of the roles of pure compounds from a crude extract in nano synthesis is interesting to phytochemistry and plant enthusiasts. However, for cost consideration. a literature search of the chemical profile of the plant species and purchase of chemically synthesized compounds for use in synthesis is mostly employed.

2.2 Phytochemicals in metallic nanoparticles synthesis

The exploration of alternatives route to the synthesis of metallic nanoparticles used for nanomedicines lead to the discovery of plant species and phytochemicals having reducing potentials for metal ions. The phytochemicals for metallic nano synthesis could be derived from the whole plant material or specific parts such as the leaves, stem, flowers, seed, or root. Generally, plant species and their phytochemicals could be nano-active or inactive by showing potential for the bio-reduction of metal ions. As discussed in the introduction section, the ability of phytochemicals to reduce metal ions is usually first determined by the SPR of the bio-reduced metal measured by recording the UV-Vis scan ranging from 300 to 800 nm [15]. The frequency of SPR absorption depends on the metal ion involved, nanoparticle size, shapes, aggregation, and crystallinity of the nanoparticles. Generally, AuNPs exhibit SPR at 500–600 nm, AgNPs show absorption at 400–500 nm, PdNPs and PtNPs at 300–400 nm, and CuNPs (280–330 nm) [23, 24]. **Table 1** shows the SPR range, color changes and other properties of some reported PM-NPs. Measurement of SPR of the metal ion is followed by confirmatory analysis through physicochemical characterization by one or more of; energy-dispersive X-ray spectroscopy (EDX), X-ray diffraction (XRD), Fourier transformed infra-red spectroscopy (FTIR), high-resolution transmission electron microscope (HRTEM), dynamic light scattering (DLS), thermogravimetry analysis etc. Several scientific reports have shown that nano-active plants consist of secondary metabolites rich in flavonoids and polyphenols [32, 33]. PM-NPs synthesis begins with the selection of potential plant species. This is usually through phytochemical screening of the plant species to identify the presence of secondary metabolites known with capping/reducing potentials of metal ions. The screening may be done through spectrophotometric assay for polyphenol and flavonoid content of the total extract or by reacting the total extract with specific reagents that give a characteristic color change to secondary metabolites [20, 34]. Del Buono et al. [20] showed

Plant	Phytochemical	PM-NPS	SPR (nm)	Color change	Size (nm)	Morphology	Bioactivities	Reference
Leucosidea sericea	Procyanidin dimers and *Leucosidea sericea* total extract	Ag	424–432	Yellowish to brownish	2–7	c, s, t	Antioxidant, antibacterial, and enzyme inhibition (α-Amylase and α-Glucosidase)	Badeggi et al. [6]
Lemna minor (Duckweeed)	Total extract of duckweed	Zn	369	NA	11.7	c, s,	Maize Growth, Chlorophyll Content, Carotenoids, Anthocyanin and MDA	Buono et al. [20]
Terminalia mantaly	Total extract of *Terminalia mantaly*	Au	500–600	Pale yellow to ruby red	22.5–43	c, s,	Anticancer (Caco-2, MCF-7, and HepG2)	Majoumouo et al. [25]
Aspalathus linearis	Total extract, and two pure chalcone compounds (helichrysetin and helichrysin)	Au	536 (Total extract), 540 (Helichrysin), and 546 (Helichrysetin)	Pale yellow to ruby red	12.5 (Total extract), 6.7 (Helichrysin) and 2.02 (Helichrysetin)	c, s, h, t	Enzyme inhibition (α-glucosidase and α-amylase) glucose uptake (HEK293 kidney cells) and cytotoxicity (HaCaT cells)	Omolaja et al. [23]
Curcuma longa	Total extract C. longa extract	Cu	524	yellow to brick brown	5–20	a	Antibacterial (*B. subtilis* (+ve) and *E. coli* (−ve)	Jayarambabu et al. [26]
Eucalyptus robusta Sm	Leave extract	Fe	280	Appearance of black color	0.8	c, a, s	antioxidant and antimicrobial (*Pseudomonas aeruginosa*, *Escherichia coli*, *Staphylococcus aureus* and *Bacillus subtilis*)	Vitta et al., [27]
Peganum harmala	Seed alkaloid fraction	Pt, Pd, and Pt-Pd respectively	269 (pt), 279 (Pd)	Yellow to dark brown	20.3 (Pt) 22.5(Pd), 33.5 (Pt-Pd)	c, s	Antioxidant and anticancer (A549 and MCF-7)	Fahmy et al. [28]
Averrhoa bilimbi	Fruit extract	Sn	280–290		3.08	c, s	Antioxidant and antimicrobial (*Klebsiella aerogenes*, *S. aureus*)	Nisha Elizabeth and Venkat Kumar [29]

Plant	Phytochemical	PM-NPS	SPR (nm)	Color change	Size (nm)	Morphology	Bioactivities	Reference
Citrus limon and *Citrus paradise* (grapefruits)	Fruit and peel extracts	Se	395	Pale yellow to brick red	4462	a, p	Antibacterial (*E. coli*, *M. luteus*, *B. subtilis* and *Klebsiella pneumoniae*)	Alvi et al. [30]
Viola betonicifolia	Leaves extract	Mn	NA	Yellowish green to brownish	10.5 ± 0.85 nm	c, s	Antibacterial (*K. pneumoniae* and *S. aureus*), antifungal (*Ammophilus fumigatus*), *Trichoderma harzianum*, and *A. flavus*, biofilm inhibition, anticancer (MCF-7) antioxidannt, and cytobiocompatibility (hMSC)	Lu et al. [31]

a, agglomerated; c, crystalline; h, hexagonal; s, spherical; t, triangular; p, polydisperse; NA, Not applicable.

Table 1.
Physicochemical properties of some plant-mediated metallic nanoparticles.

how the screening of *L. minor* (duckweed) plant for phenolic compound could be done through ultra-high-pressure liquid chromatography quadrupole-time-of-flight mass spectrometry (UHPLC-ESI/QTOF-MS). Once a plant is found to be nano-active, its optimum concentration for bio-reduction of the metal ion is determined using different concentrations on a pilot scale. The optimum concentration is then upscaled in the final synthesis. Two major synthesis procedures for PM-NPs have been reported. These procedures are the addition of phytochemicals to a stirring solution of a metal salt and the addition of a metal salt solution to a solution of phytochemicals on stirring. Akinfenwa et al. reported a procedure in which optimum concentration (10 mL of aqueous 5% each) of green rooibos extract, and aspalathin compound were added to a 90 mL of 1 mM heated solution of respective gold and silver precursors while stirring at 70°C. A similar procedure for zinc oxide nanoparticles where 25 mL of aqueous 1% mango seed extract was added to 75 mL of 10 mM zinc nitrate solution maintained at 30°C in an orbital shaker was described by Rajeshkumar et al. [35, 36]. Contrarily, different concentrations of the metal precursor are added to a predetermined concentration of phytochemicals. Thipe et al. explained this method in which 100 μL of 0.1 M $NaAuCl_4$ (in deionized water) was added to an aqueous solution of 2 mg resveratrol, a phytochemical from grapefruit in 6 mL of deionized water and the reaction mixture was stirred at room temperature and allowed to stir overnight to achieve optimal capping. Also, Elbagory et al. previously reported a method, for large-scale screening of plants with microtitre-plate. Through this method, the authors determined the optimum concentration for the biosynthesis of gold nanoparticles for further scale-up [36–38]. PM-NPs can also be synthesized by mixing plant extract clear solution with a predetermined concentration of the metal solution and boiling the above mixture at desired time and temperature while mixing. The choice of synthesis procedure depends on the hands-on experience of researchers and literature reports. In all cases, visual color change from the initial solution to the final is regarded as the first evidence for synthesis.

2.3 Mechanism and factors affecting the biogenic synthesis of metallic nanoparticles

Detailed literature report on the mechanism of biogenic synthesis using plant phytochemicals is still lacking although there exists extensive discussion on the roles of chemical functional groups present in plant phytochemicals. From a typical experience, proposed mechanism may proceed through a four-step reaction [16, 39, 40]. The first step which begins with the ionization of the metal in an aqueous solution and trapping of the metal ion on the surface of the phytochemical (dispersion medium) involves electrostatic interaction between the metal ions and the surface charge of phytochemicals. In the second step, attracted ions at the surface of the dispersion medium are reduced to form metal nuclei (nucleation) through electron transfer. In the case of an extract rich in polyphenol or other biomolecules, a metal-polyphenol framework is formed in the process. The continuous reduction and nucleation of ions result in the growth and accumulation of stabilized NPs. In the final stage, accumulated NPs are capped by plant phytochemicals which leads to the growth termination of NPs [24, 36, 41, 42]. Beyond the mechanism, researchers are also concerned about the size, shape, distribution of NPs, and time required in the synthesis. Hence, several factors that affect these properties have been identified and discussed. Factors such as the concentration and nature of plant extract, metal precursor concentration, temperature, time, and pH play significant roles in the properties (shape, size, and crystallinity) of the synthesized PM-NPs [42–44].

The potential of any plant species for bio-reduction of metal ion for synthesis depends on the contents of its active biomolecules and their concentrations in the plant extract. Plant extracts with high concentrations of biomolecules like proteins, phenolics, and flavonoids are known to greatly enhanced PM-NPs formation. Literature report shows that polyphenols and flavonoids are one of the most frequently reported phytochemicals responsible for plant-mediated metallic nanoparticle synthesis [33]. *Hibiscus sabdariffa* is a tea plant rich in polyphenols, anthocyanine compounds, and flavonoids including hibiscic acid, ascorbic acid luteolin, gallic acid, chlorogenic acid, caffeic acid, protocatechuic acid, eugenol, quercetin, delphinidin-3-sambubioside, delphinidin-3-glucoside, cyanidin-3-sambubioside and cyanidin-3-glucoside [45]. Soto-Roblesa et al. studied the effect of the different concentrations (1%, 4%, and 8%) of *H. sabdariffa* extract on the biogenic synthesis of ZnO NPs and their application in the photocatalytic degradation of methylene blue dye. The result showed that the formation of smaller (5–12 nm) ZnO-NPs of uniform shapes and photocatalytic degradation of methylene blue was achieved with the highest (8%) concentration of *H. sabdariffa* extract compared to the 1%, and 4% concentration [41, 46].

The ratio of the concentration of the metal substrate to plant extract plays a key role in biogenic synthesis. Generally, a concentration range of 1–10 mM of the metal substrate is employed in the synthesis [47]. Within this range, a lower concentration of metal precursor takes longer reaction time but is beneficial for the formation of monodispersed spherical shape and non-aggregated nanoparticles during the nucleation and growth process. The use of a high concentration of the metal precursor has the advantage to speed up the reaction rate but often leads to polydisperse and amorphous NPs due to the aggregation of a large number of nuclei [10]. This is supported by a study conducted by Vitta et al. on the synthesis of iron nanoparticles from an aqueous extract of *Eucalyptus robusta* Sm using different concentrations of iron salt concentrations; 1 mM, 5 mM, and 0.1 M $(NH4)_2\ FeSO_4.6H_2O$ (Mohr's Salt). The authors showed that the size distribution of FeNPs formed was dependent on the concentration of the metal substrate used, as the 1 mM concentration of Fe precursor gave average size distribution of 0.8 nm smaller than the size obtained with 5 mM and 0.1 M concentrations. Interestingly, this agrees with the report of the particle size distribution for Cu-NPs biosynthesized from *A. indica* leaf extract by Nagar and Devra [48]. Nagar and Devra, attributed an increase in particle size from 48.01 nm to 78.51 nm due to increased concentration of $CuCl_2$ substrate from 6×10^{-3} mol/L to 7.5×10^{-3} mol/L, respectively [27, 48].

The effect of temperature is noticeable mainly on the nucleation and NPs growth time, and the size distribution of PM-NPs. Most PM-NPs are formed between 30 and 90°C (working temperature) depending on the metal substrate used and their electrochemical potentials [49]. The working temperature is important for the excitation of electrons and interaction with phytochemicals for the bio-reduction of metals. In an experiment to study the effect of different temperatures (70, 75, 80, 85, and 90°C) on the size of the biosynthesized silver nanoparticle, Liu et al. [50] explained that high working temperatures have a positive effect on the nucleation kinetics constant k_1 and growth kinetics constant k_2. Nucleation rate constant k_1 was found to increases slightly when temperature was raised from 70 to 80°C, while it rises sharply when the temperature exceeds 80–90°C . To further the effect of temperature on the nucleation and growth of NPs, Jayachandran et al. [51] showed temperature dependence in the biosynthesis of zinc oxide nanoparticles using *Cayratia pedata* leaf extract for the immobilization of glucose oxidase enzymes. From the selected working temperature of 55, 65, and 75°C. It was noted that ZnO NPs were formed with a working temperature of 65°C while no synthesis occurred at 55 and 75°C respectively.

The role of pH maintained for reaction synthesis is also a key factor to be considered for controlled synthesis. Different literature studies attributed the effect of pH on the stability, morphology, optical and bioactivity of the synthesized PM-NPs [39, 52, 53]. A comparative stability and biological activity (cytotoxicity and antimicrobial) study of AgNPs synthesized with citrate, and green tea by Bélteky et al. [52] reveals that both NPs are more stable in alkaline and neutral media than in acidic medium with higher stability for green tea AgNPs. Additionally, synthesis in alkaline and neutral pH tend to produce smaller size particles than in acidic pH [9]. Also, the effects of pH in synthesis reflects on the optical properties of the synthesized NPs and their therapeutic applications. The optical properties of ZnO NPs were shown to depend on the pH. This was demonstrated by the report of Thipe et al. [34] showing different colors of ZnO-NPs synthesized from orange fruit peel extract for antibacterial activities. In the study, pH values of 4.0, 6.0, and 9.0 had ivory color, pH values of 7.0 and 8.0 had burnt black color and white color was noticed at pH values of 10 and 11. With Pt NPs synthesis via orange peel extract, Karim et al. [54] illustrated different size distribution and crystallinity properties for pH 3, 5, 9, 11, and 13. The report shows a reduction in sizes (2 nm – 1.8 nm), agglomeration, and uniform spherical shapes of NPs as pH increase from pH 3 through pH 11. pH 11 was noted with the smallest particle size (1.8 nm) and uniformity (which could be standard for the biosynthesis of Pt-NPs) while at pH 13 particles became agglomerated.

3. Therapeutic applications of phyto-metallic nanoparticles

New innovative technologies to improve the development of novel therapeutics and treatment of disease conditions using combined synergies between synthetic and natural products have been of interest to many nanomedicine experts. Extracts from plant has a long history of usage as traditional herbal medicine, as prototype in the formulation of synthetic drugs and in the detoxification of heavy metal accumulation in the body [55, 56]. The pharmacological potentials of many plants are closely associated to the major phytochemicals they contain. For instance, plants rich in polyphenols are often indicated as therapeutic agents for complications resulting from oxidative stress. A systematic formulation of these therapeutic agents as nano drugs for use in biomedicine is gaining acceptance due to the rising preference for natural product-based therapies and their versatilities [32]. The bioinorganic network provided by PM-NPs and their biocompatibility make them suitable as a drug over conventional speciality drugs with many side effects. Hence, their roles in therapeutic applications such as in bactericidal, anticancer, antioxidants, antidiabetic, and photothermal treatment (PTT) are briefly discussed below.

3.1 Bactericidal effects

The antibacterial potentials of various PM-NPs have been extensively discussed by different researchers given the resistance of bacterial to the first generation and conventional antibiotic drugs. Most reports show that the activities of the PM-NPs are attributed to the binding and interaction of PM-NPs on surfaces of bacterial cell membranes. The interaction inhibits synthesis of new cell membrane thus preventing cell growth. Also, the binding of PM-NPs could disrupt the energy transduction process, and cause damage to the cell membrane through the generation of reactive oxygen species leading to the eventual death of bacterial cells [57–59]. The efficacies of the bactericidal effect of PM-NPs have been tested both as a singular drug and in combination treatment with conventional antibiotic drugs [9]. Among metals and metal compounds, Ag ions and Ag-based compounds are the most pernicious on microorganisms showing strong antibacterial properties [60]. Hence, PM- AgNPs are the most studied for the inhibition of a broad spectrum of both gram-negative (G-ve) and gram-positive (G + ve) bacteria species including *Pseudomonas aeruginosa*, *Staphylococcus aureus*, *Bacillus cereus*, *Salmonella enterica*, *Escherichia coli* and *Serratia marcescens* [60]. This was recently demonstrated by Balachandar et al., [61] for antibacterial activities of AgNPs mediated by *Glochidion candolleanum* (GC) leaf against *Bacillus subtilis, Listeria monocytogens* and *S. aureus* (G+ve)*; E. coli, P. aeruginosa, and S. enterica* (G−ve). The study shows a general inhibition for all bacteria based on the concentration of GC-AgNPs used. The highest maximum inhibition zone (12.2 mm diameter) due to GC-AgNPs was observed against S. enterica and for P. aeruginosa *(*11.8 mm diameter), *L. monocytogens* (11.0 mm), *S. aureus* (10.9 mm) and 10.8 mm for both *B. subtilis and E. coli* in the descending order [61]. A similar result was presented by Nishanthi et al., [9], which showed a synergistic antibacterial activity of rind extract of *Garcinia mangostana* biosynthesized Au, Ag, and Pt NPs in combination with commercial antibiotics (penicillin G (2.0 μg), methicillin (5.0 μg), vancomycin (30 μg), gentamycin (50 μg), streptomycin (10 μg), ciprofloxacin (5.0 μg), azithromycin (30 μg) and clotrimoxazole (25 μg) and tested against *Staphylococcus* sp., and *Bacillus* sp. (G+ve) and *Pseudomonas* sp. and *Klebsiella* sp. (G+ve). The authors found that AgNPs, displayed relatively higher antibacterial activity when compared to AuNP and PtNPs against all the tested pathogens [9]. In addition to Ag, Au, and Pt NPs, biosynthesized ZnO-NPs, and FeNPs using extracts of orange fruit peel, and *E. robusta* Sm, respectively have been reported [27, 62]. ZnO/Zn $(OH)_2$ NPs are known to exhibit photocatalysis when illuminated with UV light producing reactive oxygen species (ROS), such as superoxide anion ($O_2^{\bullet-}$) or hydroxyl radical ($OH^{\bullet}$) which may cause disruption of the electron transport chain and or oxidative stress in the microbial cell membrane [59]. ZnO-NPs were shown to exhibit strong antibacterial activities towards *E. coli* and *S. aureus* at a concentration of 0.025 mg/mL without UV radiation after 8 h of incubation [62]. Vitta et al. [27] recorded no significant differences ($p > 0.05$) for *E. Coli* with Fe-NPs (positive control), while for *S. aureus, P. aeruginosa and B. subtilis*, statistically significant differences ($p < 0.05$) were found as there was an increase in the values of inhibition zone as the size of the nanoparticles diminished [62]. Also reported is, the biogenic synthesis and antimicrobial evaluation of Cu-NPs and Mg-NPs from leaf extracts of strawberry and *Viola betonicifolia (L.)* respectively. Bayat et al. [63] noted that the Cu-NPs like their Ag-NPs counterpart displayed a bactericidal effect on *P. aeruginosa* at an effective concentration (EC50) of 2.2 mg/mL. Interestingly, the authors indicated a higher antibacterial activity of Ag-NPs over Cu-NPs due to a greater minimum bactericidal concentration (MBC) of Cu-NPs (5 mg/mL) than for MBC of Ag-NPs (0.01 mg/mL). According to Lu et al. [31], synthesized *Viola betonicifolia*-MnO_2 NPs displayed higher antimicrobial and reductions in colony forming unit for *K. pneumoniae* (4.14 ± 0.03 $\log_{10}$ reduction), and *S. aureus* (4.65 ± 0.07 $\log_{10}$ reduction) respectively, than the value obtained from commercially available Mn-NPs [31].

Among other metals that have received considerable attention for the biosynthesis of PM-NPs as antimicrobial agents are titanium (TiO_2-NPs), palladium (Pd-NPs) nickel (NiO-NPs), selenium (Se-NPs), and tin (SnO_2-NPs) [30, 64–67]. Amanulla and Sundaram [64] reported that at dose concentrations from 6.75 to 50 mg/mL, TiO_2-NPs synthesized from orange peel extract showed bactericidal effects on *S. aureus, E. coli, and P. aeruginosa*. In a similar report, Pd–NPs (±3 nm) obtained via *Garcinia pedunculata roxb* leaf extract was observed to elicit antimicrobial activity against *Cronobacter sakazakii* AMD 04 at dose concentration from 0.39 mM and 0.52 mM [66]. From the report of Srihasam et al. [65], biogenic synthesis and antimicrobial activities of NiO-NPs from the leaf extract of stevia plant were implicated in significant bactericidal effects on *E. coli, Streptococcus pneumoniae, and B. subtilis* at a concentration of 200 μg/mL. Furthermore, Alvi et al. [30] demonstrated that Se-NP synthesized from *Citrus paradisi and Citrus limon* for antimicrobial activities against *E. coli, M. luteus, B. subtilis, and K. pneumoniae* show significant activities against all the bacterial pathogens when compared with the standard antibiotic Ciprofloxacin. SnO_2-NPs are also promising PM-NPs that have been found effective against microbes. SnO_2-NPs from fruit extract of *Averrhoa bilimbi* showed satisfying inhibitory activities against *S. aureus* and, *K. aerogenes* while from *Saraca indica* flowers were effective in inhibiting the growth of *E. coli* [67, 68].

3.2 Anticancer effects

The applications of metallic nanoparticles in the treatment of cancer cells from different organs such as breast, cervical, colon, ovarian, and lung cancers have attracted more interest in biomedicine. In this regard, MNPs act as nanocarriers due to the available large surface area for the attachment of a large number of vectors for targeted delivery at designated sites [69]. This method has been reported to improve the efficacy of anti-cancer drugs over traditional drugs which are sometimes identified and removed from the systemic circulation by the liver and the spleen [16, 36]. Usually, studies on anticancer potentials of nanoparticles require both in vitro and in vivo (animal and clinical) studies. In vitro, studies are mostly reported using conventional MTT and apoptosis assay to determine cell viability via cytoprotection, and extrinsic cell death respectively. In the experiment, a decrease in cell metabolic activities is implicated in the induced secretion of reactive oxygen species by cells due to biogenic NPs which gradually leads to the death of cellular components [70]. The values of half maximal inhibitory concentration IC50 are used to determine the cell behaviors when treated with PM-NPs. The in vivo experiment is a follow-up and confirmation of in vitro, although limited studies of in vivo antitumor studies are reported [71]. Because MNPs are functionalised and stabilized for selective delivery at specific sites, biodistribution and degradation of the MNPs may take much circulation time in systemic circulation resulting in metal accumulation and particle-induced toxicity of mammalian cells and tissues. Different biodegradable materials that help with bio-clearance such as plant phytochemicals, polysaccharides such as dextran and chitosan, polyvinyl alcohol (PVA), phosphorylcholine-based copolymers and so forth have been suggested for coating of MNPs [15, 72, 73].

Addressing the drawback of drug residence time and biodistribution MNPs as a potential anticancer agent in mammalian cells, Akinfenwa et al. [36] showed that phytochemicals from *Aspalathus linearis* plant from PM-NPs could serve to reduce, capped, and increase bio-clearance of biosynthesized AgNPs and AuNPs. From the results of the study, the authors highlighted that AgNPs are more efficacious showing antiproliferation effects on human hepatocellular carcinoma (HepG2), and neuroblastoma (SH-SY5Y) cells in vitro than AuNPs. In a complementary report by Alharbi and Alsubhi [74], in vitro anticancer activity of AgNPs prepared using fruit extract of *Azadirachta indica* on pneumocyte lung tumor (A549) cells show that all treatments involving biosynthesized AgNPs only, and AgNPs in combination with cisplatin were more toxic to A549 cells than with the extracts and cisplatin only. A remarkable combination of both in vitro and in vivo anticancer experiments of biogenic AgNPs was demonstrated by Kabir et al. [70]. The authors concluded that Ag/AgCl-NPs synthesized from *Geodorum densiflorum* rhizome extracts inhibit human cancer cell proliferation in vitro and also for Ehrlich ascites carcinoma (EAC) cell growth in vivo. From the in vitro experiment, the authors found that samples treated with *G. densiflorum*-Ag/AgCl-NPs induced apoptosis in glioblastoma stem cells (GSCs), pancreatic cancer (BxPC-3) and breast cancer (MCF-7) cells. While in vivo, there was inhibition of up to 60 and 95% of EAC-mice cell growth at the doses of 2 and 4 mg/kg/day after intraperitoneal treatment respectively. Noteworthy is the overall increase in mice life span by 75% compared to EAC-bearing control mice which gives credence to the efficacy of PM-NPs. A prototype drug for chemotherapeutic treatment of cancer is Cisplatin, a platinum coordination complex found with most antitumor properties although with some side effects such as neurotoxicity, ototoxicity and renal impairment. Formulation of the complex analogue for cancer therapy through biosynthesis is a welcome idea to overcome its side effects. In the study of the anticancer effects of PtNPs synthesized from black cumin seed extract, Aygun et al. [75] found that PtNPs were efficient against the proliferation of MDA-MB-231 breast and HeLa cervical cancer lines (IC50: 36.86 μg/mL and 19.83 μg/mL, respectively). The result agrees with a previous result by Chuang et al. [76] of biogenic PtNPs from peppermint leave extract which showed a decrease in the viability of HTC 116 colon cancer cells at an IC50 value of 20 μg/mL. Similarly, dose-dependent anticancer activities of biogenic Au NPs (inhibition of MCF-7 cells by 70.2%), CuO NPs (decrease breast cancer MDA-MB-231 viability and increase ROS at IC50: 20 μg/mL), MnONPs (cytotoxic against MCF-7 cells at 120 μg/mL concentration), ZnO NPs (toxicity on both A549 and MOLT4 cells and a size reduction in A549 tumor), NiO NPs (toxicity against HepG2 cancer cells, IC_{50}: 37.84 μg/mL) and Se NPs (decrease in cell viability at a concentration of 2–6 μg Se/mL) have been described as promising anticancer agents through different studies [31, 77–82].

It is believed that combining two metals in biosynthesized as bimetallic NPs (BMNPs) often synergize their effects with more biological potentials than their monometallic counterpart. As anticancer agents, biogenic BMNPs have been reported with improved activities over their respective single metals [5]. Several reports supported this hypothesis. Tamimi et al. [83] report showed that chemically synthesized bimetallic Ag-Cu NPs showed significant toxicity on MCF-7 cells at 10 μg/mL concentration than the separate AgNPs (20 μg/mL), and Cu-NP (showed no toxic effects). Hence, the hypothesis should be true for biogenic BM-NPs. Fahmy et al. [28] in their report on green synthesis of Pt-Pd NPs using *Peganum harmala* L. seed alkaloids revealed that bimetallic Pt–Pd NPs exhibited significant cytotoxic activities against A549 (IC50; 8.8 μg/mL) and, MCF-7 (IC_{50}; 3.6 μg/mL) cells when compared to the individual metals, and carboplatin standard (IC_{50}; 23 and 9.5 μg/mL,). In the study, relatively lower anticancer activities of each mono MNPs of Pt NPs and PdNPs for A549 and MCF-7 were noted at IC_{50}; 10.9 μg/mL and 6.7 μg/mL, and IC_{50}; 31 μg/

mL and 10.8 μg/mL, respectively [28]. A similar study by Athinarayanan et al. [84] revealed that bimetallic Pt-Cu NPs synthesized with catechin extract of green tea induced cell death in human cervical cancer dells (SiHa) cells with increasing the dose and time with IC_{50} value of the Pt-Cu NPs ranged from 32 to 35 g/mL for 24 and 48 h. All these reports of BMNPs from plants portend higher efficacy in therapeutics within the limit of acceptable toxicity.

3.3 Antioxidants potentials

Over-expression of free radicals is regarded as a lead factor to a variety of diseases including cancer, hyperglycemia, hyperlipidemia, inflammation, and so forth via apoptosis signaling pathways and oxidative stress on cellular components. Free radicals such as reactive nitrogen species (RNS; nitrogen dioxide, NO_2), reactive sulfur species (RSS; hydrogen sulphide, H_2S), reactive chlorine species (RCS; hypochlorous acid, HOCl), and most especially reactive oxygen species (ROS; superoxide, ($O_2^{\bullet -}$), are pro-oxidants which induce cellular oxidative stress [85]. Natural antioxidants produced by cells help to check on the excess production of ROS by scavenging activity. However, when the ROS level becomes higher than the natural antioxidant activities of cells, the use of natural antioxidants from plants or food rich in antioxidants to mitigate oxidative stress becomes necessary. Interestingly, compounds of the phenolic and flavonoid classes especially vitamins A, C, and E, quercetin, and rutin have been shown to attenuate ROS-induced oxidative stress with different radical-based and metal-related assays [86, 87]. Methods used to evaluate free radical scavenging includes 1,1-diphenyl-2-picryl-hydrzyl (DPPH) radical, 2,2-azino-bis-3-ethylbenzothiazoline-6-sulfonic acid (ABTS) radical, ferric reducing antioxidant power (FRAP), nitric oxide (NO), superoxide ($O_2^{\bullet -}$), hydroxyl ($OH^{\bullet}$) cupric reducing antioxidant capacity (CUPRAC), Trolox equivalent antioxidant capacity (TEAC) assays etc. [88]. It is proposed that during the interaction, the metal ions in PM-NPs scavenge free radicals by electron transfer and proton loss thus inhibiting oxidative DNA damage. Hence, the therapeutic potentials of PM-NPs are largely associated with the phenolics, and antioxidant properties of the phyto-reducing agent and the metal ions used during synthesis. According to Vera et al. [89], the antioxidant activities of plant phytochemicals are a strong indication of the synthesis efficiency of MNPs and their biomedical applications. Antioxidant-functionalized NPs are reported to show comparative advantages over antioxidants due to their high permeability and stability during membrane trafficking. Different PM-NPs have demonstrated nano-antioxidant activities which serve as an alternative route to the use of conventional synthetic antioxidant therapy. For example, *Rhazya stricta* plant (Apocynaceae), used in folk medicine in the Middle East and Indian subcontinent is reported with antioxidant activities. When functionalized as AgNPs, exhibited superior two-folds antioxidant activity measured at 75.16% ± 0.04 over the plant extract (43.12% ± 2.1) [90]. It is also documented that the antioxidant activity of biosynthesized MnO_2 NPs using leaves extract of *Viola betonicifolia* on linoleic acid peroxidation was higher (84.94 ± 0.77%) than for the leaves extract and a little less than that of ascorbic acid (90.57 ± 1.21%) standard [31]. When the inhibition of DPPH radicals of FeNPs from an aqueous extract of *E. robusta* leaves was compared to the extract alone, Vitta et al. [27] showed that FeNPs were significantly more potent ($p < 0.01$) than the extract alone, showing IC50 values of 81.63 ± 11.75 μg/mL and 423.14 ± 73.27 μg/mL respectively [27]. Overall, the improved antioxidant activities of PM-NPs have opened research interest in phyto-nano therapy for the repair of damaged cellular macromolecules (like proteins, lipids, DNA) and regulation of cellular functions from degenerative and pathological ailments such as aging, cancer, diabetes, and neurodegenerative diseases.

3.4 Antidiabetic application

Type 2 diabetes is a result of hepatic dysfunction and insufficient insulin secretion from pancreatic β-cells for glucose uptake accounts for over 90% the worldwide diabetic patients [91]. Being a global disease, a lot of efforts on developing therapeutic agents and advocacies to reduce the increasing number of undiagnosed and diabetes are being explored. Synthetic drugs such as sulfonylureas, metformin, acarbose are among current antidiabetic drugs. In addition to these synthetic drugs, natural products from plants general and plant-derived phytochemicals such as chalconaringenin 2′-*O*-β-*D*-glucopyranoside and aureusidin 6-*O*-β-*D*-glucopyranoside, obtained from flower extract of *H. arenarium* have been reported [55]. Despite the several *in vivo, in vitro,* and clinical studies, till date, a complete cure for treatment of diabetes is yet to be achieved. Recently, several nanomedicine research are focusing on nanoscale drugs by combining the synergies between potential antidiabetic phytochemicals and metallic nanoparticles. In this regard, Ul Haq et al. [92] showed that plant mediated silver nanoparticles from *Taverniera couneifolia* (*TC*-AgNPs) significantly improved Alloxan-induced diabetic Wistar rats. The effects of *TC*-AgNPs on the lipid, liver, and kidney profiles of treated rats (10 mg/kg body weight) were observed with lowered blood glucose levels, and improvements in the lipid, liver, and renal profiles after treatment with *TC*-AgNPs. A similar therapeutic application of ZnO NPs synthesized from *Amygdalus scoparia* stem bark extract showed significant antidiabetic potentials on treated rat (30 mg/kg). As reported, a higher level of insulin and lower AST, ALT, lower blood glucose, higher levels of IR, GluT2, and GCK expression and lower TNFα expression were recorded when compared with the STZ induced diabetic rats [93]. These reports together give insight into the biomedical and therapeutic application of phyto-metallic nanoparticles.

3.5 Photothermal and photocatalytic effects

The most important property of zerovalent metallic NPs is their electronic excitation and light scattering characters under incident laser beam. The absorption and scattering of light give rise to intense colors and interesting optical properties. These properties are used in theranostic nano probe and therapy in biomedicals such as in biosensing, tissue imaging, molecular imaging, cancer therapy etc. Photothermal therapy is less invasive using therapeutic agents like noble metals and near infrared radiation (NIR). Specifically, AuNPs are known to exhibit excellent photothermal transduction in the interconversion of heat-to-light by absorbing light in the near infrared region (NIR; 700–1400 nm). For example, NIR laser radiations delivery by AuNPs could induce stimulations in cells for necrosis, healing of open wound, in pain relief therapy etc. This property of gold and the paramagnetic property of iron metals and irradiation by NIR have been used for active targeting cancer cells, drug carrier and delivery, and removal of cancer cells [94]. The synergy between the NIR and PM-NPs (as therapeutic agents) has been explored by different researchers. Ali et al. [95] and Wang et al. [96] studies showed an enhanced cytotoxicity on lung (A549), and breast cancer (MCF-7) cell lines when treated with NIR irradiated synthesized *Cordia myxa L.* leaf extract-PtNPs and *Memecylon edule* leaf extract-graphene oxide nanoparticles (GO-NPs) respectively. The effects were induced apoptosis in MCF-7/A549 cells due to photothermal transduction (conversion of low energy into heat) by the respective PtNPs and GO-NPs.

In photocatalysis experiment, the decolorization of dyes; methyl orange, methylene blue, Eosin Y, etc. by UV light in the presence of metal catalyst which result in degradation products are usually reported. During UV irradiation of samples, the photocatalysts cause electron excitation from the Valence to the Conduction band where the material becomes chemically responsive indicated by absorption peak at λ wavelength. The difference between the valence and the conduction band is calculated as Bandgap (Eq. 1). Phytochemical-inspired metallic nanoparticles could improve the photocatalytic degradation of dyes for treatment of wastewater through reduction of conduction electrons for UV absorption spectral determination. This effect was implied by Gupta and Chundawat [97] for the photocatalytic deterioration of methyl orange by fungus mediated PtNPs for potential application in wastewater remediation.

$$E_g = h\ c_e V/\lambda \text{ or } E_g = 1240_e V/\lambda \tag{1}$$

$H = 6.626 \times 10^{-34}$ JS, $c = 3 \times 10^{8}\ ms^{-1}$, λ = absorption wavelength.

4. Cellular uptake and toxicity of PM-NPs

Cellular uptake of PM-NPs accounts for the dose of internalized NPs and their interactions in physiological media. The uptake of NPs has been investigated by different methods; quantitatively using inductively coupled plasma-mass spectrometry (ICP-MS), inductively coupled plasma-optical emission spectroscopy (ICP-OES), and inductively coupled plasma-atomic emission spectroscopy (ICP-AES). The qualitative method uses transmission electron microscopic (TEM), energy dispersive X-ray (EDX) analysis, colourimetry, and fluorescence imaging. Cellular uptake experiment is important to understand the stability of NPs and their toxicities on both normal and cancer cells. Most uptake studies discussed uptake efficiency and toxicity. Still, the exact mechanisms of cellular uptake for PM-NPs in different cells have not been well characterized [98]. It is however clear that in the process of uptake NPs used as drug carriers permeate biological barriers of the bilayer cell membrane (consisting of hydrophobic and hydrophilic layers) through a passive diffusion pathway to distribute the NPs within cells. Most studies show that permeation and assimilation of NPs depend on several factors such as surface modification/interactions between the NPs and the cells, size, shape and concentrations of NPs [99]. NPs have surface charges (positive, negative, or neutral) that attract ions of opposite charges in the cytosol for cellular uptake. Different reports have shown that positively charged NPs show better cellular uptake than neutral and negatively charged NPs [2]. This is supported by a report of high cellular uptake of AuNP synthesized from *A. linearis* by SH-SY5Y and HepG2 cells. The authors attributed the uptake of AuNPs at a concentration of 5.368 μg/mL and 3.625 μg/mL by SH-SY5Y and HepG2 cells respectively to high acidity and positive surface charges from *A. linearis* [36, 100]. Also, the nano (small) size factor of PM-NPs plays an important role in the cellular internalization of NPs. Because microorganisms and cell organelles such as DNA have nano-size ranges, they can cross the plasma membrane faster and with uniform distribution for intracellular particle trafficking. Different reports revealed that smaller-size PM-NPs ($\leq$ 40 nm) easily permeate cells than sizes greater than 50 nm. Amaliyah et al. demonstrated that the antibacterial activity of biogenic AgNP from *Piper retrofractum* Vahl fruit extract was facilitated by the entry of the small sizes (1–5 nm) AgNPs through *E. Coli* and *S. Aureus* [98, 101].

Nanotoxicological assessment of NPs has become essential before application in therapy to determine their toxicity safe level. Toxicological studies are mostly conducted in vitro using standard toxicological assays, such as the MTT assay, colony forming efficiency (CFE), lactose dehydrogenase (LDH) assays and so forth in immortalized cell lines. MTT assay measures cell viability via mitochondrial enzymatic activities, CFE checks the ability of a single cell to form a colony by survival from toxic media, and LDH assay measures the damage to the cell membrane. Every organism is made up of natural defense mechanisms which protect it from toxic exogeneous particles. Despite this, exposure to MNPs beyond a threshold could overcome the body's defense to cause hazards [102]. Toxicity by MNPs may occur during preparation or on therapeutic administration leading to cytotoxicity and oxidative DNA damage. Like cellular uptake, the toxicological behavior of NPs is dependent on their surface activities and morphological properties (size, shape, and agglomeration). It is well-known that spherical, smaller sizes and non-agglomerated NPs are more compatible and less toxic to cell membranes. Again, the toxicity of NPs is dependent on exposure concentration and the amount of the NPs that reached the targeted cells [15]. PM-NPs are considered non toxic due to the synergy that exist between metals and bioorganic networks of the reducing phytochemicals. However, PM-NPs have safe level if delivered to specific cancer cells to elicit DNA damage and cell multiplication [103]. Cellular toxicity of PM-NPs is beneficial when it displays specificity on targeted cells. The therapeutic potential of PM-NPs is lost where it is toxic against normal cells. Toxicity on cells is generally established through in vivo and in vitro experiments. Few published reports have also shown the toxic effects of PM-NPs on normal cells and these reports are briefly discussed. For example, Majoumouo et al. [25] reported in vitro cytotoxic effects of biosynthesized *Terminalia mantaly* extracts gold NPs (TM-AuNPs) investigated on cancer (Caco-2, MCF-7 and HepG2) and non-cancer (KMST-6) cell lines. According to the study, *T. mantaly* extracts showed some cytotoxicity towards the cancer cells while the TM-AuNPs exhibited more cytotoxicity on all the cells. The effects of PM-NPs concentrations on toxicity were studied in vivo on EAC cells using *G. densiflorum* -AgNP. At a higher concentration of 4 mg/kg/day, the biosynthesized inhibited growth of mice tumor cells up to 95% greater than 60% recorded on a dose of 2 mg/kg/day [70].

5. Conclusions

In summary, the synthesis of phyto-metallic nanoparticles is a paradigm shift and preferred approach to conventional environmentally unfriendly and unsustainable nanodrugs. Tremendous efforts are being made to combine the principle of green chemistry and nanotechnology for application in nanomedicine. Phytochemistry and drug development from phytochemicals functionalized on metal nanoparticles have been revisited. Factors like morphology of nanoparticles, time, temperature and pH are identified as critical in the synthesis and therapeutic applications of phyto-metallic nanoparticles. Reviews of several scientific results that compared the effectiveness of phytomediated nanoparticles with their separate phytochemicals, and separate metallic nanoparticles as potential therapeutic agents favored phytomediated nanoparticles. A more definitive conclusion on the subject will unfold with further investigations. This chapter gives current insights into the green biogenic synthesis of nanoparticles, mechanisms, therapeutic applications, uptake, and cellular toxicity.

Acknowledgements

We wish to acknowledge the Cape Peninsula University of Technology for its financial support towards the completion of this research.

Conflict of interest

The authors declare no conflict of interest.

Author details

Akeem Omolaja Akinfenwa[1,2]* and Ahmed Abdelfattah Hussein[1]

1 Department of Chemistry, Cape Peninsula University of Technology, Bellville Campus, Bellville, South Africa

2 Department of Chemical Sciences, Summit University Offa, Kwara State, Nigeria

*Address all correspondence to: oa.akeemlaja@gmail.com

References

[1] Jadoun S, Arif R, Jangid NK, Meena RK. Green synthesis of nanoparticles using plant extracts: A review. Environmental Chemistry Letters. 2021; **19**:355-374. DOI: 10.1007/s10311-020-01074-x

[2] Hiba H, Thoppil JE. Medicinal herbs as a panacea for biogenic silver nanoparticles. Bulletin of the National Research Centre. 2022;**46**:9. DOI: 10.1186/s42269-021-00692-x

[3] Swilam N, Khaled AN. Polyphenols profile of pomegranate leaves and their role in green synthesis of silver nanoparticles. Scientific Reports. 2020; **10**:14851. DOI: 10.1038/s41598-020-71847-5

[4] Hammad DM, Asaad AA. A comparative study: Biological and chemical synthesis of iron oxide nanoparticles and their affinity towards adsorption of methylene blue dye. Desalination and Water Treatment. 2021;**224**:354-366. DOI: 10.5004/dwt.2021.27053

[5] Berta L, Coman NA, Rusu A, Tanase C. A review on plant-mediated synthesis of bimetallic nanoparticles, characterisation and their biological applications. Materials. 2021;**14**:7677. DOI: 10.3390/ma14247677

[6] Badeggi UM, Badmus JA, Botha S, Ismail E, Jeanine ML, Africa CWJ, et al. Biosynthesis, characterization, and biological activities of procyanidin capped silver nanoparticles. Journal of Functional Biomaterials. 2020;**11**:66. DOI: 10.3390/jfb11030066

[7] Van Staden AB, Kovacs D, Cardinali G, Picardo M, Lebeko M, Khumalo NC, et al. Synthesis and characterization of gold nanoparticles biosynthesised from *Aspalathus linearis* (Burm.f.) R. Dahlgren for progressive macular hypomelanosis. Journal of Herbal Medicine. 2021;**2**: 100481. DOI: 10.1016/j.hermed.2021.100481

[8] Peralta-Videa JR, Huang Y, Parsons JG, Zhao L, Lopez-Moreno L, Hernandez-Viezcas JA, et al. Plant-based green synthesis of metallic nanoparticles: Scientific curiosity or a realistic alternative to chemical synthesis? Nanotechnological Environment Engineering. 2016;**1**:4. DOI: 10.1007/s41204-016-0004-5

[9] Nishanthi R, Malathi S, John Paul S, Palani P. Green synthesis and characterization of bioinspired silver, gold and platinum nanoparticles and evaluation of their synergistic antibacterial activity after combining with different classes of antibiotics. Materials Science & Engineering C. 2019;**96**:693-707. DOI: 10.1016/j.msec.2018.11.050

[10] Nguyen NTT, Nguyen LM, Nguyen TTT, Nguyen TT, Nguyen DTC, Tran TV. Formation, antimicrobial activity, and biomedical performance of plant-based nanoparticles: A review. Environmental Chemistry Letters. 2022; **20**:2531-2571. DOI: 10.1007/s10311-022-01425-w

[11] Kumar H, Bhardwaj K, Nepovimova E, Kuča K, Dhanjal DS, Bhardwaj S, et al. Antioxidant functionalized nanoparticles: A combat against oxidative stress. Nanomaterials. 2020;**10**: 1334. DOI: 10.3390/nano10071334

[12] MohamadSukri SN, Shameli K, Wong MM-T, Teow SY-T, Chew J, Ismail NA. Cytotoxicity and antimicrobial activities of plant-

mediated synthesized zinc oxide (ZnO) nanoparticles using *Punica granatum* (pomegranate) fruit peel extract. Journal of Molecular Structure. 2019;**1189**:57-65

[13] Hano C, Abbasi BH. Plant-based green synthesis of nanoparticles: Production, characterization, and applications. Biomolecules. 2022;**12**:31. DOI: 10.3390/biom12010031

[14] Khan F, Shariq M, Asif M, Siddiqui MA, Malan P, Ahmad F. Green nanotechnology: Plant-mediated nanoparticle synthesis and application. Nanomaterials. 2022;**12**:673. DOI: 10.3390/nano12040673

[15] Anu R, Krishna Y, Sheeja J. A comprehensive review on green synthesis of nature-inspired metal nanoparticles: Mechanism, application and toxicity. Journal of Cleaner Production. 2020;**272**:122880. DOI: 10.1016/j.jclepro.2020.122880

[16] Mustapha T, Misni N, Ithnin NR, Daskum AM, Unyah NZ. A review on plants and microorganisms mediated synthesis of silver nanoparticles, role of plants metabolites and applications. International Journal of Environmental Research and Public Health. 2022;**19**:674. DOI: 10.3390/ijerph19020674

[17] Selvakesavan RK, Franklin G. Prospective application of nanoparticles green synthesized using medicinal plant extracts as novel nanomedicines. Nanotechnology, Science and Applications. 2021;**14**:179-195. DOI: 10.2147/NSA.S333467

[18] Kakakhel MA, Sajjad W, Wu F, Bibi N, Shah K, Yali Z, et al. Green synthesis of silver nanoparticles and their shortcomings, animal blood a potential source for silver nanoparticles: A review. Journal of Hazardous Materials Advances. 2021;**1**:100005. DOI: 10.1016/j.hazardv.2021.100005

[19] Singh AV, Jahnke T, Xiao Y, Wang S, Yu Y, David H, et al. Peptide-induced biomineralization of tin oxide (SnO_2) nanoparticles for antibacterial applications. Journal of Nanoscience and Nanotechnology. 2019;**19**(9):5674-5686. DOI: 10.1166/jnn.2019.16645

[20] Del Buono D, Di Michele A, Costantino F, Trevisan M, Lucini L. Biogenic ZnO nanoparticles synthesized using a novel plant extract: Application to enhance physiological and biochemical traits in maize. Nanomaterials. 2021;**11**:1270. DOI: 10.3390/nano11051270

[21] Azmir J, Zaidul ISM, Rahman MM, Sharif KM, Mohamed A, Sahena F, et al. Techniques for extraction of bioactive compounds from plant materials: A review. Journal of Food Engineering. 2013;**117**(4):426-436. DOI: 10.1016/j.jfoodeng.2013.01.014

[22] Alvand ZM, Rajabi HR, Mirzael A, Masoumiasl A. Ultrasonic and microwave assisted extraction as rapid and efficient techniques for plant mediated synthesis of quantum dots: Green synthesis, characterization of zinc telluride and comparison study of some biological activities. New Journal of Chemistry. 2019;**43**:1512-15138. DOI: 10.1039/C9NJ03144H

[23] Omolaja AA, Pearce B, Omoruyia SI, Badmus JA, Ismail E, Marnewick JL, et al. The potential of chalcone-capped gold nanoparticles for the management of diabetes mellitus. Surfaces and Interfaces. 2021;**25**:101251. DOI: 10.1016/j.surfin.2021.101251

[24] Narendra Kumar HK, Chandra Mohana N, Nuthan BR, Ramesha KP, Rakshith D, Geetha N, et al. Phyto-

mediated synthesis of zinc oxide nanoparticles using aqueous plant extract of *Ocimum americanum* and evaluation of its bioactivity. SN Applied Sciences. 2019;**1**:651. DOI: 10.1007/s42452-019-0671-5

[25] Majoumouo MS, Sharma JR, Sibuyi NRS, Tincho MB, Boyom FF, Meyer M. Synthesis of biogenic gold nanoparticles from *Terminalia mantaly* extracts and the evaluation of their in vitro cytotoxic effects in cancer cells. Molecules. 2020; **25**:4469. DOI: 10.3390/molecules25194469

[26] Jayarambabu N, Akshaykranth A, Rao TV, Rao KV, Kumar RR. Green synthesis of Cu nanoparticles using Curcuma longa extract and their application in antimicrobial activity. Materials Letters. 2020;**259**:126813. DOI: 10.1016/j.matlet.2019.126813

[27] Vitta Y, Figueroa M, Calderon M, Ciangherotti C. Synthesis of iron nanoparticles from aqueous extract of *Eucalyptus robusta* Sm and evaluation of antioxidant and antimicrobial activity. Materials Science for Energy Technologies. 2020;**3**:97-103. DOI: 10.1016/j.mset.2019.10.014

[28] Fahmy SA, Fawzy IM, Saleh BM, Issa MY, Bakowsky U, Azzazy HME-S. Green synthesis of platinum and palladium nanoparticles using *Peganum harmala* L. seed alkaloids: Biological and computational studies. Nanomaterials. 2021;**11**:965. DOI: 10.3390/nano11040965

[29] Sunny NE, Kumar V. Biogenesis, characterization and bioefficacy of tin oxide nanoparticles from *Averrhoa bilimbi* fruit extract. International Journal of Recent Technology and Engineering. 2019;**8**:10309-10315. DOI: 10.35940/ijrte.D4523.118419

[30] Alvi GB, Iqbal MS, Ghaith MMS, Haseeb A, Ahmed B, Qadir MI. Biogenic selenium nanoparticles (SeNPs) from citrus fruit have anti-bacterial activities. Scientifc Reports. 2021;**11**:4811. DOI: 10.1038/s41598-021-84099-8

[31] Lu H, Zhang X, Khan SA, Li W, Wan L. Biogenic synthesis of MnO_2 nanoparticles with leaf extract of *Viola betonicifolia* for enhanced antioxidant, antimicrobial, cytotoxic, and biocompatible applications. Frontiers in Microbiology. 2021;**12**:761084. DOI: 10.3389/fmicb.2021.761084

[32] Chopra H, Bibi S, Singh I, Hasan MM, Khan MS, Yousafi Q, et al. Green metallic nanoparticles: Biosynthesis to applications. Frontiers in Bioengineering and Biotechnology. 2022;**10**:874742. DOI: 10.3389/fbioe.2022.874742

[33] Guo Y, Sun Q, Fu-Gen W, Dai Y, Chen X. Polyphenol-containing nanoparticles: Synthesis, properties, and therapeutic delivery. Advanced Materials. 2021;**33**:2007356. DOI: 10.1002/adma.202007356

[34] Thipe VC, Amiri KP, Bloebaum P, Karikachery AR, Khoobchandani M, Katti KK, et al. Development of resveratrol-conjugated gold nanoparticles: Interrelationship of increased resveratrol corona on anti-tumor efficacy against breast, pancreatic and prostate cancers. International Journal of Nanomedicine. 2019;**14**:4413-4428. DOI: 10.2147/IJN.S204443

[35] Latif MS, Abbas S, Kormin F, Mustafa MK. Green synthesis of plant-mediated metal nanoparticles: The roles of polyphenols. Asian J. pharm. Clinical Research. 2019;**12**(7):75-84. DOI: 10.22159/ajpcr.2019.v.12i7.33211

[36] Akinfenwa AO, Abdul NS, Docrat FT, Marnewick JL, Luckay RC, Hussein AA. Cytotoxic effects of Phytomediated silver and gold nanoparticles synthesised

from rooibos *(Aspalathus linearis)*, and Aspalathin. Plants. 2021;**10**:2460. DOI: 10.3390/plants10112460

[37] Rajeshkumar S, Parameswari RP, Sandhiya D, Al-Ghanim KA, Nicoletti M, Govindarajan M. Green synthesis, characterization and bioactivity of *Mangifera indica* seed-wrapped zinc oxide nanoparticles. Molecules. 2023;**28**: 2818. DOI: 10.3390/molecules28062818

[38] Elbagory AM, Cupido CN, Meyer M, Hussein AA. Large scale screening of southern African plant extracts for the green synthesis of gold nanoparticles using microtitre-plate method. Molecules. 2016;**21**:1498. DOI: 10.3390/molecules21111498

[39] Jameel MS, Abdul Aziz A, Dheyab MA. Green synthesis: Proposed mechanism and factors influencing the synthesis of platinum nanoparticles. Green Processing and Synthesis. 2020;**9**: 386-398. DOI: 10.1515/gps-2020-0041

[40] Keerthiga N, Anitha R, Rajeshkumar S, Lakshmi T. Antioxidant activity of cumin oil mediated silver nanoparticles. Pharmacognition Journal. 2019;**11**(4): 787-789. DOI: 10.5530/pj.2019.11.125

[41] Soto-Roblesa CA, Luquea PA, Gómez-Gutiérreza CM, Navaa O, Vilchis-Nestorb AR, Lugo-Medinac E, et al. Study on the effect of the concentration of *Hibiscus sabdariffa* extract on the green synthesis of ZnO nanoparticles. Results in Physics. 2019; **15**:102807. DOI: 10.1016/j.rinp.2019.102807

[42] Letchumanan D, Sok SPM, Ibrahim S, Nagoor NH, Arshad NM. Plant-based biosynthesis of copper/copper oxide nanoparticles: An update on their applications in biomedicine, mechanisms, and toxicity. Biomolecules. 2021;**11**:564. DOI: 10.3390/biom11040564

[43] Keijok WJ, Pereira RHA, Alvarez LAC, Prado AR, daSilva AR, Ribeiro J, et al. Controlled biosynthesis of gold nanoparticles with *Cofea arabica* using factorial design. Scientific Reports. 2019; **9**:16019. DOI: 10.1038/s41598-019-52496-9

[44] Nadaroglu H, Gungor AA, Ince S, Babagi A. Green synthesis and characterisation of platinum nanoparticles using quail egg yolk. Spectrochimica Acta Part A: Molecular and Biomolecular Spectroscopy. 2017; **172**:43-47. DOI: 10.1016/j.saa.2016.05.023

[45] Huang HC, Chang WT, Wu YH, Yang BC, Xu MR, Lin MK, et al. Phytochemicals levels and biological activities in *Hibiscus sabdariffa* L. were enhanced using microbial fermentation. Industrial Crops and Products. 2022;**176**: 114408. DOI: 10.1016/j.indcrop.2021.114408

[46] Singha M, Thrimawithana T, Shuklaa R, Benu A. Extraction and characterization of polyphenolic compounds and potassium hydroxycitrate from *Hibiscus sabdariff*. Future Foods. 2021;**4**:100087. DOI: 10.1016/j.fufo.2021.100087

[47] Bisht B, Pancholi D, Pande V, Dandapat A. A green approach to synthesize Au nanoplates using *Morus indica* L. fruit extract and their superior activities in catalysis and surface enhanced Raman scattering. Vegetos. 2021;**34**:867-875. DOI: 10.1007/s42535-021-00245-7

[48] Nagar N, Devra V. Green synthesis and characterization of copper nanoparticles using *Azadirachta indica* leaves. Materials Chemistry and Physics.

2018;**213**:44-51. DOI: 10.1016/j.matchemphys.2018.04.007

[49] Singh R, Hano C, Nath G, Sharma B. Green biosynthesis of silver nanoparticles using leaf extract of *Carissa carandas* L. and their antioxidant and antimicrobial activity against human pathogenic bacteria. Biomolecules. 2021; **11**(2):299. DOI: 10.3390/biom11020299

[50] Liu H, Zhang H, Wang J, Wei J. Effect of temperature on the size of biosynthesized silver nanoparticle: Deep insight into microscopic kinetics analysis. Arabian Journal of Chemistry. 2020;**13**:110-119. DOI: 10.1016/j.arabjc.2017.09.004

[51] Jayachandran A, Aswathy TR, Nair AS. Green synthesis and characterization of zinc oxide nanoparticles using *Cayratia pedata* leaf extract. Biochemistry and Biophysics Reports. 2021;**26**:100995. DOI: 10.1016/j.bbrep.2021.100995

[52] Bélteky P, Rónavári A, Igaz N, Szerencsés B, Tóth IY, Pfeiffer I, et al. Silver nanoparticles: Aggregation behavior in biorelevant conditions and its impact on biological activity. International Journal of Nanomedicine. 2019;**14**:667-687. DOI: 10.2147/IJN.S185965

[53] Wei S, Wang Y, Tang Z, Hu J, Su R, Lin J, et al. A size-controlled green synthesis of silver nanoparticles by using the berry extract of *Sea Buckthorn* and their biological activities. New Journal of Chemistry. 2020;**44**(22):9304-9312

[54] Karim N, Rubinsin N, Burukan M, Kamarudin S. Sustainable route of synthesis platinum nanoparticles using orange peel extract. International Journal of Green Energy. 2019;**16**:1518-1526. DOI: 10.1080/15435075.2019.1671422

[55] Akinfenwa AO, Sagbo IJ, Makhaba M, Mabusela WT, Hussein AA. *Helichrysum* genus and compound activities in the management of diabetes mellitus. Plants. 2022;**11**:1386. DOI: 10.3390/plants11101386

[56] Mehrandish R, Rahimian A, Shahriary A. Heavy metals detoxification: A review of herbal compounds for chelation therapy in heavy metals toxicity. Journal of Herbmed Pharmacology. 2019;**8**(2):69-77. DOI: 10.15171/jhp.2019.12

[57] Srikar SK, Giri DD, Pal DB, Mishra PK, Upadhyay SN. Green synthesis of silver nanoparticles: A review. Green and Sustainable Chemistry. 2016;**6**:34-56. DOI: 10.4236/gsc.2016.61004

[58] Das B, Dash SK, Mandal D, Ghosh T, Chattopadhyay S, Tripathy S, et al. Green synthesized silver nanoparticles destroy multidrug resistant bacteria via reactive oxygenspecies mediated membrane damage. Arabian Journal of Chemistry. 2017;**10**:862-876. DOI: 10.1016/j.arabjc.2015.08.008

[59] Tortella G, Rubilar O, Fincheira P, Pieretti JC, Duran P, Lourenço IM, et al. Bactericidal and Virucidal activities of biogenic metal-based nanoparticles: Advances and perspectives. Antibiotics. 2021;**10**:783. DOI: 10.3390/antibiotics 10070783

[60] Hamad A, Khashan KS, Hadi A. Silver nanoparticles and silver ions as potential antibacterial agents. Journal of Inorganic and Organometallic Polymers and Materials. 2020;**30**(12):4811-4828. DOI: 10.1007/s10904-020-01744-x

[61] Balachandar R, Navaneethan R, Biruntha M, Kumar KKA, Govarthanan M, Karmegam N. Antibacterial activity of silver nanoparticles phytosynthesized from *Glochidion candolleanum* leaves.

Materials Letters. 2022;**311**:131572. DOI: 10.1016/j.matlet.2021.131572

[62] Thi TUD, Nguyen TT, Thi YD, Thi KHT, Phan BT, Pham KN. Green synthesis of ZnO nanoparticles using orange fruit peel extract for antibacterial activities. RSC Advances. 2020;**10**: 23899-23907. DOI: 10.1039/D0RA04926C

[63] Bayat M, Zargar M, Chudinova E, Astarkhanova T, Pakina E. In vitro evaluation of antibacterial and antifungal activity of biogenic silver and copper nanoparticles: The first report of applying biogenic nanoparticles against *Pilidium concavum* and *Pestalotia* sp. Fungi. Molecules. 2021;**26**:5402. DOI: 10.3390/molecules26175402

[64] Amanulla AM, Sundaram R. Green synthesis of TiO_2 nanoparticles using orange peel extract for antibacterial, cytotoxicity and humidity sensor applications. Materials Today Processing. 2019;**8**:323-331. DOI: 10.1016/j.matpr.2019.02.118

[65] Srihasam S, Thyagarajan K, Korivi M, Lebaka VR. Mallem SPR.: Phytogenic generation of NiO nanoparticles using *Stevia* leaf extract and evaluation of their In-vitro antioxidant and antimicrobial properties. Biomolecules. 2019;**10**(1):89. DOI: 10.3390/biom10010089

[66] Hazarica M, Borah D, Bora P, Silva AR, Das P. Biogenic synthesis of palladium nanoparticles and their applications as catalyst and antimicrobial agent. PLoS One. 2017;**12**:e0184936. DOI: 10.1371/journal.pone.0184936

[67] Vidhu K, Philip D. Biogenic synthesis of SnO_2 nanoparticles: Evaluation of antibacterial and antioxidant activities. Spectrochimica Acta. Part A, Molecular and Biomolecular Spectroscopy. 2015;**134**: 372-379. DOI: 10.1016/j.saa.2014.06.131

[68] Sunny NE, Kumar VS. Biogenesis, characterization and bioefficacy of tin oxide nanoparticles from *Averrhoa Bilimbi* fruit extract. International Journal of Recent Technology and Engineering (IJRTE). 2019;**8**:4. DOI: 10.35940/ijrte.D4523.118419

[69] Behzadi S, Serpooshan V, Tao W, Hamaly MA, Alkawareek MY, Dreaden EC, et al. Cellular uptake of nanoparticles: Journey inside the cell. Chemical Social Review. 2017;**46**:4218-4244. DOI: 10.1039/c6cs00636a

[70] Kabir SR, Islam F, Asaduzzaman AKM. Biogenic silver/silver chloride nanoparticles inhibit human cancer cells proliferation in vitro and Ehrlich ascites carcinoma cells growth in vivo. Scientific Reports. 2022;**12**:8909. DOI: 10.1038/s41598-022-12974-z

[71] Ferreira LAB, Garcia-Fossa F, Radaic A, Duran N, Favaro WJ, de Jesus MB. Biogenic silver nanoparticles: In vitro and in vivo antitumor activity in bladder cancer. European Journal of Pharmaceutics and Biopharmaceutics. 2020;**151**:162-170. DOI: 10.1016/j.ejpb.2020.04.012

[72] Alharbi NS, Alsubhi NS, Felimban AI. Green synthesis of silver nanoparticles using medicinal plants: Characterization and application. Journal of Radiation Research and Applied Sciences. 2022;**15**:109-124. DOI: 10.1016/j.jrras.2022.06.012

[73] Chenthamara D, Subramaniam S, Ramakrishnan GS, Krishnaswamy S, Essa MM, Lin F-H, et al. Therapeutic efficacy of nanoparticles and routes of administration. Biomaterials Research. 2019;**23**:20. DOI: 1186/s40824-019-0166-x

[74] Alharbi NS, Alsubh NS. Green synthesis and anticancer activity of silver nanoparticles prepared using fruit extract of *Azadirachta indica*. Journal of Radiation Research and Applied Sciences. 2022;**15**:335-345. DOI: 10.1016/j.jrras.2022.08.009

[75] Aygun A, Gülbagca F, Ozer LY, Ustaoglu B, Altunoglu YC, Baloglu MC, et al. Biogenic platinum nanoparticles using black cumin seed and their potential usage as antimicrobial and anticancer agent. Journal of Pharmaceutical and Biomedical Analysis. 2020;**179**:112961. DOI: 10.1016/j.jpba.2019.112961

[76] Chuang Y, Minghua W, Junde Z, Qiang C. Biosynthesis of peppermint leaf extract polyphenols capped nano-platinum and their in-vitro cytotoxicity towards colon cancer cell lines (HCT 116). Materials Science and Engineering C. 2017;**77**:1012-1016. DOI: 10.1016/j.msec.2017.04.020

[77] Sonkusre P. Specificity of biogenic selenium nanoparticles for prostate cancer therapy with reduced risk of toxicity: An *in vitro* and *in vivo* study. Frontiers in Oncology. 2020;**9**:1541. DOI: 10.3389/fonc.2019.01541

[78] Chabattula SC, Gupta PK, Tripathi SK, Gahtori R, Padhi P, Mahapatra S, et al. Anticancer therapeutic efficacy of biogenic Am-ZnO nanoparticles on 2D and 3D tumor models. Materials Today Chemistry. 2021;**22**:100618. DOI: 10.1016/j.mtchem.2021.100618

[79] Zughaibi TA, Mirza AA, Suhail M, Jabir NR, Zaidi SK, Wasi S, et al. Evaluation of anticancer potential of biogenic copper oxide nanoparticles (CuO NPs) against breast cancer. Journal of Nanomaterials. 2022;**2022**:1-7. DOI: 10.1155/2022/5326355

[80] Abbasi BA, Iqbal J, Mahmood T, Ahmad R, Kanwal S, Afridi S. Plant-mediated synthesis of nickel oxide nanoparticles (NiO) via *Geranium wallichianum*: Characterization and different biological applications. Materials Research Express. 2019;**6**(8):0850a7. DOI: 10.1088/2053-1591/ab23e1

[81] Viswanathan S, Palaniyandi T, Kannaki P, Shanmugam R, Baskar G, Rahaman AM, et al. Biogenic synthesis of gold nanoparticles using red seaweed *Champia parvula* and its anti-oxidant and anticarcinogenic activity on lung cancer. Particulate Science and Technology. 2023;**41**(2):241-249. DOI: 10.1080/02726351.2022.2074926

[82] El-Deeb NM, Khattab SM, Abu-Youssef MA, Badr AM. Green synthesis of novel stable biogenic gold nanoparticles for breast cancer therapeutics via the induction of extrinsic and intrinsic pathways. Scientific Reports. 2022;**12**(1):11518. DOI: 10.1038/s41598-022-15648-y

[83] Al Tamimi S, Ashraf S, Abdulrehman T, Parray A, Mansour SA, Haik Y, et al. Synthesis and analysis of silver–copper alloy nanoparticles of different ratios manifest anticancer activity in breast cancer cells. Cancer Nano. 2020;**11**:13. DOI: 10.1186/s12645-020-00069-1

[84] Athinarayanan J, Periasamy VS, Alshatwi AA. Eco-friendly synthesis and characterization of platinum-copper alloy nanoparticles induce cell death in human cervical cancer cells. Process Biochemistry. 2016;**51**(7):925-932. DOI: 10.1016/j.procbio.2016.04.006

[85] Burgos-Morón E, Abad-Jiménez Z, Martínez de Marañón A, Iannantuoni F, Escribano-López I, López-Domènech S, et al. Relationship between oxidative stress, ER stress, and inflammation in

type 2 diabetes: The Battle continues. Journal of Clinical Medicine. 2019;**8**: 1385. DOI: 10.3390/jcm8091385

[86] Azeem A, Hanif M, Mahmood K, Ameer N, Chughtai FRS, Abid U. An insight into anticancer, antioxidant, antimicrobial, antidiabetic, and anti-inflammatory effects of quercetin: A review. Polymer Bulletin. 2023;**80**:241-262. DOI: 10.1007/s00289-022-04091-8

[87] Akbari B, Baghaei-Yazdi N, Bahmaie M, Akbari FM. The role of plant-derivednatural antioxidants in reduction of oxidative stress. BioFactors. 2022; **2022**:1-23. DOI: 10.1002/biof.1831

[88] Muteanu IG, Apetrei C. Analytical methods used in determining antioxidant activity: A review. International Journal of Molecular Sciences. 2021;**22**:3380. DOI: 10.3390/ijms22073380

[89] Vera J, Herrera W, Hermosilla E, Diaz M, Parada J, Seabra AB, et al. Antioxidant activity as an indicator of the efficiency of plant extract-mediated synthesis of zinc oxide nanoparticles. Antioxidants. 2023;**12**:784. DOI: 10.3390/antiox/12040784

[90] Rahman H, Rauf A, Khan SA, Ahmad Z, Alshammari A, Alharbi M, et al. Green synthesis of silver nanoparticles using *Rhazya stricta* decne extracts and their anti-microbial and anti-oxidant activities. Crystals. 2023;**13**: 398. DOI: 10.3390/cryst13030398

[91] Alam F, Islam MA, Kamal MA, Gan SH. Updates on managing type 2 diabetes mellitus with natural products: Towards antidiabetic drug development. Current Medicinal Chemistry. 2018;**25** (39):5395-5431

[92] Ul Haq MN, Shah GM, Menaa F, Khan RA, Althobaiti NA, Albalawi AE, et al. Green silver nanoparticles synthesized from *Taverniera couneifolia* elicits effective anti-diabetic effect in Alloxan-induced diabetic Wistar rats. Nanomaterials. 2022;**22**(12):1035. DOI: 10.3390/nano12071035

[93] Jobie FN, Ranjbar M, Moghaddam AH, Kiani M. Green synthesis of zinc oxide nanoparticles using *Amygdalus scoparia Spach* stem bark extract and their applications as an alternative antimicrobial, anticancer, and anti-diabetic agent. Advanced Powder Technology. 2021;**32**(6):2043-2052. DOI: 10.1016/j.apt.2021.04.014

[94] Gupta N, Malviya R. Understanding and advancement in gold nanoparticle targeted photothermal therapy of cancer. Biochimica et Biophysica Acta (BBA): Reviews on Cancer. 1875;**2021**(1): 188532. DOI: 10.1016/j.bbcan.2021.188532

[95] Ali SA, Ahmed MM, Yousif HK. Novel photothermal therapy using platinum nanoparticles in synergy with near-infrared (NIR) against human breast cancer (MCF-7) cell line. Results in Chemistry. 2022;**4**:100591. DOI: 10.1016/j.rchem.2022.100591

[96] Wang C, Wang X, Chen Y, Fang Z. In-vitro photothermal therapy using plant extract polyphenols functionalized graphene sheets for treatment of lung cancer. Journal of Photochemistry and Photobiology B: Biology. 2020;**204**: 111587. DOI: 10.1016/j.jphotobio.2019.111587

[97] Gupta K, Chundawat TS. Bio-inspired synthesis of platinum nanoparticles from fungus *Fusarium oxysporum*: Its characteristics, potential antimicrobial, antioxidant and photocatalytic activities. Materials Research Express. 2019;**6**:1050d6. DOI: 10.1088/2053-1591/ab4219

[98] Wu M, Guo H, Liu L, Liu Y, Xie L. Size-dependent cellular uptake and localization profiles of silver nanoparticles. International Journal of Nanomedicine. 2019;**14**:4247-4259. DOI: 10.2147/IJN.S201107

[99] Naz S, Gul A, Zia M. Toxicity of copper oxide nanoparticles: A review study. IET Nanobiotechnology. 2020;**14** (1):1-13. DOI: 10.1049/iet-nbt.2019.0176

[100] Nakamura H, Watano S. Direct permeation of nanoparticles across cell membrane: A review. KONA Powder and Part Journal. 2018;**35**:49-65. DOI: 10.14356/kona/201811

[101] Amaliyah S, Sabarudin A, Masruri M, Sumitro SB. Characterization and antibacterial application of biosynthesized silver nanoparticles using *Piper retrofractum* Vahl fruit extract as bioreductor. Journal of Applied Pharmaceutical Science. 2022;**12**(03): 103-114. DOI: 10.7324/JAPS.2022.120311

[102] Sani A, Cao C, Cui D. Toxicity of gold nanoparticles (AuNPs): A review. Biochemistry and Biophysics Reports. 2021;**26**:100991. DOI: 10.1016/j. bbrep.2021.100991

[103] Andleeb A, Andleeb A, Asghar S, Zaman G, Tariq M, Mehmood A, et al. A systematic review of biosynthesized metallic nanoparticles as a promising anti-cancer-strategy. Cancers. 2021;**13**: 38. DOI: 10.3390/cancers13112818

Chapter 6

Efficacy of Cobalt-Incorporated Mesoporous Silica towards Photodegradation of Azodyes and Its Kinetic Study for Advanced Application

Krishna Vaddadi, Nookaraju Muralasetti and Naginami Naidu

Abstract

Along with MCM-41, cobalt-incorporated mesoporous silica (Co-MCM-41) has been created. Powder X-ray diffraction, scanning electron microscopy, and nitrogen adsorption-desorption studies were used to describe the materials. It has been discovered that the Co-MCM-41 has less surface area (SBET, $m^2\ g^{-1}$), pore volume ($cc{\cdot}g^{-1}$), and pore size (Å) than the MCM-41. The SEM-EDAX analysis has also unmistakably demonstrated the existence of the appropriate elements in the materials. The photoactivity was significantly impacted by the extremely distributed Co^{3+} species present on the MCM-41 structure. A theoretical loading of 3.5 wt% permitted an AO_7 degradation percentage of about 70% for the samples that were simply treated with Co. Increased Co^{3+} inactive species, such as clusters or $-Co_2O_3$ nanoparticles, are present at higher loadings, but the photoactivity is not noticeably increased. By using the Kubelka-Munk function to the UV-Vis DRS results, it was discovered that the band gap (eV) in the Co-MCM-41 was also substantially smaller than in its parent template. The Alizarin Red S dye was successfully photodegraded employing the materials as photocatalysts, and pseudo-first order kinetics was carried out using the Langmuir-Hinshelwood kinetic model. The necessary experimental setups were all optimised.

Keywords: mesoporou silica, MCM-41, heterogeneous catalyst, photocatalyst, Alizarin Red S

1. Introduction

A growing focus on heterogeneous catalysts has emerged in recent years due to economic and environmental factors. These catalysts typically have low cost, high reactivity, environmental friendliness, high selectivity, easy setup, and recoverability of catalysts [1]. Mesoporous MCM-41 materials have come to light as highly stable compounds with a significant surface area. They have been widely used as catalysts or catalyst supports in a variety of processes. This was explained by the reaction mixture's higher reusability and straightforward recoverability [2]. Due to their increased oxidative or acidic character, the introduction of metal ions including Ti^{4+}, Al^{3+}, Co^{3+}, and Fe^{3+} into the framework has demonstrated improved catalytic activity [3].

Due to contaminated ground water and dangerous industrial effluents, the entire world is currently experiencing environmental issues [4, 5]. These highly coloured effluents harm the environment when they are dumped into water systems because they impede light penetration and hinder aquatic life's ability to photosynthesize [6, 7]. Intense colour is imparted by the presence of dyes in the effluent at very low concentrations (1 mg L^{-1}) but it is discovered that they are harmful to the environment [8, 9]. In order to effectively remove colour from waste fluids, physical or chemical procedures must be used [10]. The majority of dyes used on an industrial basis are derivatives of azo, anthraquinone, indigo, triphenylmethane, xanthene, and others [11, 12]. Due to their advantageous qualities—bright colour, easy application, and low energy consumption—these dyes are widely utilised in the textile industry. They are, however, typically the most poisonous and mutagenic substances found in nature [13, 14].

One of the anthraquinone class of dyes, Alizarin Red S (ARS), is widely used in the textile, woven fabric, wool, and cotton industries [15–17]. The paint, plastics, leather, and cosmetics sectors all utilise anionic dye extensively [18]. However, when industrial effluents are released into aquatic environments at amounts exceeding what is permitted, the aquatic life is negatively impacted [19, 20]. It was shown that traditional aerobic digestion techniques were ineffective at degrading these resistant compounds [21]. Photo catalysis has become a green technique for gathering solar energy and degrading organic pollutants due to the global energy crisis and environmental challenges [22–25]. ARS was selected as the test molecule to undergo photodegradation in the presence of visible light as a result.

MCM-41 and its metal-included derivatives have a wide range of photocatalytic uses, so cobalt metal ion (Co^{+3}) was successfully inserted into the structure of MCM-41. In order to research the photodegradation of ARS under ideal experimental settings, such as effect of photocatalyst, effect of photocatalyst dose, effect of dye concentration, and effect of pH, the materials were characterised and used as photocatalysts. To identify the active species participating in the photodegradation phenomenon, a scavenger experiment was done in addition to these research. To assess how well the outcomes matched each other, the kinetic research was carried out.

2. Results and discussion

2.1 Powder XRD studies

Figure 1 displays the XRD patterns ($2° \leq 2\theta \leq 10°$) of the MCM-41 and Co-MCM-41 samples. The patterns only display one low-angle peak for the d100 plane, which corresponds to the mesophase at a value of 2 approximately 2.2o (d-spacing: 32.54623 A, wall thickness: 2.712 A). This is typical of MCM-41's long-range hexagonal structure [26]. Planes that mirror the characteristics of mesoporous nature as in MCM-41. Co-diffraction MCM-41's pattern has a lesser intensity of the low angle peak than MCM-41, which suggests that the metal ions are obstructing the structure that directs the template's action in the materials' regular ordering.

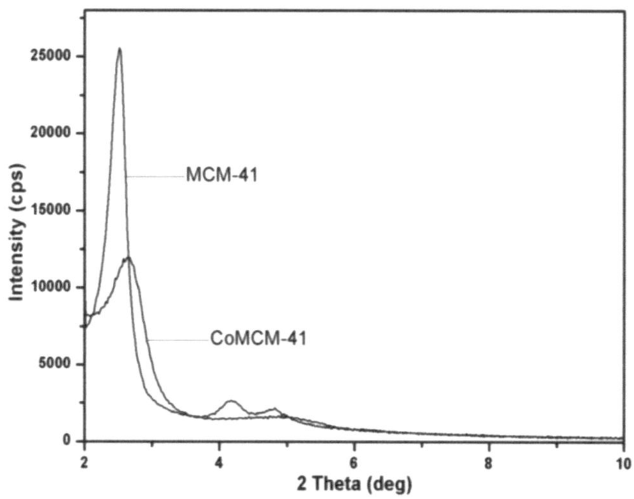

Figure 1.
Powder XRD patterns of MCM-41 and CoMCM-41.

The diffraction planes d110 (d-spacing = 19.48761 A, wall thickness = 4.531 A), and d200 (16.95948 A, wall thickness = 5.207 A), which reflect the hexagonal array of MCM-41, are responsible for the less intense and broader peaks in the 2 of 4.0°–5.5°. Three diffraction peaks suggest that the mesopores are ordered crystallographically. The reason for the low value of 2 is primarily the template's long carbon chain, which was employed to synthesise MCM-41 [27]. Co-MCM-41 materials, in contrast, exhibit one large peak at 2 = 2.5o, which corresponds to the mesoporous phase, and two succeeding, less intense peaks at (110) and (200) crystal planes, which mirror the mesoporous characteristic of MCM-41. Co-diffraction MCM-41's pattern has a lesser intensity of the low angle peak than MCM-41, which suggests that the metal ions are obstructing the structure that directs the template's action in the materials' regular ordering.

2.2 Nitrogen adsorption-desorption studies

It was discovered that the synthetic materials followed a standard type-IV adsorption isotherm without hysteresis. This demonstrates how mesoporous these materials are [28, 29]. By using the BET (Brunauer, Emmet, and Teller) method to determine the specific surface area of the materials from adsorption isotherms, it can be demonstrated that the insertion of metal ions reduces the materials' surface area. This might be explained by metal ions filling part of the pores. By using the BJH (Barrett-Joyner-Halenda) method, the pore size and pore volume of the materials are assessed. **Table 1** lists the textural characteristics of MCM-41 and Co-MCM-41 materials.

2.3 SEM-EDAX studies

Figures 2 and **3** show the SEM-EDAX micrographs of MCM-41 and Co-MCM-41, respectively. All of the materials have spherical morphologies similar to those of MCM-41, according to SEM micrographs of the materials. The functionalization of materials with metal ions (Co^{+3}) was also validated by EDAX analysis. It was discovered that adding metal ions to the framework has no effect on the morphology of the materials.

Material	S_{BET} (m^2g^{-1})	Pore size (Å)	Pore volume (cc g^{-1})
MCM-41	1023.50	17.20	0.28
CoMCM-41	698.22	16.90	0.22

Table 1.
Textural characteristics of the materials.

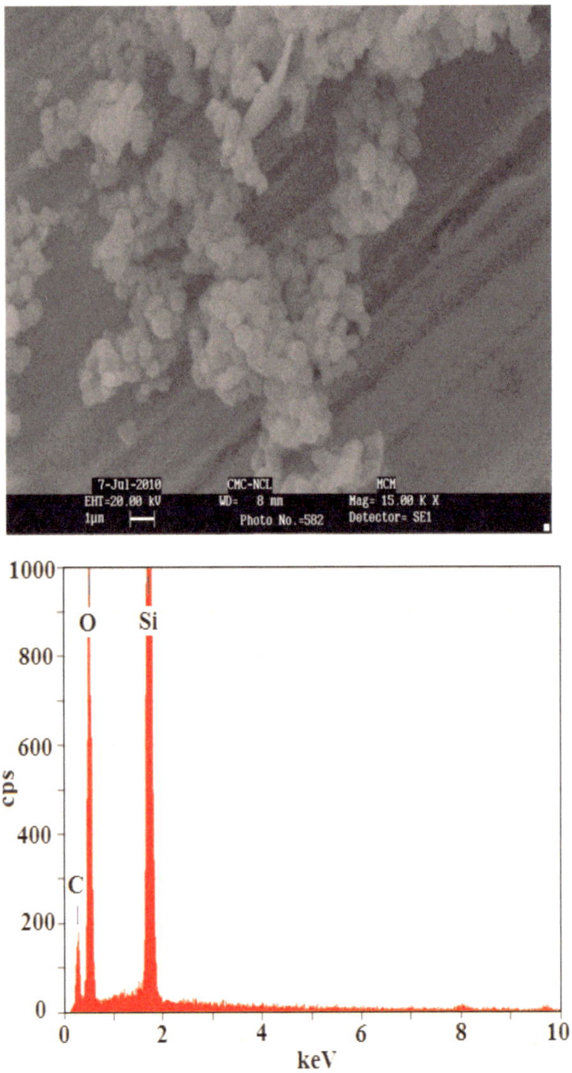

Figure 2.
SEM-EDAX micrographs of MCM-41.

2.4 UV-Vis diffuse reflectance spectra & Kubelka-Munk function curve

The UV-Vis diffuse reflectance spectra were taken in order to comprehend the coordination between the Cobalt and MCM-41. Pure MCM-41 was found to lack a

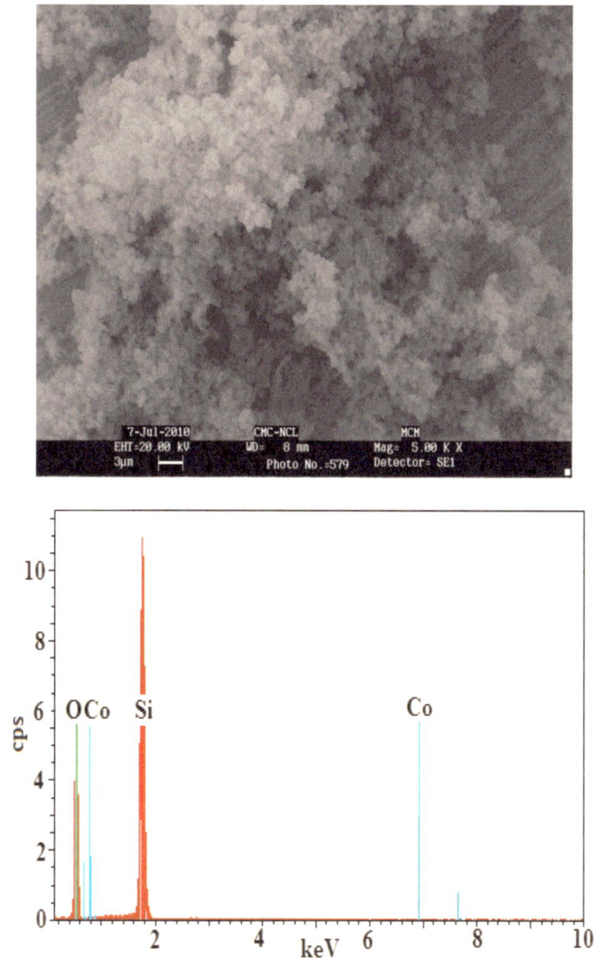

Figure 3.
SEM-EDAX micrographs of Co-MCM-41.

distinctive absorption peak in the 200–800 nm range, which suggests that it was not sensitive in the UV-Visible range [30]. However, a strong absorption peak was seen in Co-MCM-41 in the 400–450 nm (430 nm) range, which is consistent with the octahedral geometry of the Co^{+3} crystal field transition.

$^5T_{2g}(D)\rightarrow{}^5E_g(D)$ $[(t_{2g}{}^4e_g{}^2)\rightarrow(t_{2g}{}^3e_g{}^3)]$. By using the Kubelka-Munk (KM) function, the produced mesoporous materials have also been described for their band gap values (in electron volts, eV). **Figure 4** showed the findings and a plot of the band gap energy values (eV) vs. the modified Kubelka-Munk function [F(R)hv]2 [31]. In MCM-41 and Co-MCM-41 materials, the band gap was discovered to be 2.9 eV and 2.7 eV, respectively. It was clear from the analysis of these results that immobilising Cobalt caused a sizable reduction in the band gap as well as proper coordination of the Cobalt (+3) ion in the zeolite framework. A necessary component of the phenomena of photocatalysis, the photocatalytic activity in the visible light can be improved by the smaller band gap [31].

Adsorption studies: Adsorption isotherm experiment was carried out in the dark to investigate the maximum ARS adsorption capacity on the surface of the MCM-41 and Co-MCM-41 materials. The following statement [32] inevitable factor for the

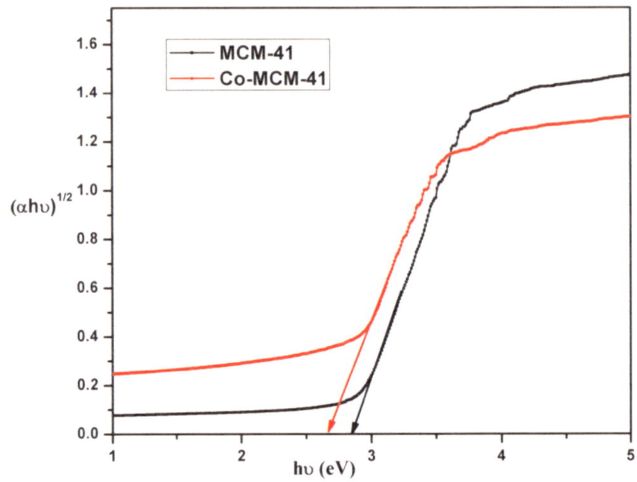

Figure 4.
KM function curve of MCM-41 and Co-MCM-41 materials.

phenomena of photocatalysis [31] was used to determine the amount of ARS absorbed per unit mass of the adsorbent at time, t (min).

$$q_t = \frac{(Co - Ct)V}{m.} \tag{1}$$

Co and Ct are the dye solution concentrations (mol/L) before and after adsorption, respectively. V is the dye solution's volume (L) in the photoreactor, and m is the photocatalyst's mass (g).

Figure 5 shows the amount of ARS adsorption as a function of time. In the photoreactor, it was seen that a fast adsorption developed after 15 min of contact. However, the adsorption propensity started out very mildly and only became saturated after 15 min. It demonstrates that the addition of Cobalt to the MCM-41 framework increased the dye's tendency to be removed. Based on these findings, the equilibration time was optimised to be 15 min with both mesoporous materials in darkness, and the same was fixed as the equilibration duration for additional research.

Effect of photocatalyst: The synthesised materials were used to investigate the impact of deterioration on the 2.92×10^{-5} M aqueous solutions of ARS. The results are depicted in **Figures 6** and 7. Co-MCM-41 (10 mg) achieved about 98% degradation efficiency in 90 min, whereas MCM-41 only destroyed 68% of ARS during the same period (**Figure 6**). This might be because Co-MCM-41 material has a higher capacity for absorbing light due to its smaller pore size and volume, which could boost its activity for efficient photocatalytic degradation.

Studying the effect of Co-MCM-41 on ARS absorbance, it was found that the intensity of absorbance significantly decreased within 90 min, as shown in **Figure 7**. The dye's efficient degradation may be to blame for the decrease in ARS absorption.

Effect of photocatalyst dose: **Figure 8** shows the results of research on the effect of photocatalyst (Co-MCM-41) on ARS degradation. From 2 mg to 12 mg of catalyst were added to 100 mL of 2.92×10^{-5} M ARS solution. The findings indicated that increasing the catalyst dose increased the degradation efficiency, and that increasing the catalyst dose over 10 mg had no effect on the degradation. Because there are more active sites on the

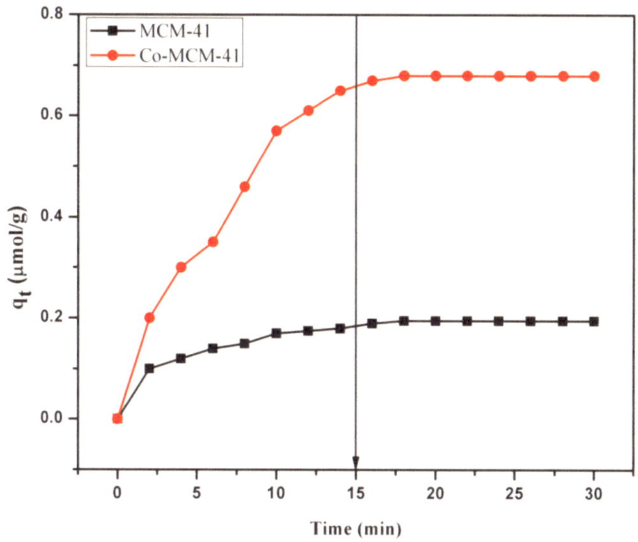

Figure 5.
Adsorption studies of ARS with MCM-41 and Co-MCM-41 materials. Catalyst load = 10 mg, [ARS] = 2.92 × 10^{-5} M, ARS volume = 100 mL.

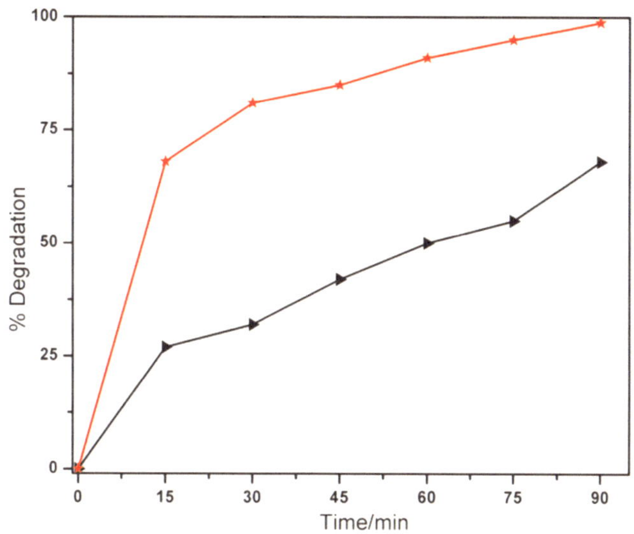

Figure 6.
Effect of photocatalyst on degradation of ARS with the photocatalysts (each 10 mg). (triangles) *MCM-41;* (stars) *Co-MCM-41.*

surface of the catalyst as catalyst concentration increases, the rate of dye adsorption also increases. High catalyst doses, however, cause opacity in aqueous solutions and limit visible light penetration while slowing the rate of breakdown. Therefore, a catalyst dose of 10 mg/100 mL of ARS solution was determined to be optimal for the tests.

Effect of dye concentration: By adjusting the dye's concentration from 1.16 × 10^{-5} M to 4 × 10^{-5} M, the impact of initial dye concentration on the photocatalytic degradation of ARS was calculated. According to **Figure 9**, the percentage degradation efficiency was high at 2.92 × 10^{-5} M concentration and subsequently declined as

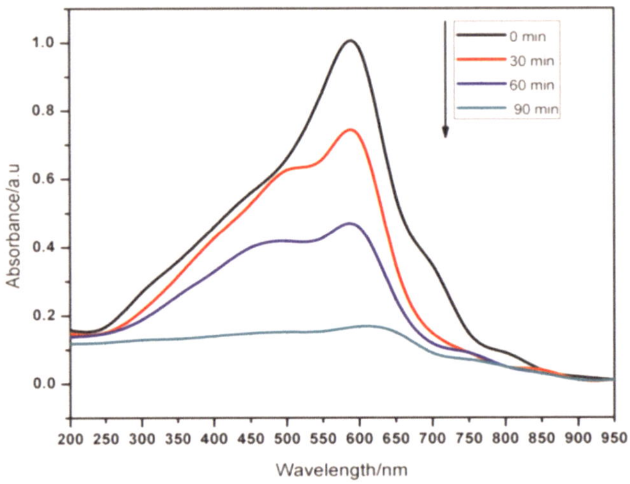

Figure 7.
Effect of Co-MCM-41 (10 mg) on absorbance of ARS at λ $_{max}$ = 597 nm with respect to time (min).

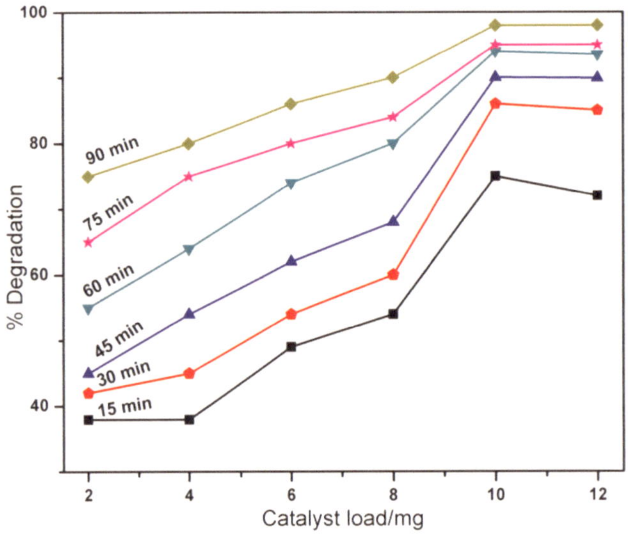

Figure 8.
Effect of catalyst load on degradation of ARS with Co-MCM-41 (10 mg). (squares) *15 min;* (pentagon) *30 min;* (left in triangles) *45 min;* (right in triangles) *60 min;* (stars) *75 min;* (diamonds) *90 min.*

concentrations increased. The dye's rate of adsorption on the active sites of the photocatalyst increases with increasing dye concentration while concurrently decreasing the dye's adsorption inclination. Additionally, excessive dye concentrations reduce the number of photons and their adsorption on the catalyst's surface, which ultimately reduces the degradation efficiency [33].

Effect of pH: At different pH levels, the effluents discharged by the textile industry include hazardous colours. As a result, the effect of the dye medium's pH on its degradation was assessed by altering pH values between 2.0 and 12.0 for a constant dye concentration (2.92×10^{-5} M) and catalyst load (10 mg). The breakdown efficiency of ARS increased at pH 4.0 and peaked at pH 6.0, as shown in **Figure 10**. The

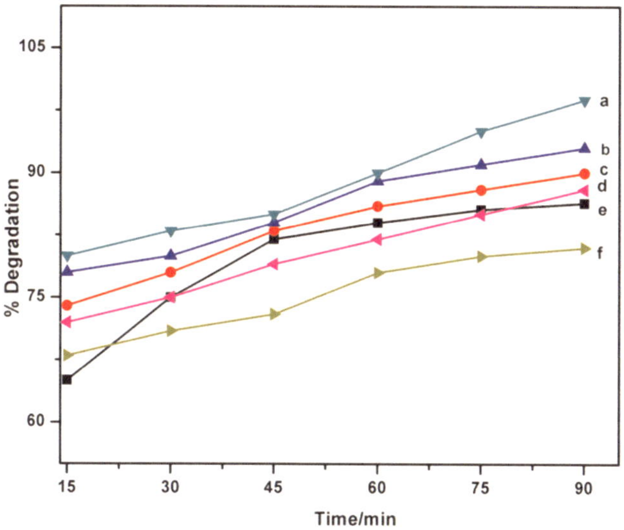

Figure 9.
Effect of dye concentration on degradation of ARS with Co-MCM-41. Catalyst load = 10 mg/100 mL of ARS. (a = 2.92×10^{-5} M, b = 2.33×10^{-5} M, c = 1.75×10^{-5} M, d = 3.5×10^{-5} M, e = 1.16×10^{-5} M, f = 4×10^{-5} M).

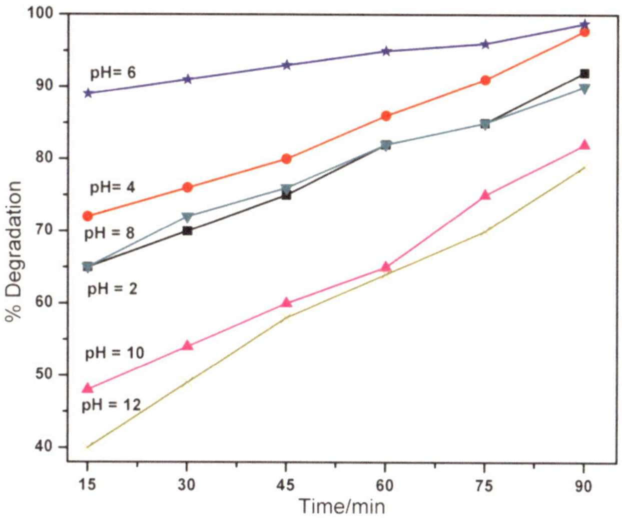

Figure 10.
Effect of pH on degradation of ARS with Co-MCM-41. [Dye] = 2.92×10^{-5} M, catalyst load = 10 mg/100 mL of ARS.

effectiveness of the degradation has diminished when pH levels have increased to 12.0. Since it controls the charge on the particle surface, a solution's pH is a crucial variable in photocatalytic reactions. Therefore, at pH levels below 6.8, the negatively charged dye molecules and surface of Co-MCM-41 may interact chemically strongly and adsorb. At a pH of between 4.0 and 6.0, this situation may have caused ARS dye to breakdown. Repulsion between the negatively charged dye and the catalyst surface may account for the extremely low degradation rate at higher pH levels.

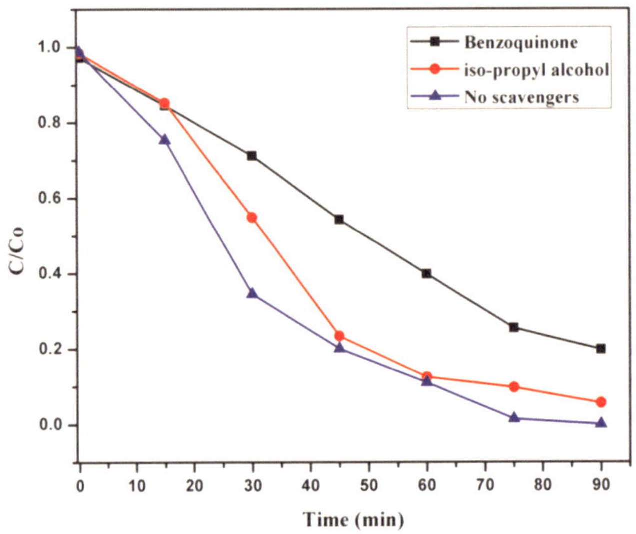

Figure 11.
Variation of C/Co against time (min) in the scavenger experiment. (squares)*: Benzoquinone;* (circles)*: Isopropyl alcohol;* (upward triangles)*: No scavengers.*

Role of scavengers in the photocatalysis of ARS: Studies on the photocatalytic degradation of ARS were conducted in both the absence and presence of scavengers. **Figure 11** illustrates how the dye's C/Co varied as a function of time (in minutes).

Because it prevents the release of a significant amount of superoxide radicals (O2•), benzoquinone limits the degradation of dye (78.5%). The elimination of ARS (91.2%) from its aqueous solution, which primarily regulates the activity of the hydroxyl radicals (OH•), is also slightly constrained by the presence of iso-propyl alcohol [34]. In contrast, when these scavengers were absent, the degradation efficiency was over 98.8%. Superoxide radical and hydroxyl radical were therefore the two main active species in the redox process of the current investigation.

Plausible mechanism for photocatalytic degradation of ARS: The electrons in the photocatalyst's valence band would excite into the conduction band and generate electron-hole pairs when the dye solution and photocatalyst were exposed to visible light. The degradation efficiency would be decreased as a result of these charged pairs recombining either directly or through the catalyst surface. A good photocatalyst with increased visible light absorption (reduced band gap, Eg) might transport the electron to oxygen, generating the active superoxide radical ion, and therefore limit the tendency to recombine. According to the UV-Vis DRS investigation, Cobalt added MCM-41 has a lowered bang gap of 2.7 eV, which would support the visible light driven photocatalytic activity. The electron can be excited from the valence to conduction band by the photocatalyst when it interacts with visible light and proceeds in the manner described above. In subsequent steps, the superoxide radical ions might produce hydroxyl radicals, which ultimately efficiently degraded the ARS dye (**Figure 12**).

Kinetic analysis: The kinetic study was performed for the degradation of ARS dye using Langmuir-Hinshelwood kinetic model [35].

$$r = dC/dt = kKC/(1 + KC) \tag{2}$$

$$\text{Co-MCM-41} + h\upsilon \rightarrow \text{Co-MCM-41}\ (h^+ + e^-)$$
$$O_2 + e^- \rightarrow O_2^{-\bullet}$$
$$O_2^{-\bullet} + H^+ \rightarrow HO_2^{\bullet}$$
$$2HO_2^{\bullet} \rightarrow O_2 + H_2O_2$$
$$H_2O_2 + O_2^{-\bullet} \rightarrow OH\bullet + OH^- + O_2$$
$$H_2O_2 + h\upsilon \rightarrow 2OH\bullet$$
$$O_2^{-\bullet}/OH\bullet + \textbf{ARS} \rightarrow \text{degradation products}$$

Figure 12.
Photocatalytic degradation mechanism of ARS using Co-MCM-41.

On neglecting the value of K_C in the denominator ($K_C < <1$) and integrating with respect to time t, the above equation accords to the pseudo-first order equation.

$$ln\,\frac{Co}{C} = kKt = k_{(\text{appt})} \tag{3}$$

where Co denotes the starting concentration and C denotes the current concentration of the ARS solution. K is the adsorption coefficient of the ARS dye onto the photocatalyst, and t. k is the rate constant.

For the rate of degradation using MCM-41 and Co-MCM-41, respectively, the rate constants (k) were estimated as 1.526×10^{-2} min^{-1} (2.543×10^{-4} s^{-1}) and 9.20×10^{-2} min^{-1} (15.33×10^{-4} s^{-1}). Compared to the MCM-41, the rate constant has grown six times with the Co-MCM-41. It demonstrates how adding Cobalt to MCM-41 increases the rate of reaction.

3. Experimental

Materials: Cetyltrimethylammonium bromide [CH_3-$(CH_2)_{15}$-$N(CH_3)_3Br$, CTMAB], tetraethylorthosilicate [$(C_2H_5O)_4Si$, TEOS], Cobalt acetate [Co $(CH_3COO)_2.2H_2O$], Ethyl alcohol, were procured with AR grade quality (99% pure) from Sigma Aldrich and Merck. Alizarin Red S (Merck) with molecular formula, $C_{14}H_7NaO_7S$ (M.Wt = 342.26 g.mol^{-1}, λ_{max} = 597 nm). Scavenger experiments were carried out using benzoquinone and iso-propyl alcohol. None of the compounds were further purified before usage. Utilising double distilled water, all solutions needed for the experimental work were created.

Synthesis of MCM-41: MCM-41 A straightforward room-temperature co-precipitation technique was used to create the material [26]. In a typical synthesis, a clear, homogenous solution was created by dissolving 2.40 g of the surfactant CTMAB in 50.00 mL of distilled water and stirring continuously. This homogenous solution received 76.00 mL of ethyl alcohol and 13.00 mL of 25 weight percent aqueous ammonia while being stirred. The aforementioned mixture received a dropwise addition of 10.00 mL of TEOS. The solution turned milky and a gel was formed due to the hydrolysis of TEOS. The resultant mixture was stirred for about 2 h to completely hydrolyze TEOS. White precipitate thus formed was centrifuged, washed consecutively with distilled water and methanol. The product was dried overnight at 110°C. Solid product thus obtained was calcined at 550°C in air atmosphere for 5 h to remove the trapped surfactant.

Synthesis of Cobalt incorporated MCM-41 (Co-MCM-41): The in-situ approach was used to create MCM-41 with metal incorporation. These were made using the same procedure as MCM-41, with the exception that the addition of the appropriate amounts (Si/M ion ratio of 100:1, M = metal) of the corresponding metal precursors was made 20 min after the addition of TEOS. Using cobalt acetate as a precursor material, cobalt (Co) integrated MCM-41 materials were created.

Characterisation: Step scanning on the Rigaku D/MAX-2500 diffractometer (Rigaku Co., Japan) with Cu-K -radiation (k = 0.15406 nm), operated at 40 kV and 100 mA, was used to analyse the produced MCM-41 and its Cobalt doped derivative. A scanning electron microscope model Philips XL 30 ESEM was used to capture SEM images of the materials (FEI-Philips Company, Hillsboro). The surface areas of the catalysts were measured using a Quantachrome Nova 2000e surface area and pore size analyser by nitrogen adsorption-desorption in a liquid nitrogen atmosphere (77 K). With the help of the Single Monochromator UV-2600 (optional ISR-2600Plus, up to 1400 nm), UV-Vis diffuse reflectance spectra were captured.

Photocatalytic measurements: The efficiency of the photocatalysts in the photodegradation of ARS was examined using a UV-Vis Spectrophotometer (UV-2550, Shimadzu, wavelength range: 180–1100 nm). With a known weight (10 mg) of the photocatalyst per 100 mL of a variety of diluted ARS solutions, adsorption-desorption equilibrium tests were carried out. The chosen equilibration period was 15 min. The solutions were exposed to a 300 watt tungsten halogen lamp in a photocatalytic chamber with an electrical supply to conduct the photocatalytic study. A UV-Vis spectrophotometer was used to evaluate the translucent solution after centrifuging 5 mL aliquots at regular intervals of 5 min. Eq. (4) was used to get the percentage of dye degradation.

$$\text{Photocatalytic degradation\%} = (\text{Co} - \text{Ct})/\text{Co} \times 100 \qquad (4)$$

where C_0 and C_t are the initial concentration and concentration of the dilute ARS solution at a time interval, t respectively.

Scavenger experiment: The purpose of the in-situ scavenger experiment was to look for active species produced by the efficient photocatalyst during the photocatalytic breakdown of ARS. Scavengers were added to the photocatalytic process to capture holes, or superoxide radicals ($O_2•$) and hydroxyl radicals (OH•), such as benzoquinone (1 mmol/L) and iso-propyl alcohol (0.1 mmol/L).

4. Applications

There are many benefits to using this Co-MCM-41 heterogeneous catalyst in sophisticated oxidation processes [36]. Co-MCM-41 mesoporous materials have been synthesised, and their high specific surface area, well-ordered mesoporous structure, large surface area, and flexible framework make it easy to incorporate metal active sites in silica materials [37, 38]. These materials also make good dispersions for reaction and product molecule molecules. The host material's adsorption characteristics are crucial for the purification of biomolecules. According to a publication, MCM-41 with Co inserted is a strong contender for use as a high surface area adsorbent for biological compounds like amino acids [39]. This has undergone testing as a method of regulated drug release for medications like ibuprofen [40]. These heterogeneous catalyst Co-MCM-41 based mesoporous materials have successfully immobilised or absorbed a number of proteins, including cytochrome C [41], lysozym [42], and trypsin [43], and their activities have been investigated [44]. Well-ordered mesoporous silica nanoparticles have recently been demonstrated to be useful as cell markers by Lin et al. [45].

5. Conclusions

The derivative of MCM-41 with cobalt was created and described. Under ideal circumstances, the materials have produced effective results in the photodegradation of the anthroquinone dye Alizarin Red S. The fact that Co-MCM-41 degrades more effectively than MCM-41 is proof of the metal's influence on the microporous material's framework when serving as a photocatalyst.

Acknowledgements

The author is thankful to Sophisticated analytical instrumentation facility (SAIF), India and National Institute of Technology, Warangal (NIT-W), India for providing the characterisation results. Also thankful to the Management of NS Raju Institute of Technology, Visakhapatnam, India for providing the funding and laboratory facilities.

Author details

Krishna Vaddadi[1], Nookaraju Muralasetti[2]* and Naginami Naidu[1]

1 Department of Basic Sciences and Humanities, N.S. Raju Institute of Technology, Sontyam, Andhra Pradesh, India

2 Aditya College of Engineering and Technology, Surampalem, Andhra Pradesh, India

*Address all correspondence to: mnookarajuphd@gmail.com

References

[1] Dodd J, Schwender CF, Moore JB Jr., Ritchie DM, Gray-Nunez Y, Loughney D, et al. Design and discovery of RWJ 22108–a novel bronchoselective calcium channel blocker. Drug Design Discovery. 1998;**15**:3

[2] Vicini P, Geronikaki A, Incerti M, Busonera B, Poni G, Kabras CA, et al. Synthesis and Biological Evaluation of Benzo[d]isothiazole, Benzothiazole and Thiazole Schiff Bases. Bioorganic and Medicinal Chemistry. 2003;**11**:4785

[3] Bergbreiter DE, Newcomb M. Asymmetric Synthesis. In: Morrison JD, editor. Vol. 2A. Academic Press; p. 243

[4] Rubin AJ. Aqueous-Environmental Chemistry of Metals. Ann Arbor , MI, USA: Science Publishers; 1974

[5] Mercier L, Pinnavaia TJ. Heavy Metal Ion Adsorbents Formed by the Grafting of a Thiol Functionality to Mesoporous Silica Molecular Sieves: Factors Affecting Hg(II) Uptake, Environmental Science & Technology. 1998;**32**:2749

[6] Clarke EA, Anliker R. Organic dyes and pigments. In: Handbook of Environmental Chemistry, Anthropogenic Compounds. Vol. 3. New York: Springer-Verlag; 1980

[7] Robinson T, Mcmullan G, Marchant R, Nigam P. Remediation of dyes in textile effluent: a critical review on current treatment technologies with a proposed alternative. Bioresource Technology. 2001;**77**:247

[8] Nigam P, Armour G, Banat IM, Singh D, Marchant R. Physical removal of textile dyes from effluents and solid-state fermentation of dye-adsorbed agricultural residues. Bioresource Technology. 2000;**72**:219

[9] Naeem A, Abdul H, Safia A. Physicochemical characterization and Bioremediation perspective of textile effluent, dyes and metals by indigenous Bacteria.Journal of Hazardous Materials. 2009;**164**:322

[10] Sharma S, Kaur A. Various Methods for Removal of Dyes from Industrial Effluents - A Review. Indian Journal of Science and Technology. 2018; **11**(12):1

[11] Fu Y, Viraraghavan T. Removal of Congo Red from an aqueous solution by fungus *Aspergillus niger*. Advanced Environmental Research. 2002;**7**:239

[12] Baptista MS, Indig GL. Effect of BSA Binding on Photophysical and Photochemical Properties of Triarylmethane Dyes. The Journal of Physical Chemistry B. 1998;**102**:4678

[13] Wu J, Eitman MA, Law SE. Evaluation of membrane filtration and ozonation processes for treatment of reactive-dye wastewater. Journal of Environmental Engineering. 1998;**12**:272

[14] Ozdemir O, Armagan B, Turan M, Celik MS. Comparison of the adsorption characteristics of azo-reactive dyes on mesoporous minerals. Dyes and Pigments. 2004;**62**:49

[15] Parra S, Sarria V, Moloto S, Periner P, Pulgarin C. Photocatalytic degradation of atrazine using suspended and supported TiO2. Applied Catalysis. B, Environmental. 2004;**51**:107

[16] Talib TH, AlDamen MA, Alani RR. Modeling of Advanced Photo Oxidation of Alizarin Red-S Dye Using Tio2 as Photo Catalyst. American Chemical Science Journal. 2014;**4**(6):918

[17] Kaur S, Sharma S. Sushil Kumar Kansal. Synthesis of ZnS/CQDs nanocomposite and its application as a photocatalyst for the degradation of an anionic dye, ARS. Superlattices and Microstructures. 2016;**98**:86

[18] Navas CS, Reboredo MM, Granados DL. Comparative Study of Agroindustrial Wastes for their use in Polymer Matrix Composites. Procedia Materials Science. 2015;**8**:778

[19] Schiavello M. Heterogeneous Photocatalysis. England: John Wiley & Sons; 1997

[20] Kansal SK, Lamba R, Mehta SK, Umar A. Photocatalytic degradation of Alizarin Red S using simply synthesized ZnO nanoparticles. Materials Letters. 2013;**106**:385

[21] Kansal SK, Kaur N, Singh S. Photocatalytic Degradation of Two Commercial Reactive Dyes in Aqueous Phase Using Nanophotocatalysts. Nanoscale Research Letters. 2009;**4**:709

[22] Hwang SD, Kim SW, Lee WK, Kim EJ, Hahn SH. Visible Light Photocatalytic Activity of N-Ion Implanted TiO2 Thin Films Prepared by Oblique Incident Electron-Beam Evaporation Method. Journal of Nanoscience and Nanotechnology. 2013;**13**:7059

[23] Corma A. From Microporous to Mesoporous Molecular Sieve Materials and Their Use in Catalysis. Chemical Reviews. 1997;**97**:2373

[24] Ciesla U, Schuth F. Ordered mesoporous materials. Microporous and Mesoporous Materials. 1999;**27**:131

[25] Kraushaar B, Hooff JHCV. A test reaction for titanium silicalite catalysts. Catalysis Letters. 1989;**2**:43

[26] Rajini A, Nookaraju M, Reddy IAK, Venkatathri N. Vanadium dodecylamino phosphate: A novel efficient catalyst for synthesis of polyhydroquinolines. Chemical Papers. 2013;**68**(2):170

[27] Rouquerol F, Rouquerol J, Sing K. Adsorption by Powders & Porous Solids. San Diego: Academic Press; 1999

[28] Ferrini C, Kowenhoven HW. Modified Zeolites for Oxidation Reactions. Studies in Surface Science and Catalysis. 1990;**55**:53

[29] Petrini G, Cesana A, De Alberti G, Geroni F, Leofanti G, Padovan M, et al. Selective oxidations with hydrogen peroxide and titanium silicalite catalyst. Studies in Surface Science and Catalysis. 1991;**68**:761

[30] Shen S, Chen J, Koodali RT, Hu Y, Xiao Q, Zhou J, et al. Activation of MCM-41 mesoporous silica by transition-metal incorporation for photocatalytic hydrogen production. Applied Catalysis B. 2014;**150**:138

[31] Addamo M, Augugliaro V, Paola AD, Garcia-Lopez E, Loddo V, Marci G, et al. Preparation and photoactivity of nanostructured TiO2 particles obtained by hydrolysis of TiCl4. Colloid Surface A. Physciochemical EngineeringAspects. 2005;**265**:23

[32] Sekhar RS, Douglas SP. Graphene oxide-nano-titania composites for efficient photocatalytic degradation of indigo carmine. Journal of the Chinese Chemical Society. 2018;**65**:1423

[33] Blasco T, Corma A, Navarro MT, Pariente JP. Synthesis, characterization, and catalytic activity of Ti-MCM-41 structures. Journal of Catalysis. 1995;**156**:65

[34] Liu T, Wang L, Lu X, Fan J, Cai X, Gao B, et al. Comparative study of the

photocatalytic performance for the degradation of different dyes by ZnIn2S4: adsorption, active species, and pathways. RSC Advances. 2017;**22**(7): 12292

[35] Sivakumar S, Ranga Rao V, Nageswara Rao G. Efficient Photocatalytic Degradation of Alizarin Red S by Silver-Impregnated Zinc Oxide. Proceedings of the National Academy Science. 2013;**83**:309

[36] Anipsitakis GP, Stathatos E, Dionysiou DD. Heterogeneous activation of oxone using Co3O4. The Journal of Physical Chemistry. B. 2005;**109**: 13052-13055

[37] Pradhan AC, Paul A, Rao GR. Sol-gel-cum-hydrothermal synthesis of mesoporous Co-Fe@Al2O3–MCM-41 for methylene blue remediation. Chemical Science. 2017;**129**:381-395

[38] Zhao Q, Zhou X, Ji M, Ding H, Jiang T, Li C, et al. Stability and textural properties of cobalt incorporated MCM-48 mesoporous molecular sieve. Applied Surface Science. 2011;**257**: 2436-2442

[39] O'Connor AJ, Hokura A, Kisler JM, Shimazu S, Stevens GW, Komatsu Y. Amino acid adsorption onto mesoporous silica molecular sieves. Separation and Purification Technology. 2006;**48**:197

[40] Vallet-Regi M, Ramila A, Del Real RP, Perez-Pariente J. A new property of MCM-41: Drug delivery system. Chemistry of Materials. 2001;**13**:308

[41] Muñoz B, Rámila A, Pérez-Pariente J, Dýáz I, Vallet-Regý M. MCM-41 organic modification as drug delivery rate regulator. Chemistry of Materials. 2003;**15**:500

[42] Gimon-Kinsel ME, Jimenez VL, Washmon L. Mesoporous molecular sieve immobilized enzymes. Studies in Surface Science and Catalysis. 1998;**117**:373

[43] Vinu A, Murugesan V, Hartmann M. Adsorption of lysozyme over mesoporous molecular sieves MCM-41 and SBA-15: Influence of pH and aluminum incorporation. The Journal of Physical Chemistry. B. 2004;**108**:7323

[44] Fu F-M, DeOliveira DB, Trumble WR, Sarkar HK, Singh BR. Secondary structure estimation of proteins using the amide-III region of Fourier-transform infrared-spectroscopy – Application to analyze calcium binding-induced structural changes in casequestrin. Applied Spectroscopy. 1994;**48**(11):1432

[45] Tope AM, Srivinas N, Kulkarni SJ, Jamil K. Mesoporous molecular sieve (MCM-41) as support material for microbial cell immobilization and transformation of 2,4,6-trinitrotoluene (TNT): A novel system for whole cell immobilization. Journal of Molecular Catalysis B. 2001;**16**:17

Chapter 7

Iron Oxide Nanoparticles: A Mighty Pioneering Diagnostic Tool But Is It Really Safe for Carcinoma and Neurodegenerative Diseases?

V. Sandhiya, Selvaraja Elumalai, K. Dhunmati, C.N. Nalini, Ganesh Mani, Senthilkumar Balakrishnan and Ubaidulla Uthumansha

Abstract

Iron oxide nanoparticles have been used in medicine for around 90 years, and this time has demonstrated their versatility, therapeutic efficacy, and safety. The primary constituents of iron oxide nanoparticles (IONs) are either magnetite (FeO Fe2O3) or maghemite (-Fe2O3). The most major clinical application of IONs is based on MRI. To detect cancers and age-related diseases, IONs are being used in medical diagnostic imaging. The two IONs with the best clinical repute are Resovist and Feridex IV. In addition to being used to detect cancers, IONs are also adapted as gastrointestinal negative contrast agents and as slow-release iron supplements to treat iron deficiency anemia. With IONs exposed to alternating magnetic fields, targeted imaging and thermal energy production are both feasible. Radiation therapy, immunotherapy, or chemotherapy be facilitated by the effects of heat. A growing number of IONs are being studied in therapeutic settings as nanotechnology develops swiftly. How IONs are used in biomedicine is determined by their interaction with the human immune system.

Keywords: iron oxide, nanoparticles, MRI, diagnostic tool, role in cancer, neurodegenerative safety

1. Introduction

Due to their capacity to give anatomical (mainly dimensions) and functional characteristics of solid tumors and their environs, imaging biomarkers are becoming more crucial in cancer research. The characteristics of metabolism, tissue water diffusion, perfusion, chemical composition, and hypoxia are among those that PET, CT, and MRI may measure. Anatomical and functional information (physiological and pathophysiological) are only available with MRI, making it special. Noninvasive imaging probes like nanoparticles (NPs) have a lot of potential in this field of study because they can be made to carry and release anticancer medications into the target tissue while also functioning as diagnostic tools by utilizing the physical and chemical properties of their constituents (or moieties) [1–4]. While further acting as tools for diagnosis. Clinical trial imaging biomarker-based response criteria ought to aid in directing early choices and reducing the likelihood that patients would get needless treatment. Size-based response assessment is typically ineffective in detecting responses in patients who are experiencing either cytostasis or pseudoprogression because it is frequently insensitive to early biological alterations. These situations are typically seen with innovative target therapy, where the cancer response is more variable than with cytotoxic drugs. Biological changes such apoptosis, necrosis, cystic degeneration, intralesional hemorrhage, edoema, and immune cell infiltration happen quickly after the start of treatment (up to 12 weeks later). It is possible that anatomical imaging would not be able to identify them, which could affect clinical outcomes. Judgment. Many of these modifications can be seen on MRI and may serve as preliminary therapeutic response indications. As a result, there is an urgent demand for particular MRI biomarkers for cutting-edge treatments.

This study will concentrate on one of the most fascinating uses of NPs in cancer imaging, specifically their role in the early evaluation of immunotherapy efficacy and their capacity to change macrophage polarization.

A cutting-edge therapeutic strategy called immunotherapy works by inducing an immunological response in cancer cells. The recruitment of immune cells to the tumor site, which may be accompanied by a decline in tumor growth, is a sign of an early response to immunotherapy. NPs' propensity to be internalized by inflammatory cells in vivo is correlated with their ability to act as diagnostic agents [5]. Their capacity to be internalized by various cells, both in vitro and in vivo, has been utilized for a variety of purposes throughout the previous 20 years.

As will be briefly stated in the first half of this study, the ability of iron oxide NPs to penetrate cells, including stem cells, can enable MRI detection of inflammatory cell recruitment and provide information on the fate of the cells when transplanted into living beings. Applications in cancer immunotherapies will be highlighted in the sections that follow. The final section of the study will focus on magnetic particle imaging (MPI), a cutting-edge tomographic imaging technique that uses iron oxide nanoparticles (NPs) as tracers and describes how iron oxide NPs can be directed toward lesions. MPI is anticipated to have a significant diagnostic role in cancer immunotherapy due to its high sensitivity.

2. Contrast-enhancing iron oxide nanoparticles for cellular imaging

Several iron-based MR contrast agents were created for MRI in the middle of the 1990s. They were referred to as ferrites, magnetites, ferumoxides, or superparamagnetic iron oxides (SPIOs) because they were often made up of tiny (30–200 nm) clusters of iron-containing crystals that formed single magnetic domains. Iron-based MR contrast agents are referred to as T2-relaxing contrast agents because they have higher transverse relaxivity and r2/r1 ratios than Gd chelates. They can also have a considerable effect on the T2 relaxation time since they considerably increase the inhomogeneity of the static magnetic field outside of their immediate neighborhood. On T2 weighted pictures taken close to the iron, iron oxide NPs therefore cause a signal attenuation (commonly referred to as the "blooming effect") [6].

Iron oxide nanoparticles (NPs) have been suggested as liver-specific contrast agents due to their size and affinity for collection by the reticuloendothelial system of the liver following intravenous injection [7]. Due to the variety of cells' ease of internalization, iron oxide nanoparticles (NPs) have been employed extensively during the past 20 years to identify and track cells administered as therapy for various disorders. A detailed summary of the experiment's approach, states that NPs are given to the medium for cell growth, maybe coupled with transfection agents. In terms of cellular iron content and cell survival, the ideal experimental parameters, such as incubation period, iron oxide NP concentration, and transfection agent addition, are identified.

Using MRI, the cells are tracked in vivo after being injected into the recipient's body [8]. The fate of many cell types, including stem cells [9–11], pancreatic islets [12, 13], dendritic cells [14], and even exosomes generated from stem cells [15, 16], has been investigated using this approach in a number of preclinical studies. Benefits and limitations of the approach have been demonstrated in preclinical research. The benefits of MRI include its high sensitivity, which can even detect single cells [17, 18], as well as its outstanding anatomical detail, which clarifies cell homing and allows transferability to the clinical setting [19].

The main drawbacks include the inability to differentiate between live and dead cells, the fact that MRI's signal void does not quantitatively report on the number of cells, label dilution due to in vivo cell replication, and the removal of iron oxide NPs that were previously approved for use as MRI contrast agents in clinical settings. Kostevšek et al. [20] provides information on the most recent list of iron oxide (IO)-based contrast agents that have undergone clinical studies or received approval for use as MRI contrast agents as well as specifics on their intended purpose and current market position.

Another possibility is that SPIOs are absorbed by cells in vivo, where circulating monocytes that can enter tumors and transform into macrophages phagocytose iron oxide NPs that have been injected into the circulation. Consequently, immune cell recruitment in malignancies as well as in other organs and tissues can be detected using MRI. In a recent study, Kirschbaum et al. [21] have used high-field MRI to map inflammatory infiltrates in an experimental multiple sclerosis model using iron oxide NPs for cell tracking. They discovered an association between NP absorption and the innate immune cells-only disease's clinical severity. Their research opens the door for more accurate clinical and diagnostic treatment of a range of inflammatory diseases. in addition to therapeutic oversight [22]. Similar techniques have been applied in organ transplant experimental models, where the recruitment of macrophages is one indicator of transplant rejection. Additionally, studies have been done in clinical settings. In a recent clinical investigation, myocardial edoema and macrophage inflammation have been successfully visualized in patients who suffered myocardial infarction, utilizing T2 mapping and Ultrasmall SPIO-enhanced T2 MRI. The study concludes by showing that the technology can offer a noninvasive way to detect and track tissue inflammatory macrophage activity in the heart [23]. It is common practice to use iron oxide NPs to detect macrophages in solid tumors. This is because iron oxide NPs are not antibody-conjugated and can be administered directly into the vein and detected using a conventional 1 H radiofrequency coil and a T2 weighted sequence since they are primarily taken up by phagocytic cells like macrophages and Kupffer cells. To detect the spatial distribution of tumor-associated macrophages (TAMs) and quantify the amount of iron deposition, it is possible to collect a T2 map using a multi-gradient echo sequence. As an alternative, quantitative susceptibility mapping can be used to gauge the change in susceptibility brought about by the treatment by the contrast substance. Both of these methods have a linear correlation with the concentration of iron oxide NPs.

3. NPs' function in oncolytic virotherapy

Oncolytic virotherapy infects tumors with viruses, which kills cancer cells. Only attacking cancer cells is a capability of many distinct virus types. Along with this underlying effect, there is also significant inflammation in the cancer microenvironment. The tumor primarily targets the virus with the recruitment of inflammatory cells. The production of cancer-associated antigens as a result of virus-mediated cell lysis, however, may trigger an immune response that targets the tumor, such as by activating macrophages and T cells. The latter produce cytokines that actively stimulate the production of new immune cells as well as cancer cells. Both innate and adaptive immune responses produce an immunological memory that works in conjunction with the oncolytic action of the viruses to prevent cancer from coming back. Oncolytic virotherapy's effectiveness has been evaluated in a large number of preclinical and clinical investigations, mostly in patients with melanoma and brain tumors. Oncolytic virotherapy-induced intratumoral inflammation can be found using MRI. The effectiveness of oncolytic virotherapy may be monitored, virus sites can be indirectly identified and quantified, and new therapeutic virus strains can be improved utilizing 19F MRI [24] and iron oxide NPs [25].

Perfluorocarbon nanoemulsions (PFC) and 19F MRI were used by Weibel and colleagues [26] to establish a longitudinal, noninvasive monitoring of intratumoral inflammation during oncolytic virotherapy. By comparing in vivo and ex vivo 19F/1H MRI with histology, the authors demonstrated the potential of this imaging modality for the localization of the host immune response and for sentinel lymph node detection. Tumor viral colonization significantly altered the 19F signal distribution and intensity in solid tumors as well as in the nearby lymph nodes. Compared to virally infected tumors, which only displayed 19F-positive hot patches along the tumor margin, the mock-infected tumors had a uniform distribution of both the 19F signal and CD68 + -macrophages. The population of CD68+ macrophages displayed a similar pattern of distribution. According to our research, PFC NPs are more likely than intratumoral TAMs to detect circulating immune cells that enter the tumor after viral infection.

4. Magnetic particle imaging is a recent development in imaging technology

A new imaging method called magnetic particle imaging (MPI) has just been developed to find iron oxide nanoparticles (NPs). Some MRI flaws, such as poor specificity (caused by other low signal regions in MRI, such as hemorrhagic regions or those containing air) and challenging quantification, can be resolved with MPI. High tracer specificity is made feasible by leveraging MPI's direct detection of iron oxide NPs, which offer positive contrast without any underlying background signal from biological tissues. Iron (Fe) concentrations of 550 pg./L in vitro and 7.8 ng Fe in vivo, as well as detection limits as low as 1.1 ng Fe, have all been demonstrated. A static gradient field with a single, field-free parameter identifies signals in MPI. (FFR), which could be a line or a point. Then, using the particles already present in the FFR, a signal is produced by an oscillating magnetic field. Raster scanning the FFR throughout the entire field of vision produces images. Outside of the FFR, superparamagnetic particles are still fully magnetized and do not increase the signal [27].

Because MPI directly detects SPIO magnetization, the signal is very dependent on the SPIO tracer's physical characteristics. For NPs to be suitable for MPI, they must possess the following three characteristics: superparamagneticity, susceptibility to magnetic saturation, and a nonlinear magnetic curve.

Since many SPION agents for MRI have the aforementioned characteristics, their prospective application in MPI has been looked at. Resovist®, a previously developed MRI contrast agent for the liver, was used to produce the most efficient MPI [20]. Magnetic Insight, Inc. has unveiled VivoTrax®, a carboxydextran-coated iron oxide NP formulation, with the same reference standard. It is intriguing to note that MPI can locate the clinically approved ferumoxytol rapidly [28]. Other research teams are working to develop new MPI tracers as it was established that neither Resovist® nor VivoTrax® was the optimal MPI solution.

There are numerous MPI biological applications that are now being studied. Early cancers can be identified using MPI by utilizing the tumor's enhanced permeability and retention (EPR) impact [29, 30]. Due to the leaky capillaries with big holes that result in the EPR phenomena, tumor tissue is an excellent target for therapy with nanomedicines and nanosized contrast agents. As a result, MPI plays a crucial clinical role in the early detection of cancer.

Cell tracking is one of MPI's oldest and most promising applications due to its superior tissue penetration, absence of background noise, and high degree of sensitivity, which enables it to identify as little as 200 tagged cells. Recently, mesenchymal stem cells (MSCs) tagged with clinically relevant feromuxytol NPs have been found by Nejadnik et al. [31] using MPI. A noteworthy achievement for the therapeutic application of MPI technology was the precise in vivo identification and quantification of ferumoxytol-labeled stem cells.

5. Alzheimer's disease (AD)

The Alzheimer Association attributes 60–80% of dementia cases to Alzheimer's disease (AD), a degenerative brain illness. Depending on the stage of the illness, apathy, depression, decreased communication, disorientation, poor judgment, difficulties swallowing and walking, and behavioral changes are some of the characteristics that evolve to make doing daily tasks difficult [32, 33]. Age, genetics, and sex are some of the factors that influence how long it takes for a continuum of these symptoms to emerge, according to current estimates, and the COVID-19 pandemic has seen a 16% increase in the number of deaths (Alzheimer's Association, 2021). Amyloid-beta (Aβ) and tau protein buildup is linked to the progression of cognitive deterioration in AD. Beta-secretase and gamma-secretase sequentially cleave the amyloid precursor protein (APP), resulting in the formation of Aβ. Thus, the aggregation of Aβ produces hazardous oligomers for the neurons. Tau, on the other hand, is produced by alternative splicing from the soluble protein isoforms of the microtubule-associated protein tau (MAPT) gene. The damage to brain circuits and cognitive impairment in AD has been linked to a number of functional interactions between Aβ and tau. In the neuropathology of Alzheimer's disease, there is a loss of neurons and atrophy in the temporofrontal cortex, which results in inflammation and the deposition of amyloid plaques, an abnormal cluster of protein fragments, and tangled bundles of fibers. As a result, there is an increase in the presence of monocytes and macrophages in the cerebral cortex, and it also activates the microglial cells in the parenchyma [34–36].

6. Pathophysiology

The primary neuropathologic symptoms of AD include extracellular amyloid plaques, intracellular NFTs, synaptic deterioration, and neuronal death. Without detecting granulovacuolar degeneration in the hippocampus or amyloid buildup in blood vessels (congophilicangiopathy), the diagnosis can be made. According to the "amyloid cascade" idea, amyloid plaques interfere with synaptic transmission and trigger a series of following processes that eventually cause cell death.

7. Amyloid plaques

Even though amyloid plaques can be categorized into different groups depending on their structure, all types of -amyloid protein (A) are present in them. When APP is degraded proteolytically by - and -secretase, an amino acid peptide known as A is produced. The primary results of this cleavage are A1–40 and A1–42. The development of amyloid oligomers and fibrils, which come together to form amyloid plaques, is predisposed by a relative excess of A1–42. Since the generation, processing, and/or trafficking of amyloid is connected to the proteins encoded by APP, PS1, PS2, SorL1, and ApoE, this suggests that amyloid plays a significant role in the etiology of AD [37].

8. Neurofibrillary tangles (NFT)

Tau, a protein associated with microtubules, is required for normal neuronal development and axonal expansion. However, frontal association cortices, the lateral parietotemporal area, and the mesial temporal lobe (especially the hippocampus) neurons are where hyperphosphorylated tau protein aggregates are most frequently found as helical filamentous NFT. The relationship between the density and distribution of tau NFT and the symptoms and severity of AD dementia emphasizes the critical role of NFT in AD pathology.

9. Loss of neurons and synapses

Synapse loss and neuronal cell death have a similar distribution to NFT. Due to the death of neurons in the nucleus basalis of Meynert, acetylcholine (Ach), a neurotransmitter linked to memory, is decreased in typical AD. Most current therapies seek to remedy this cholinergic deficit. Serotonin and norepinephrine deficits result from the loss of neurons in the brainstem's locus ceruleus and median raphe. Dysfunctional serotonergic and adrenergic activity in the brain is probably the root cause of dysphoria and insomnia in AD [38, 39].

10. The metal ion theory

Metal dyshomeostasis has a role in the development and pathophysiology of illnesses, including as cancer and neurodegenerative disorders. A number of these compounds are employed in clinical studies. Ionosphere and metal chelators are well-known modulators of transition metal homeostasis. Other medications besides the metal-binding ones can also target the homeostasis of transition metals. The balance of redox transition metals, primarily copper (Cu), iron (Fe), and other trace metals, is changing, according to current findings. In AD, their brain levels are observed to be elevated. Other neurological diseases also involve copper, manganese, aluminum, and zinc. The cholinergic theory Acetyl-cholinesterase inhibitors (AChEIs) and the impact of apo-lipo-protein E (APOE) genotype in Alzheimer's disease patients. The AchEI drugs are the cornerstone of AD treatment, and the most significant risk factor for AD is the APOE genotype [36].

11. Role of iron in AD

Iron is the most abundant transition element on Earth and one of the most important minerals in the body. It plays an indispensable role in many physiological and pathological processes of the body. Iron homeostasis is even more crucial in the brain to maintain its normal function. Iron dyshomeostasis within the brain can cause oxidative stress and inflammatory responses, leading to cell damage and finally neurological diseases. Ferroptosis, a programmed cell death process associated with iron dysregulation, has been supposed to be linked to neurological diseases, especially neurodegenerative diseases. Till date, it is impossible to explain AD with a single pathological path. Currently, metal dyshomeostasis in AD has been extensively studied. Studies have found intracellular iron deposition even before the formation of senileplaques and neurofibrillary tangles (NFT), and ferroptosis is proposed to be one of key causes of neuronal loss in AD patients [38].

12. Iron metabolism in healthy and Alzheimer's disease brain

About 48% of the iron in the body is bound to hemoglobin and is involved in oxygen transport in the body. About 17% of the iron is found as the cofactor in proteins to carry out functions in several crucial biological processes such as the tricarboxylic acid cycle, oxidative phosphorylation, DNA synthesis and repair, and iron homeostasis. In the brain, iron is involved in myelination, neurotransmitter synthesis, and antioxidant enzyme function, and its entry and exit are tightly regulated by a variety of molecules. Aging, inflammation, and oxidative stress, which disturb the functions of molecules involved in iron metabolism, present as the main contributors to iron dyshomeostasis. Iron transport across blood-brain barrier in the brain, transferrin receptor 1 (TfR1), responsible for the strict control of the level of iron transported into the brain, is expressed on the luminal side of the brain microvascular endothelial cells (BMECs) and the blood-cerebrospinal fluid barrier. After circulation, a complex was formed (holo-Tf) by iron with transferrin (Tf), it binds to TfR1 on the surface of the BMECs, followed by entry into the BMECs via clathrin-mediated endocytosis. $Fe3+$ detaches from Tf in the acidic environment of the endosome and is reduced to $Fe2+$ by six-transmembrane epithelial antigen of prostrate 3(STEAP3) or duodenal cytochrome b (DCYTB), both of which are metalloreductases. It then enters the cytoplasm via divalent metal transporter 1 (DMT1). $Fe2+$ in BMECs can then enter the brain by the secretion of ferroportin 1 (FPN1), followed by the oxidation by extracellular ceruloplasmin (Cp) or hephaestin. Non-transferrin-bound iron (NTBI) can cross the blood-brain barrier (BBB) via receptor-mediated transcytosis after binding to heavy-chain ferritin (H-ferritin;Lf). It was also reported that Lf increased in the brains of aged individuals and those with AD, allowing large amounts of non-Tf-bound iron to enter the brain. Iron transport and storage within the brain neuronal iron metabolism TfR1 is highly expressed on the surface of neurons, and similar to BMECs, iron enters neurons via clathrin-mediated endocytosis of holo-Tf/TfR1 and exits the endosomes in the form of reduced $Fe2+$ via DMT1. NTBI can also enter neurons in a DMT1-dependent manner independent of Tf. Cellular prion protein (PrPC) is abundantly expressed on the surface of neuronal membranes. It functions as a ferrireductase partner for DMT1, mediating $Fe2+$ uptake in the plasma membrane in the form of complex PrPc/DMT1. PrPC knockout in mice can lead to iron deficiency in brain and uptake increase of holo-Tf. By comparing the brain tissues of juvenile, adult, and aged rats that had the pathological features of AD, it was found that DMT1 abnormally increased with age. They supposed that DMT1 may be one of the main reasons why the iron concentration in the brain gradually increases with age. Some $Fe2+$ undergo normal metabolism in the cytoplasm of neurons, while some are stored in ferritin in the form of nontoxic $Fe3+$; when neurons are low in iron, ferritin can be degraded by lysosomes to release the stored iron to meet the physiological needs of the neurons. Ferritin is positively correlated with iron overload and is found deposited in senile plaques in the AD brain. It had been shown that there was an age-dependent increase in ferritin in the brain, probably a contributor to the iron overload in aged and AD brains. Autopsy studies of AD patients have revealed that mitochondrial ferritin is upregulated. Ferritin in the cerebrospinal fluid (CSF) of AD patients has been shown significantly increased, which is negatively correlated with cognitive decline and hippocampal atrophy in AD. Additionally, iron can enter mitochondria to form iron ferroptosis and Alzheimer's disease sulfur cluster and participate in the process of aerobic respiration. Regarding the transport of excess iron out of neurons, FPN1 is the only known iron exporter to date. Both Cp and hephaestin (Heph) can oxidize $Fe2+$ and facilitate FPN1to export iron, so the FPN1/Cp and FPN1/Heph are the main iron efflux pathways. Decrease of any of these three export proteins can induce iron retention and consequently the memory impairment. It was reported that FPN1 was downregulated in the brains of AD patients and triple-transgenic AD mouse models; thus, excessive iron could not be excreted normally, initiating intracellular iron deposition. Since Cp is a crucial partner of FPN1 to oxidize $Fe2+$ before it is excreted by FPN1, the dysfunction of Cp serves as an upstream event of iron retention, which has been found in AD. Noteworthily, both of amyloid precursor protein (APP) and tau, which are the substrates of the AD hallmarks in pathological condition, are crucial for neuronal iron efflux. APP is defined as a metalloprotein involved in iron homeostasis. With the assistance of soluble tau protein, APP is transported to the cell membrane where it stabilizes FPN1 and facilitates the efflux of iron. APP or tau knockdown can lead to abnormal FPN1 function and the inability of neuronal iron to flow out normally, resulting in neuronal iron overload. APP with the pathogenic Italian mutation A673V is more prone to be cleaved by β-secretase to produce Aβ1–42, impeding its support of FPN1 and thus increasing iron retention. Because of the continuous cleavage of APP and hyperphosphorylation of tau in AD brain, the iron efflux was hindered in neurons. Glial support for neuronal iron metabolism glial cells help to maintain the iron availability at a safe level in neurons. Astrocytes and microglia respond during iron overload or deficiency in order to maintain neuronal iron homeostasis. As a buffer pool, astrocytes express abundant TfR1 and DMT1, which facilitates taking up of both holo-Tf and NTBI from the abluminal side of BMECs and the brain interstitium, precisely regulating the iron concentration in neurons. Microglia also express TfR1 and reduce iron toxicity by promoting the influx of excess iron (for storage in ferritin) via the TfR1/DMT1 pathway. Microglia and astrocytes are capable of releasing ferritin carrying Fe3+ to supplement the iron deficiency or to support oligodendrocytes for myelination or remyelination. Iron is essential for myelination in oligodendrocytes, which are the most iron-rich cell type in the brain. TfR1 is absent in oligodendrocytes, while H-ferritin is the main source of iron for oligodendrocyte by interaction with T-cell immunoglobulin mucin domain 2 (TIM2). Noteworthily, when iron is overloaded, oligodendrocytes provide an antioxidant defense for neurons by secreting H-ferritin, scavenging extracellular extra iron [40–46].

13. Impact of iron overload on Alzheimer's disease pathology

Currently, the involvement of iron in the early pathology of AD has been well accepted since the discovery of the link between dysregulation of brain iron homeostasis and AD pathogenesis in 1953. In the preclinical stage of AD, there is significant abnormal iron elevation in cortical, hippocampal, and cerebellar neurons while much severe in the cortex and hippocampus, the main brain areas affected by AD [47]. The iron overload in the brain is corresponding to the severity of AD lesions and the rate of cognitive decline. It is also proposed that hippocampal iron deposition could be the predictor of the rate of cognitive decline caused by Aβ. Iron overload drives a series of events, including glial activation, formation of Aβ plaque and tau tangles, and even neuronal loss, pushing the progress of the disease and accelerating cognitive decline. Iron interaction with Aβ plaques and neurofibrillary tangles iron accumulation was demonstrated to accelerate senile plaque deposition and the production of neurofibrillary tangles [48, 49]. Autopsy evidence and magnetic resonance imaging analysis provide evidence that there are a large amount of iron deposition not only in and around senile plaques but also in the sites of cortical tau accumulation, indicating the potential cross talk of iron with both of senile plaques and neurofibrillary tangles. Perturbations in iron homeostasis is one of key players in Aβ deposition. High intracellular iron concentration enhances the interaction of IRP/IRE, inducing APP upregulation. Furthermore, the enzymes that cleave APP named α- and β-secretase are tightly balanced and modulated by furin. More β-secretase is activated when α-secretase is suppressed by furin impairment in the condition of excessive iron. Upregulated APP is cleaved by more β-secretase to Aβ40/42, accelerating the Aβ deposition [50, 51]. Meanwhile, APP can no longer assist FPN1, resulting in impaired iron efflux and aggravated iron deposition. Some researchers have even proposed that Aβ is nontoxic in the absence of redox metals and that aggregation of Aβ requires the involvement of metals. Soluble Aβ binds to Fe3+ when extracellular iron increases so as to remove excess iron, but it is difficult to dissociate them after they interact; Aβ can promote the reduction of Fe3+ to Fe2+, and the reactive oxygen species (ROS) released during this process allow Aβ to be deposited more easily and rapidly, forming more senile plaques. The interactions of iron with APP and Aβ greatly increase the formation rate and degree of senile plaques. Therefore, some researchers believe that iron deposition should be included in the "Aβ cascade hypothesis" of AD. Iron can also interact with tau. Reduced soluble tau in the brain of AD patients increased brain iron deposition by suppressing FPN1 activity. On the contrary, a diet high in iron can lead to cognitive decline in mice, increased abnormal tau phosphorylation in neurons, and abnormal expression of insulin pathway-related proteins. Insulin supplementation can reduce iron-induced phosphorylation of tau, indicating that iron deposition may lead to tau hyperphosphorylation by interfering insulin signaling. In vivo research has found that iron can be involved in tau hyperphosphorylation by activating the cyclin-dependent kinase 5 (CDK5)/P25 complex and glycogen synthase kinase3β (GSK-3β). Excessive intracellular Fe2 + −induced production of oxygen free radicals can also promote tau hyperphosphorylation by activating the extracellular. Ferroptosis and Alzheimer's disease signal-regulated kinase 1/2 (Erk1/2) or mitogen-activated protein kinase (MAPK) signaling pathways. Glial activation and neuroinflammation has been demonstrated to be a prominent characteristic of AD pathology. Microglial are highly reactive cells responding to increased iron levels in the brain. When iron level increases in brain, microglia become activated, with soma volume increased and process length decreased. Iron may activate microglia through proinflammatory cytokines mediated by the nuclear factor-κB (NF-κB). After activated, they express more ferritin to scavenge the extracellular iron, resulting in intracellular iron retention, increased TNFα expression, and finally infiltrated with Aβ-plaques. Activated microglia also secret Lf, which can interact with APP, promoting the Aβ formation. Conversely, formation of Aβ induces more IL-1β expression in microglia in the environment of elevated iron, exacerbating the proinflammatory effects. Astrocytes are highly resistant to metal-induced toxicity within the brain as the critical cell type in maintaining a balanced extracellular environment and supporting the normal functioning of neurons. In the environment of high iron, astrocytes respond with a significant increase in glutathione, catalase, and manganese superoxide dismutase levels to resist the oxidative stress. They show less impairment by iron than neurons and oligodendrocytes. But later, the astrocytes were found activated with increased glial fibrillary acidic protein (GFAP). Activated astrocytes release inflammatory mediators and induce oxidative stress, which facilitate the formation of Aβ and tau tangles and hinder Aβ clearance. Iron overload induces oxidative stress and neuronal loss; iron toxicity is largely based on Fenton chemistry [52–54].

14. Conclusion

Along with the increasing importance of novel cancer immunotherapies in the fight against cancer and their translation from preclinical research to clinical practice, there is an increase in the demand for noninvasive imaging techniques that can measure macrophage responses. Although there are other imaging methods available, including PET, Gd-enhanced MRI, and 19F MRI, using MRI with superparamagnetic iron oxide contrast agents is probably the most promising. Theranostic properties, magnetic gradient actuation forces for transport to the target, and multimodal imaging capacity (MRI-MPI) are a few of the main advantages of these NPs. Additionally, despite the fact that clinical development of SPIOs has been stopped, a number of contrast agents, such as Resovist® and Feromuxytol, are still available.

As a redox-active transition metal, iron is a key player during the process of oxidative stress. Elevated iron promotes the production of ROS, which further depletes the cellular antioxidant GSH and promotes lipid peroxidation, finally triggering ferroptosis and neuronal loss.

As previously mentioned, oxidative stress, protein aggregation, and iron buildup all have a positive feedback loop where one factor encourages the other. By inducing iron buildup, oxidative stress, or protein aggregation, iron oxide nanoparticles

(IONPs) can turn on this loop. Additionally, IONPs could cause the neurons to undergo apoptotic cellular death. IONPs may cause neurodegeneration given the roles that iron buildup, oxidative stress, protein aggregation, and apoptosis play in neurodegenerative disorders. However, IONPs' properties, such as size, shape, concentration, surface charge, type of coating, and functional groups, have an impact on how toxic they are. Therefore, taking into account the properties of IONPs is crucial when applying them to the CNS.

Author details

V. Sandhiya[1], Selvaraja Elumalai[2], K. Dhunmati[3], C.N. Nalini[3], Ganesh Mani[4], Senthilkumar Balakrishnan[5] and Ubaidulla Uthumansha[6]*

1 Faculty of Pharmacy, Sree Balaji Medical College and Hospital, BIHER, Chennai, India

2 Granules Pharmaceuticalsinc, Chantilly, VA, USA

3 Department of Pharmaceutical Chemistry, C.L. Baid Metha College of Pharmacy, Chennai, India

4 Department of Quality Assurance, Channabasweshwar Pharmacy College, Latur, Maharashtra, India

5 JKKMMRFs Annai JKK Sampoorani Ammal College of Pharmacy, Tamilnadu, India

6 Department of Pharmaceutics, Crescent School of Pharmacy, B.S. Abdur Rahman Crescent Institute of Science and Technology, India

*Address all correspondence to: ubaidnkl@gmail.com

References

[1] Bazak R, Houri M, El Achy S, Kamel S, Refaat T. Cancer active targeting by nanoparticles: A comprehensive review of literature. Journal of Cancer Research and Clinical Oncology. 2015;**141**:769-784. DOI: 10.1007/s00432-014-1767-3

[2] Alqaraghuli HGJ, Kashanian S, Rafipour R. A review on targeting nanoparticles for breast cancer. Current Pharmaceutical Biotechnology. 2019;**20**:1087-1107. DOI: 10.2174/1389201020666190731130001

[3] Nag OK, Delehanty JB. Active cellular and subcellular targeting of nanoparticles for drug delivery. Pharmaceutics. 2019;**11**:543. DOI: 10.3390/pharmaceutics11100543

[4] Vannucci L, Falvo E, Failla CM, Carbo M, Fornara M, Canese R, et al. In vivo targeting of cutaneous melanoma using an melanoma stimulating hormone-engineered human protein cage with fluorophore and magnetic resonance imaging tracers. Journal of Biomedical Nanotechnology. 2015;**11**: 81-92. DOI: 10.1166/jbn.2015.1946

[5] Neuwelt A, Sidhu N, Hu C-AA, Mlady G, Eberhardt SC, Sillerud LO. Iron-based superparamagnetic nanoparticle contrast agents for MRI of infection and inflammation. American Journal of Roentgenology. 2015;**204**:W302-W313. DOI: 10.2214/AJR.14.12733

[6] Yang R, Sarkar S, Yong VW, Dunn JF. In vivo MR imaging of tumor-associated macrophages: The next frontier in cancer imaging. Magnetic Resonance Insights. 2018;**11**:1178623X18771974. DOI: 10.1177/1178623X18771974

[7] Morana G, Salviato E, Guarise A. Contrast agents for hepatic MRI. Cancer Imaging. 2007;7:S24-S27. DOI: 10.1102/1470-7330.2007.9001

[8] Barrow M, Taylor A, Murray P, Rosseinsky MJ, Adams DJ. Design considerations for the synthesis of polymer coated iron oxide nanoparticles for stem cell labelling and tracking using MRI. Chemical Society Reviews. 2015;**44**:6733-6748. DOI: 10.1039/C5CS00331H

[9] Hoehn M, Küstermann E, Blunk J, Wiedermann D, Trapp T, Wecker S, et al. Monitoring of implanted stem cell migration in vivo: A highly resolved in vivo magnetic resonance imaging investigation of experimental stroke in rat. Proceedings of the National Academy of Sciences USA. 2002;**99**:16267-16272. DOI: 10.1073/pnas.242435499

[10] Neri M, Maderna C, Cavazzin C, Deidda-Vigoriti V, Politi LS, Scotti G, et al. Efficient In vitro Labeling of human neural precursor cells with superparamagnetic iron oxide particles: Relevance for In vivo cell tracking. Stem Cells. 2008;**26**:505-516. DOI: 10.1634/stemcells.2007-0251

[11] Rosenberg JT, Yuan X, Grant S, Ma T. Tracking mesenchymal stem cells using magnetic resonance imaging. Brain Circulation. 2016;**2**:108-113. DOI: 10.4103/2394-8108.192521

[12] Jirák D, Kríz J, Herynek V, Andersson B, Girman P, Burian M, et al. MRI of transplanted pancreatic islets. Magnetic Resonance in Medicine. 2004;**52**:1228-1233. DOI: 10.1002/mrm.20282

[13] Marzola P, Longoni B, Szilagyi E, Merigo F, Nicolato E, Fiorini S, et al.

In vivo visualization of transplanted pancreatic islets by MRI: Comparison between In vivo, histological and electron microscopy findings. Contrast Media & Molecular Imaging. 2009;**4**:135-142. DOI: 10.1002/cmmi.274

[14] Martelli C, Borelli M, Ottobrini L, Rainone V, Degrassi A, Russo M, et al. In vivo imaging of lymph node migration of MNP- and 111In-Labeled dendritic cells in a transgenic mouse model of breast cancer (MMTV-Ras). Molecular Imaging and Biology. 2012;**14**:183-196. DOI: 10.1007/s11307-011-0496-0

[15] Busato A, Bonafede R, Bontempi P, Scambi I, Schiaffino L, Benati D, et al. Labeling and magnetic resonance imaging of exosomes isolated from adipose stem cells. Current Protocols in Cell Biology. 2017;**75**:3-44. DOI: 10.1002/cpcb.23

[16] Bonafede R, Turano E, Scambi I, Busato A, Bontempi P, Virla F, et al. ASC-exosomes ameliorate the disease progression in SOD1(G93A) murine model underlining their potential therapeutic use in human ALS. International Journal of Molecular Sciences. 2020;**21**:3651. DOI: 10.3390/ijms21103651

[17] Afridi MJ, Ross A , Liu X, Bennewitz MF, Shuboni DD, Shapiro EM. Intelligent and automatic In vivo detection and quantification of transplanted cells in MRI. Magnetic Resonance in Medicine. 2017;**78**:1991-2002. DOI: 10.1002/mrm.26571

[18] Shapiro EM, Skrtic S, Sharer K, Hill JM, Dunbar CE, Koretsky AP. MRI detection of single particles for cellular imaging. Proceedings of the National Academy of Sciences USA. 2004;**101**:10901-10906. DOI: 10.1073/pnas.0403918101

[19] Bulte JWM, Daldrup-Link HE. Clinical tracking of cell transfer and cell transplantation: Trials and tribulations. Radiology. 2018;**289**:604-615. DOI: 10.1148/radiol.2018180

[20] Kostevšek N, Cheung CCL, Serša I, Kreft ME, Monaco I, Franchini MC, et al. Magneto-liposomes as MRI contrast agents: A systematic study of different liposomal formulations. Nanomaterials. 2020;**10**:889. DOI: 10.3390/nano10050889

[21] Kirschbaum K, Sonner JK, Zeller MW, Deumelandt K, Bode J, Sharma R, et al. In vivo nanoparticle imaging of innate immune cells can serve as a marker of disease severity in a model of multiple sclerosis. Proceedings of the National Academy of Sciences USA. 2016;**113**:13227-13232. DOI: 10.1073/pnas.1609397113

[22] Kanno S, Wu Y-JL, Lee PC, Dodd SJ, Williams M, Griffith BP, et al. Macrophage accumulation associated with rat cardiac allograft rejection detected by magnetic resonance imaging with Ultrasmall superparamagnetic iron oxide particles. Circulation. 2001;**104**:934-938. DOI: 10.1161/hc3401.093148

[23] Stirrat CG, Alam SR, MacGillivray TJ, Gray CD, Dweck MR, Raftis J, et al. Ferumoxytol-enhanced magnetic resonance imaging assessing inflammation after myocardial infarction. Heart. 2017;**103**:1528-1535. DOI: 10.1136/heartjnl-2016-311018

[24] Zhang L, Xiao S, Kang X, Sun T, Zhou C, Xu Z, et al. Metabolic conversion and removal of manganese ferrite nanoparticles in RAW264.7 cells and induced alteration of metal transporter gene expression. International Journal of Nanomedicine. 2021;**16**:1709-1724. DOI: 10.2147/IJN.S289707

[25] Iv M, Samghabadi P, Holdsworth S, Gentles A, Rezaii P, Harsh G, et al. Quantification of macrophages in high-grade gliomas by using Ferumoxytol-enhanced MRI: A pilot study. Radiology. 2019;**290**:198-206. DOI: 10.1148/radiol.2018181204

[26] Aşık E, Akpınar Y, Güray NT, İşcan M, Demircigil GÇ, Volkan M. Cellular uptake, genotoxicity and cytotoxicity of cobalt ferrite magnetic nanoparticles in human breast cells. Toxicology Research. 2016;**5**:1649-1662. DOI: 10.1039/C6TX00211K

[27] Sinigaglia M, Assi T, Besson FL, Ammari S, Edjlali M, Feltus W, et al. Imaging-guided precision medicine in glioblastoma patients treated with immune checkpoint modulators: Research trend and future directions in the field of imaging biomarkers and artificial intelligence. EJNMMI Research. 2019;**9**:78. DOI: 10.1186/s13550-019-0542-5

[28] Daldrup-Link HE, Golovko D, Ruffell B, DeNardo DG, Castaneda R, Ansari C, et al. MRI of tumor-associated macrophages with clinically applicable iron oxide nanoparticles. Clinical Cancer Research. 2011;**17**:5695-5704. DOI: 10.1158/1078-0432.CCR-10-3420

[29] Shih Y-YI, Hsu Y-H, Duong TQ, Lin S-S, Chow K-PN, Chang C. Longitudinal study of tumor-associated macrophages during tumor expansion using MRI. NMR in Biomedicine. 2011;**24**:1353-1360. DOI: 10.1002/nbm.1698

[30] Simon GH, von Vopelius-Feldt J, Fu Y, Schlegel J, Pinotek G, Wendland MF, et al. Ultrasmall Supraparamagnetic iron oxide-enhanced magnetic resonance imaging of antigen-induced arthritis. Investigative Radiology. 2006;**41**:45-51. DOI: 10.1097/01.rli.0000191367.61306.83

[31] Aghighi M, Golovko D, Ansari C, Marina NM, Pisani L, Kurlander L, et al. Imaging tumor necrosis with Ferumoxytol. PLoS One. 2015;**10**:e0142665. DOI: 10.1371/journal.pone.0142665

[32] Nejadnik H, Pandit P, Lenkov O, Lahiji AP, Yerneni K, Daldrup-Link HE. Ferumoxytol can Be used for quantitative magnetic particle imaging of transplanted stem cells. Molecular Imaging and Biology. 2019;**21**:465-472. DOI: 10.1007/s11307-018-1276-x

[33] Mody VV, Cox A, Shah S, Singh A, Bevins W, Parihar H. Magnetic nanoparticle drug delivery systems for targeting tumor. Applied Nanoscience. 2014;**4**:385-392

[34] Mucke L. Alzheimer's disease. Nature. 2009;**461**(7266):895-897

[35] Scheltens P, Blennow K, Breteler MM, De Strooper B, Frisoni GB, Salloway S, et al. Alzheimer's disease. The Lancet. 2016;**388**(10043):505-517

[36] Bush AI. The metallobiology of Alzheimer's disease. Trends in neurosciences. 2003;**26**(4):207-214

[37] Younkin SG. The role of Aβ42 in Alzheimer's disease. Journal of Physiology-Paris. 1998;**92**(3-4):289-292

[38] Ayton S, Lei P, Bush AI. Metallostasis in Alzheimer's disease. Free Radical Biology and Medicine. 2013;**62**:76-89

[39] Wenk GL. Neuropathologic changes in Alzheimer's disease. Journal of Clinical Psychiatry. 2003;**64**:7-10

[40] Dikpati A, Madgulkar AR, Kshirsagar SJ, Bhalekar MR, Chahal AS. Targeted drug delivery to CNS using nanoparticles. JAPS. 2012;**2**(1):179-191

[41] Poduslo JF, Hultman KL, Curran GL, Preboske GM, Chamberlain R. Targeting vascular amyloid in arterioles of Alzheimer disease transgenic mice with amyloid beta protein antibody-coated nanoparticles. Neuropathology & Experimental Neurology. 2011;**70**:653-661

[42] Glat M, Skaat H, Menkes-Caspi N, Margel S, Stern EA. Age-dependent efects of microglial inhibition in vivo on Alzheimer's disease neuropathology using bioactive-conjugated iron oxide nanoparticles. Journal of Nanbiotechnology. 2013;**11**(32):1-12

[43] Chertok B, Mofat BA, David AE, Yu F, Bergemann C, Ross BD, et al. Iron oxide nanoparticles as a drug delivery vehicle for MRI monitored magnetic targeting of brain tumors. Biomaterials. 2008;**29**(4):487-496

[44] Weinstein JS, Varallyay CG, Dosa E, Gahramanov S, Hamilton B, Rooney WD, et al. Superparamagnetic iron oxide nanoparticles: Diagnostic magnetic resonance imaging and potential therapeutic applications in neurooncology and central nervous system infammatory pathologies, a review. Journal of Cerebral Blood Flow and Metabolism. 2010;**30**:15-35

[45] Sripetchwandee J, Wongjaikam S, Krintratun W, Chattipakorn N, Chattipakorn SC. A combination of an iron chelator with an antioxidant efectively diminishes the dendritic loss, tau-hyperphosphorylation, amyloids-β accumulation and brain mitochondrial dynamic disruption in rats with chronic iron-overload. Neuroscience. 2016;**332**:191-202

[46] Birben E, Sahiner UM, Sackesen C, Erzurum S, Kalayci O. Oxidative stress and antioxidant defense. World Allergy Organization Journal. 2012;**5**(1):9-19

[47] Palmieri B, Sblendorio V. Oxidative stress tests: Overview on reliability and use. Part I. European Review for Medical and Pharmacological Sciences. 2007;**11**(5):309-342

[48] Mexas LM, Florang VR, Doorn JA. Inhibition and covalent modifcation of tyrosine hydroxylase by 3,4-dihydroxyphenylacetaldehyde, a toxic dopamine metabolite. Neurotoxicology. 2011;**32**(4):471-477

[49] Heo HJ, Lee CY. Protective efects of quercetin and vitamin C against oxidative stress-induced neurodegeneration. Journal of Agricultural and Food Chemistry. 2004;**52**(25):7514-7517

[50] Imam SZ, Lantz-McPeak SM, Cuevas E, Rosas-Hernandez H, Liachenko S, Zhang Y, et al. Iron oxide nanoparticles induce dopaminergic damage: in vitro pathways and in vivo imaging reveals mechanism of neuronal damage. Molecular Neurobiology. 2015;**52**:913-926

[51] Zhang Y, Wang Z, Li X, Wang L, Yin M, Wang L, et al. Dietary iron oxide nanoparticles delay aging and ameliorate neurodegeneration in drosophila. Advanced Materials. 2016;**28**(7):1387-1393

[52] Kumari M, Rajak S, Singh SP, Kumari SI, Kumar PU, Murty US, et al. Repeated oral dose toxicity of iron oxide nanoparticles: Biochemical and histopathological alterations in diferenttissues of rats. Journal of Nanoscience and Nanotechnology. 2012;**12**(3):2149-2159

[53] Szalay B, Tátrai E, Nyírő G, Vezérb T, Dura G. Potential toxic efects of iron

oxide nanoparticles in *in vivo* and *in vitro* experiments. Journal of Applied Toxicology. Jun 2012;**32**(6):446-453. DOI: 10.1002/jat.1779. [Epub 2011 Dec 7]. PMID: 22161551

[54] Sonmez E, Aydin E, Turkez H, Özbek E, Togar B, Meral K, et al. Cytotoxicity and genotoxicity of iron oxide nanoparticles: An in vitro biosafety study. Archives of Biological Sciences. 2016;**68**(1):7-16

Section 3

Biological State-of-the-Art

Chapter 8

In Vitro, *In Vivo* and *Ex Vivo* Models for Toxicity Evaluation of Nanoparticles: Advantages and Disadvantages

Neeraja Revi, Oluwatosin D. Oladejo and Divya Bijukumar

Abstract

This chapter focus on existing model systems used to evaluate the toxicity of nanoparticles. We will be discussing monolayer and 3D cell based toxicity models, In vivo models like rodents and zebrafish systems. A focus will also be given on ex vivo models like chick embryos. Each toxicity model system will be discussed with its advantages and limitations. The chapter will provide critical information to students and researchers studying nanotechnology about the potential systems to check the toxicity of the nanoparticles developed in the laboratory. This can be used as a quick guide to use a model system to check toxicity based on the different type of particle with informed decisions based on its advantages and disadvantages.

Keywords: nanoparticles, toxicity, in vitro assays, precision cut slice model, organ on a chip, *Drosophila melanogaster*, *Danio rerio*, non-human primates

1. Introduction

Nanoparticles (NPs) range from size 1–100 nm [1]. They are made from various materials like polymers, liposomes, dendrimers, and metals like Zinc, Titanium, Gold, and Aluminum [2]. NPs have been found to induce toxicity through the production of reactive oxidative species (ROS). Their small size, greater surface area to volume ratio, and ability to easily penetrate tissue cells, leading to higher chemical reactivity, cause increased ROS production when introduced into the body [3–5]. Also, due to their very small size, they can be ingested through inhalation and are able to pass through biological barriers into sensitive parts of the body like the lungs, brain, heart, liver, and spleens [6, 7]. A factor that Increases their toxicity is the solubility of the nanomaterial. NPs like zinc oxide and titanium oxide have been found to elicit more toxicity than ceria and titania, which are less soluble [8]. Also, due to the diversity in the use of these NPs in various industries like food, cosmetics, agriculture, biomedical, optics, and technology, it is easy for them to be absorbed and ingested into the human system, affecting the human gut microbiota. They have been observed to accumulate in the stomach, ileum, colon, and duodenum, which poses

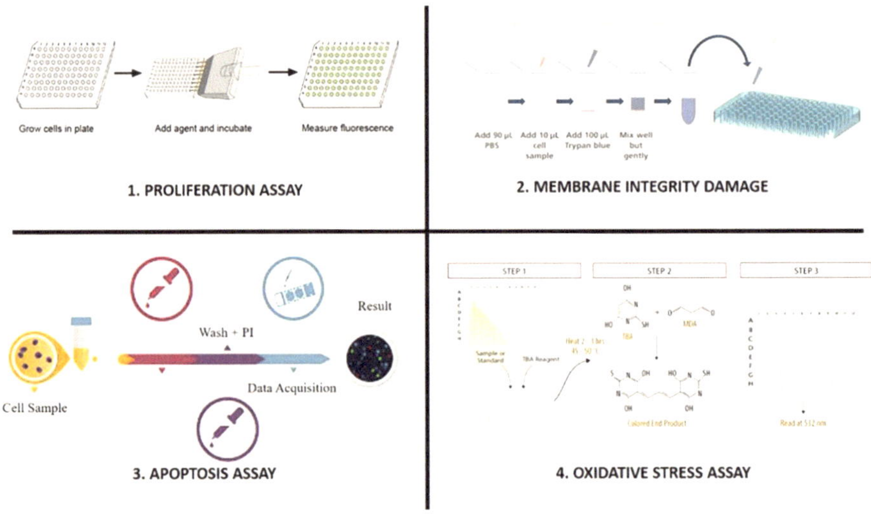

Figure 1.
Conventional methods of evaluating the cytotoxicity of the nanoparticles [12].

serious risks and concerns [9, 10]. Hence, the need to study their toxicity in human systems has become necessary [11].

The two major ways of assessing nanoparticle toxicity are *in vitro* and *in vivo* methods; *ex vivo*, although not commonly used, is another form. As shown in **Figure 1**, here are the common assays used to measure nanoparticle toxicity. With the use of these methods, there has been observed variability in results and even obstruction from the NPs themselves in the assay. In this chapter, we will be considering the various limitations and advantages of the toxicity assays commonly used in nanoparticle toxicity studies to ensure scientists have adequate knowledge and make the best decision on the assays to use for their study.

2. *In vitro* models

2.1 Proliferation assays

Proliferation assays are assays used to check for cellular metabolism in active metabolic cells. They help to ascertain the viability of cells when treated with NPs to check for the toxic effects of the particles on the cells. One of the commonly used assays is the 3-(4,5-Dimethylthiazol-2-yl)-2,5-diphenyltetrazolium bromide (MTT) assay. It is a colorimetric assay where 3-[4,5-dimethylthiazole-2-yl]-2,5 diphenyl-tetrazolium bromide (MTT) is enzymatically reduced by mitochondrial succinate dehydrogenase to formazan crystals, an insoluble product, in the mitochondria of live cells. The breakdown is spectrophotometrically measured to estimate cell viability. In this assay, there is a linear relationship between the color formed when the breakdown to formazan occurs or the absorbance when measured and the viability of the cells. It is a very sensitive and quantitative assay [13, 14]. The MTT assay proves to be advantageous because it produces results quickly under a maximum of 3 to 4 hours. It is also reproducible and does not require manipulation of the target cell [15].

The Alamar Blue Assay is also a colorimetric indicator assay characterized by the reduction of a blue, non-fluorescent dye to a pink-colored substance known as Resazurin [16]. It is both a qualitative and quantitative assay where colorimetric measurement can be taken or physical observation of color change in assessing cell viability or treatment toxicity. It is taken at a wavelength of 570 and 600 nm or 540 and 630 nm using the spectrophotometer. It is excited and emitted at wavelengths of 530–560 and 590 nm respectively [17]. Alamar Blue assay is highly sensitive, requires low cost, it's easy and safe to use, non-radioactive, and can be used for a large number of sample. It can also be used for both quantitative and qualitative analysis. Another method used to assess proliferation is the incorporation of [^{3}H] thymidine into the DNA of proliferating cells during the S phase of the cell cycle, with the use of autoradiography [18]. As the cells proliferate, new DNA strands are formed. The tritiated thymidine, a radioactive nucleoside, then enters into the new chromosomal strands as the cells divide [19]. [^{3}H] thymidine incorporation method is also widely used in immunological studies because of its high throughput and direct measurement of cell proliferation. It is also sensitive [19]. Another commonly used assay, clonogenic assay is used to assess cell survival and reproductive ability after treatment. The ability of a single cell to reproduce a colony is checked [20].

As shown in **Figure 2**, MTT assay has its limitations that affect the interpretation of data. Due to the formation of formazan, an insoluble dye, in MTT assay, other assays like XTT (2,3-bis [2-methoxy-4-nitro-5-sulfophenyl]-2H-tetrazolium-5-carboxanilide) are used. With XTT, in the presence of mitochondrial dehydrogenase enzymes, a soluble water product is produced. However, both assays also faced the limitation of reaction under other conditions like acidic media, additives found in culture media, polyphenols, the presence of NPs and O^-_2. These gives a false positive result of cell viability [22–24]. In a study carried out on Chinese hamster ovary cells (CHO-K1) after exposure to nano-TiO_2, which increases the formation of O^-_2, MTT and XTT assay were found to produce an inaccurate analysis of cell viability, because of the ability of the superoxide produced by the NPs to reduce the two salts [25]. Other factors such as cell number, the concentration of the MTT reagent, the treatment applied to the cells, and the toxic effect of the MTT itself has to be critically considered when analyzing the results from the assay. When prostate cancer cell lines (PC-3) were treated with polyethylene glycol-coated gold NPs (Au-PEG-NPs) at a concentration of 5 nm, there was an observed significant optical interference with the MTT assay. However, this effect could be minimized by washing the cells and removing the supernatant before MTT incubation and using appropriate blanks [21].

A study on the *in vitro* assessment of the toxicity of carbon based NPs HiPco single-walled carbon nanotubes (SWCNT), arc discharge SWCNT and Printex 90 carbon black NPs was done to initiate a more reliable method with the use of clonogenic assay [26]. This study was performed to avoid the interaction of carbon based NPs with the indicator dyes used in colorimetric assays, leading to inaccurate interpretations on toxicity studies [27, 28]. Graphene oxide and TiO2 NPs have also been observed to interact with the dyes [29, 30]. This assay rules out the possibility of the nanoparticle reacting with the assay itself. It is recommended as a very useful tool for testing cytotoxicity since colony number and size are taken into account. This makes it an effective differentiating tool between cell viability and cell proliferation [26]. It is also a sensitive method since colony size depicts the division rate and proliferation after the cells are treated [31].

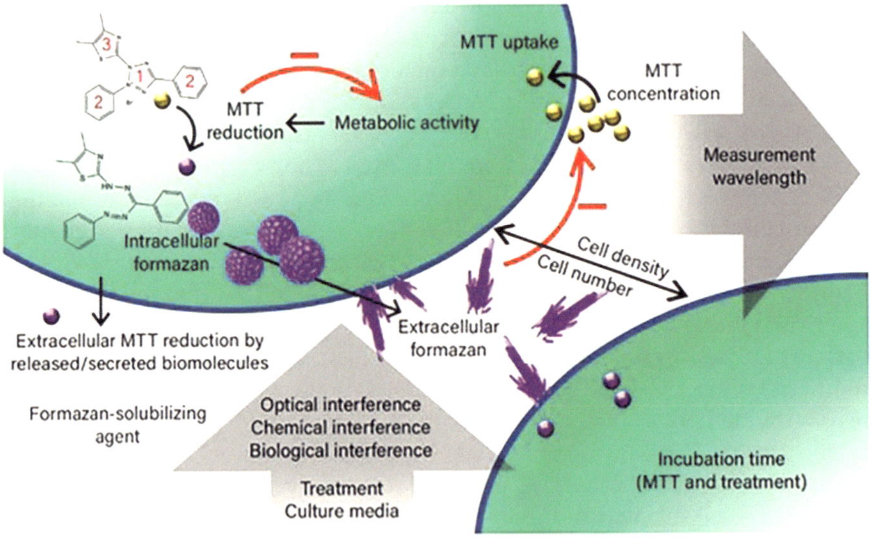

Figure 2.
Factors affecting the final optical density (OD) measurements in the MTT assay. These include the concentration of MTT reagent and the proportion that actually enters the cell, cellular metabolic activity (which is highly dependent on a multitude of variables including treatments to the cells, biological effect of culture media, cell density, and impedance of cell metabolism due to toxic effects of MTT), cell number, timing of formazan crystals extrusion (which could impede further MTT uptake), chemical interference such as abiotic reduction of MTT by culture media, the tested treatment, or released cellular content, optical interference by all the background components, time of incubating cells with MTT reagent and/or tested treatment, and ultimately the optical measurement. Chemical structure of MTT and formazan are illustrated inside the cell: MTT consists of a tetrazole ring core containing four nitrogen atoms (1) surrounded by three aromatic rings including two phenyl moieties (2) and one thiazolyl ring (3). Reduction of MTT results in disruption of the core tetrazole ring and the formation of formazan. Red arrows and the "-" sign indicate disruption of MTT reduction on the normal metabolic activity of the cells and the impeding effect of the formazan crystals (when presenting on the cell surface) on further uptake of MTT reagent by cells [21].

For Alamar Blue assay, because of its photosensitivity, it is affected by light exposure and has to be done in the dark. Cell density is also a factor that can affect the assay reading, hence cells must be cultured in high population to prevent slow growth and allow adequate dye reduction. The assay is also limited to a pH range of 7.0 and 7.4, and optimum temperature of 37°C. Longer incubation times could also lead to the reduced dye being bleached to a pink color. This quenches the fluorescence and could cause misinterpretation of the fluorescent signals. There is also the need for a positive and negative control to rule out non-specific interaction with the chemistry [17]. Specifically, with the use of nanomaterials, it is necessary to check toxicity using a combination of other assays because of the diversity and physiochemical properties of the nanomaterials, which could affect the cell and also interfere with the chemistry of the assay [32, 33].

The [^{3}H] thymidine proliferation method is toxic and also requires radioactive facilities, which can be very expensive. Also, there is the need for proper waste management of the radioactive materials used with this method [34, 35]. It has also been found to interfere with the target cells by inducing cell cycle arrest, apoptosis and fragmenting the DNA [34, 36, 37].

One of the major cons of Clonogenic assay is that it is time-consuming as it takes about 10–14 days to perform. This is quite long compared to the other assays described above which take only a few hours to a day [38].

2.2 Apoptosis assays

Apoptosis Assay is another assay carried out to evaluate the cytotoxicity of nanomaterials. It measures the extent of DNA damage and cell death when cells are treated. The release of free radicals due to oxidative stress is a pointer to DNA damage and nanomaterials have been found to elicit such reactions [39, 40]. One of the methods used to study apoptosis is the Annexin-V assay. This assay works on the principle of annexin-V binding to phosphatidylserine (PS) which is externalized on the plasma membrane due to activation of caspase-dependent pathway. In combination with Propidium Iodide (PI) which stains the nucleus indicating the last stage of cell death, both dead and apoptotic cells can be identified [41, 42]. Annexin-V and PI assays have been used to detect apoptosis in HeLa and human HepG2 hepatoma cells treated with gold and silica NPs respectively [43, 44]. Annexin-V does not penetrate the cell but only binds to phosphatidylserine on the extracellular membrane of the cell; hence it does not cause cell damage or affect the intracellular components of the cell. It is highly sensitive as it binds to only PS amidst other molecules on the cell surface. It produces bright fluorescent signals and it's easy to perform [41]. There are different options available for its labeling and also it can be performed both *in vitro* and *in vivo* [45].

Comet assay is used to determine DNA damage, single- and double-stranded DNA breaks and the mutagenicity of treatment both *in vivo* and *in vitro* [46–50]. It is one of the commonly used assay to assess the genotoxic effects of engineered nanomaterials like TiO_2, SiO_2, Zinc oxide NPs [51–55]. Damaged bases are detected when nucleoids are incubated with endonuclease III and formamidopyrimidine DNA glycosylase (FPG), specific to oxidized pyrimidine and purines, respectively [56]. Comet assay is a sensitive technique that is quick, cheap and easy to do [57]. It has been used both *in vitro* and *in vivo* and hence can be used to check the toxic effects of treatment in a particular organ or tissue and on any animal model [49, 58]. It can also be done on the first or specific site of contact.

TUNEL (terminal deoxynucleotidyl transferase biotin-dUTP nick end labeling) assay is a commonly, wisely used tool for the detection of DNA damage caused by both apoptosis, DNA fragmentation and necrosis, caused by exposure to toxic materials [59–61]. When DNA fragmentation occurs, the $3'$ hydroxyl termina becomes free. Then the enzyme terminal deoxynucleotidyl transferase incorporates labeled dUTP into the free end, which causes the staining observed in the assay [62]. It can be detected by light microscopy, fluorescence microscopy or flow cytometry [63]. TUNEL assay is fast as it can be completed within 3 hours. It has high sensitivity and versatile with the use of various techniques, as single cells can be detected using fluorescence microscopy or few cells using flow cytometry [63, 64].

Flow Cytometry is a very common and versatile technique used in the toxicity analysis of nanomaterials when applied to cells. It works on the principle of detecting fluorescently labeled cells when light beams are passed through the cell suspension. The technique analyses both the size and granularity of the cell populations, which gives a characteristic of the type of cells being studied [65]. With the use of flow cytometry, large populations can be analyzed with valid statistical results and this is done within a short period of time. This method can be used to evaluate the uptake of NPs by the cells, cell death, and expression of certain proteins [66].

However, these assays have their own set of limitations. For example, Annexin-V assay cannot be effectively used to obtain high throughput due to strong background signals when there are unbound labeled annexin-V. It can be difficult to optically image tissues when stained with annexin-v because of its short lifetime and slow diffusion in the body [67]. It can produce false positive results when it binds to negatively charged aldehyde adducts [68]. Also, Engineered NPs have been found to interfere and hinder the detection of DNA damaged from oxidation with the comet assay when ions are released when the particles are dissolved [69]. Cell toxicity, even in the absence of DNA damage, can lead to false-positive results in Comet Assay. There are also variability of protocols from the assay which can make it difficult to compare results with other laboratories, as there is no standardization [57]. For TUNEL assay, there is the limitation of non-specific staining because it labels all the free 3′ hydroxyl termini end irrespective of the cause of it. Hence, it can label non-apoptotic cells, which gives a false positive result. This means it can stain cells undergoing DNA repair, cells secretes factors causing proliferation in neighboring cells, or cells damaged through other means [70]. It is also costly. For flow cytometry, freshly prepared samples are required. If samples are kept for a long period of time before analysis, it could affect the cell properties and the analysis. Also, the use of the flow machine requires competent and skillful hands to accurately analyze and interpret the data [71]. NPs smaller than 100 nm experience low light scattering which could reduce the sensitivity of the detection [72].

2.3 Necrosis assays

Necrosis assays are generally used to check for membrane integrity of the cells after treatment. As shown in **Figure 3**, NPs can adhere to membranes causing changes in their structure and function, hence it's important to carry out these assays. Neutral Red (2-amino-3 methyl-7-dimethylaminophenazoniumchloride) uptake assay is a widely used assay used to measure cytotoxicity. The Neutral Red dye is a weak, cationic, cell permeable dye taken up by viable cells and localized in the lysosome. It permeates into the cell by nonionic passive dilution and binds to the phosphate groups found on the lysosome matrix. The dye is extracted from the viable cells and spectrophotometrically measured [74, 75]. It is a sensitive dye that measures the integrity

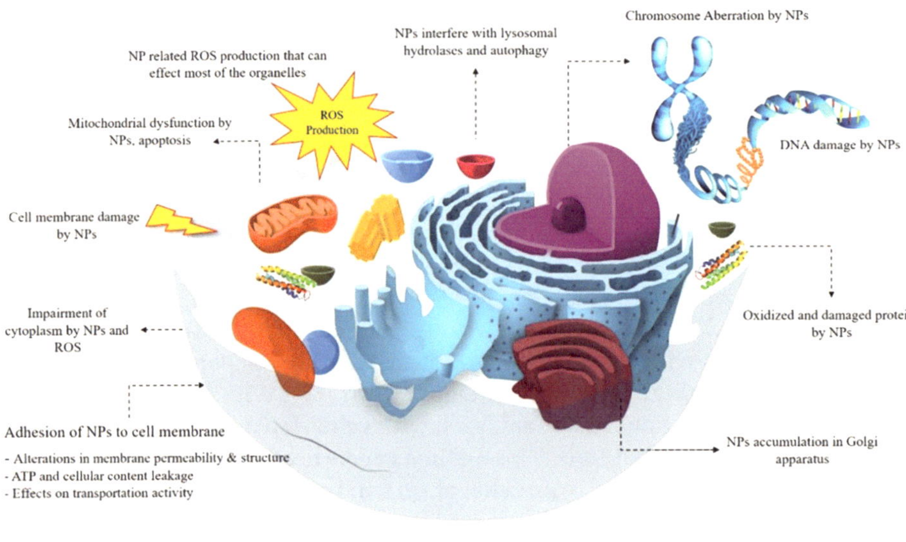

Figure 3.
The toxicity mechanisms induced by nanoparticles [73].

of the cell membrane and measures the viability of the cells [76]. The Neutral Red uptake assay is very sensitive and requires less equipment. It does not face much interference and does not have unstable reagents like other assays like MTT, XTS, etc. [77]. It is simple and can be used to detect only viable cells. It can also be done together with estimation of total protein content [78]. However, the use of Neutral Red, a fluorescent-based assay, is prone to interference from the NPs, which could lead to quenching of the dye. In a study to check the toxicity effect of silver NPs, Neutral Red, amongst other dyes, were observed to face interference from the particles [79]. Also, when applied on certain compounds that are volatile and insoluble in water, it faces problems in analysis as it mainly works when soluble in water [76]. Some chemicals could also cause the transformation of the dye into insoluble crystals, which could increase the toxicity estimation of the assay, giving a false negative result. When total protein estimation is done after the neutral red uptake assay, it could lead to reduction of the amount of protein estimated [78].

Trypan Blue Exclusion Assay is used to measure cytotoxicity of nanoparticle treatment on cells. The Trypan Blue is a dye that is absorbed by dead cells, leaving out viable cells and giving the actual number of viable cells after treatment.

Trypan Blue Exclusion Assay is very simple and quick to do, and it does not require any technical know-how. However, Trypan Blue assay is less sensitive and less reliable compared to other assays. It can be tedious and time-consuming when done on large number of samples [80]. Because of the use of hemocytometer, the possibility of making counting errors is present. Also, if the cells are not properly diluted or poorly dispersed in the counting chamber, it can contribute to the error [81]. The assay is insensitive as it cannot distinguish between live cells and cells that are gradually losing cell function. Another disadvantage of this assay is it can cause toxicity to mammalian cells [82].

2.4 Oxidative stress assays

Due to the high surface area to volume ratio of NPs, they elicit the production of reactive oxidative species (ROS) from the cells as shown in **Figure 4**, which could lead to further cellular damage. The ROS produced is measured using some of the oxidative stress assays.

2′,7′-dichlorofluorescein diacetate (DCFH-DA) is a nonionic, nonpolar fluorophore that penetrates the membrane and is sensitive to ROS. When DCFH-DA is internalized by the cell, it is enzymatically hydrolyzed by cytosolic esterases into dichlorofluorescein (DCFH), which remains in the cytosol. The nonfluorescent DCFH is then oxidized by hydroxyl radicals and cellular ROS into highly fluorescent dichlorofluorescein (DCF), which is then analyzed by flow cytometry or fluorescence microscopy [84].

Tritiated borohydride and 2,4-dinitrophenylhydrazine (DNPH) are used to measure protein carbonyl levels, which are indicative of oxidative damage. Protein carbonyls are reduced to alcohols by borohydride. This reduction is measured spectrophotometrically at wavelength absorbance of 340 nm. Also, when protein carbonyls react with DNPH a stable 2,4-dinitrophenyl (DNP) hydrazone product is generated and its absorbance is read between 360 and 390 nm [85]. DNPH is highly sensitive and specific and result from the assay can further be improved with the use of high-performance liquid chromatography (HPLC) or Western blotting [86].

Another means of assessing the production of ROS is through analysis of lipid peroxidation products like malondialdehyde (MDA) and 4-hydroxyl-2-nonenal

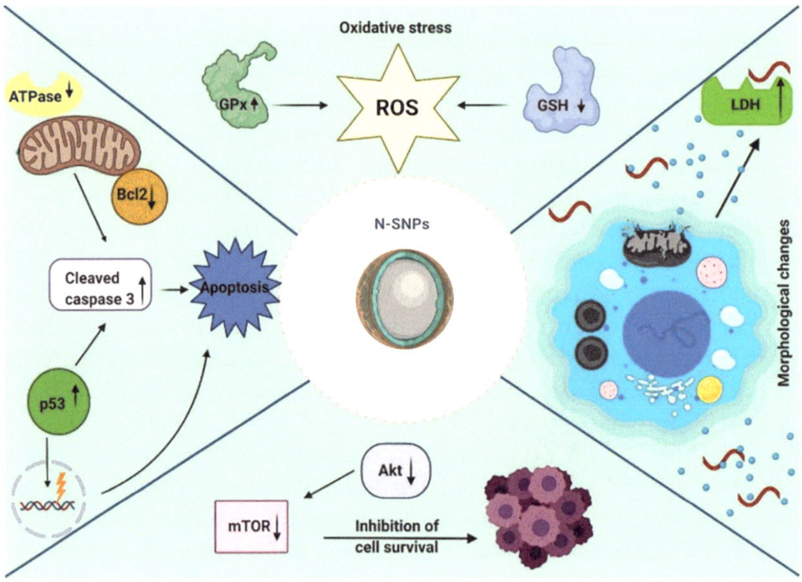

Figure 4.
Induction of oxidative stress by nanomaterials [83].

(4-HNE). MDA is formed when large polyunsaturated fatty acids (PUFAs) are broken down and during the metabolism of arachidonic acid (AA) [87]. The by-product reacts with thiobarbituric acid (TBA) at pH 3.5 to produce MDA-TBA which is measured by a spectrophotometer at wavelengths of 532 nm and fluorescently at 553 nm [88]. 4-HNE is a by-product of AA and peroxidation of PUFAs. It reacts with primary amines to form Schiff's bases and thiol. It also reacts with amino compounds to form Michael's adduct which is detected by HNE-protein adduct ELISA assays [89, 90].

DCFH-DA, However, is prone to inaccurate interpretation due to nonspecific enzymatic oxidation and photooxidation [84, 91]. With the use of DNPH, larger samples are required for the analysis [92]. The lipid peroxidation is nonspecific since TBA reacts with other molecules other than MDA, which could lead to false positive results.

3. Ex vivo models

Using *in vitro* models for studying nanoparticle mediated toxicity comes with the set of challenges as discussed in the previous section. Animal models like rodents or zebrafish embryos are ideal alternatives for studying toxicity induced by NPs, however, there are various limitations including the cost of maintenance, biosafety etc. One also needs to follow appropriate ethical guidelines and have a moral compass to ensure whether inducing pain or sacrificing the animal can be avoided by alternate studies.

Ex vivo studies where tissue slices are cultured outside the host organism, experiments on fertilized eggs, organs on chip studies are more reliant than *in vitro* studies which are often based on monolayer culture of cells. The following section will briefly describe the prominent ex vivo models.

3.1 Precision cut tissue slices

In precision cut tissue slices, tissues are collected from a model organism or human biopsy samples for testing the effects of nanoparticle treatment. This approach provides a more comparable biological scenario rather than using immortalized or primary cells cultured in monolayer. As the organ tissues provide a better biological replica of the effect of nanoparticle in the biological system as a whole. Predominantly, this method has been used to study aerosol effects or nanocarriers developed for treating lung disorders [93–96]. Using liver cut slices to study hepatotoxicity is also common [97–100].

One of the major challenges of this method involves the penetration of nanocarriers into the organ slices. In a biological system, like our body, circulatory system ensures the uptake/delivery of nanocarriers. Without this in the organ slices which also has several physical barriers makes it difficult for the passive entry of NPs. As seen in some studies, the NPs are seen around the surface of the tissue slices [101]. Besides this, sometimes slicing a tissue induces inflammation around the cut region. This is unfavorable while studying toxicity effects. An inflamed tissue could introduce bias into the experimental outcomes. Inflammation could activate cells and induce necrosis or apoptosis which will interfere with toxicity results for nanoparticle testing.

3.2 Organ-on-a-chip

A 3D culture with multiple kinds of cell population sufficiently supported with extra cellular matrix and growth factors is better than monolayer culture in terms of nanotoxicology analysis. However, these kinds of system are considered to be static and does not mimic vascular perfusion or the sheer stress in biological systems. Recently, microfluidics have been integrated to 3D culture of different population to generate micro physiological systems called organ-on-a-chip. A very simplistic model of organ-on-a-chip will have a single type of cell lined over the microfluidic channel with a continuous flow of media. In advanced models like lung-on-a chip, blood brain barrier-on-a chip or blood retinal barrier-on-a-chip model, multiple types of cells are used with microfluidic channel mimicking biological processes like breathing, blinking [102–115] etc.

These devices employ different perfusion rates and with the kind of material used, introduce mechanical strains which mimics biological circulation events like breathing, heartbeat, peristalsis, blinking, twitching etc. By using transparent materials for channel construction, the device also allows live visualization. As compared to conventional *in vitro* techniques, this provides a better physiological parallel with intricate designs to replicate organ architecture. Recently, there have been studies to develop multiple organ on a chip interconnected devices which could essentially be called body-on-chip to develop a model of whole organism rather than studying the effects on individual organs [116]. With intricate designs to carefully recreate biological systems, this could to a larger extent eliminate animal based testing.

Fluorescently labeled PEGylated gold nanoparticles were tested on tumor on a chip model which provided better insights into the circulation, elimination and uptake of nanoparticles in tumor microenvironment [117, 118]. Similarly Organ-on-a-Chip model are used in understanding the effects of shape of NP on its toxicity [119]. Understanding the effect of flow rates, especially while NP mediated targeted therapy for crossing BBB [120], the underlying changes in toxicity profile introduced by surface modifications of the particle [121] etc. are few of the recent advances in NP mediated research which has recently been revolutionized by Organ-on-a-Chip models. Specifically focusing on toxicity studies, lung-on-a-chip models have provided better insights into pulmonary toxicity of nanoparticles, specifically, TiO_2, Silica NPs [108, 122] etc. Also, the effect of NPs on crossing placental membrane has been evaluated using Organ-on-a-Chip model. In one such model, placental membrane integrity and maternal immune cell response were negatively challenged by TiO_2 NPs [123]. These models also enable the researchers to evaluate toxicity of NPs in both static and mobile conditions as opposed to static conditions in *in vitro* studies [124].

One major disadvantage of the devices is the use of precursors which adsorbs drugs while testing. This reduces the amount of drugs available for interacting with the cells of interest. The widely used material polydimethylsiloxane (PDMS) [125] is known to adsorb certain drugs which is a disadvantage to the application. Polysulphone based materials are considered to be an alternative for PDMS. However, they lack transparency and are difficult to tune the mechanical properties. If careful considerations are introduced while designing the architecture and appropriate precursor materials are chosen, organ-on-chip can be a lot desired application for nanoparticle toxicity analysis (**Figure 5**).

3.3 Chick chorioallantoic membrane (CAM) assay

Angiogenesis and neovascularization is a characteristic trait of multiple disorders including cancer and retinopathies. Preventing the formation of new leaky vasculature is thus critical in treating these disorders. NPs developed as potential therapeutic options thus need to be evaluated for their anti-angiogenic properties. Evaluating this on developing chick embryos is both cost and time effective other than the simplicity of the model [127–130]. Briefly, the assay is performed by making a hole in the eggshell and inserting a membrane which delivers the NPs of choice. At the end of study, the eggs are opened, and the blood vessels are quantified using bright field imaging. One of the major advantages of this method is its reproducibility and simplicity to conduct in small scale laboratories.

3.3.1 Ex-vivo tumor models on CAM

Cancer cells of human origin are transplanted to CAM that covers chicken embryo. After 3 days of transplantation, a tumor with host species features will be developed. The tumor features multiple cell type of human origin, with rich vasculature and extracellular matrix. In terms of tumor microenvironment and cell types, this model is a closer approximation to *in vivo* models compared to 3D organoid studies. Also, the tumor formation process is rapid and eliminates longer waiting period as compared to rodents. Also, immune compromised animals are costlier in contrast to chick embryos which poses naturally underdeveloped immune system in early developmental stages. Besides these advantages, the rich nutrients present in chick embryo encourages effective angiogenesis in the tumor model. This facilitates drug and nanoparticle testing and monitoring of reduction of micro blood vessels and other tumor characteristics (**Figure 6**).

3.3.2 Anti-angiogenesis studies for retinopathies

These are studies where implants are prepared and investigated on its ability to sustainably deliver therapeutic agents in ocular regions. Neovascularization and

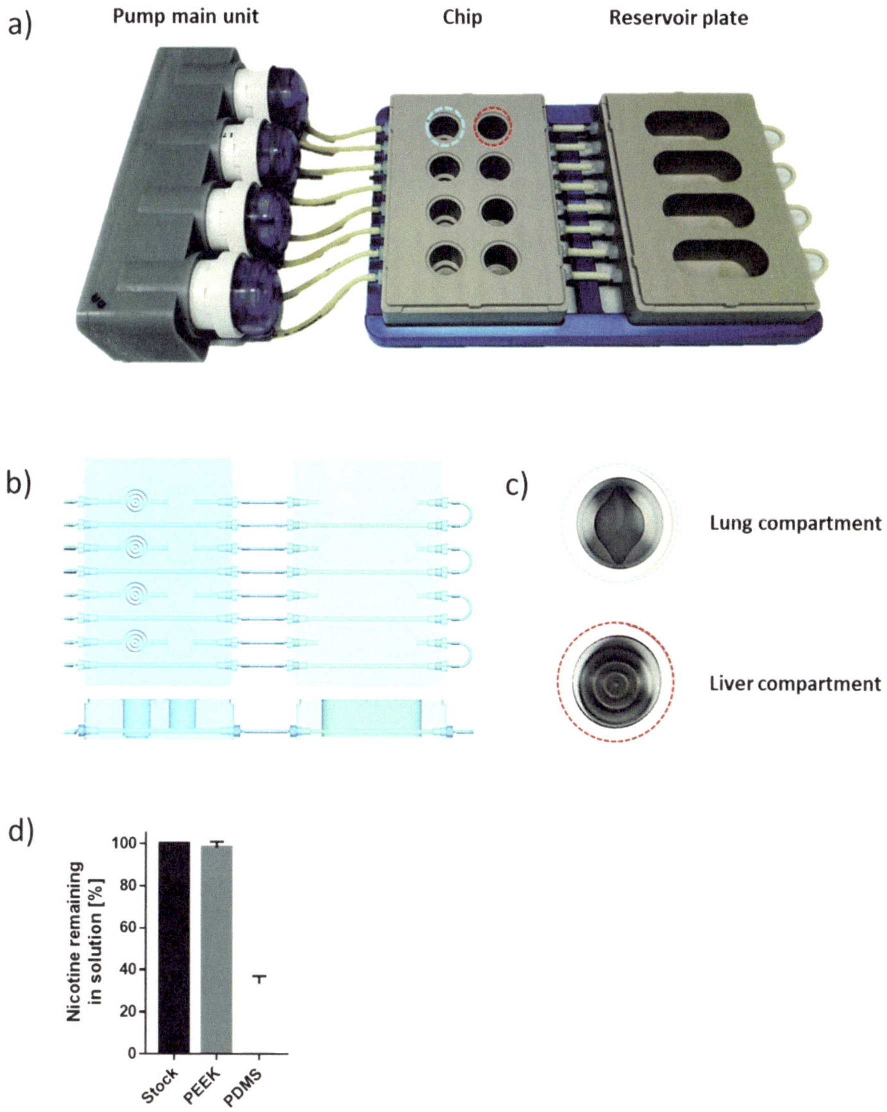

Figure 5.
The lung/liver-on-a-chip platform. (a) A photograph of the chip system comprising the pump main unit with four pump heads, the PEEK chip, and the reservoir plate. (b) a schematic view of the chip comprising four circuits is shown; each circuit includes two compartments to house the lung and liver tissues, respectively. The cross-section schemas of the plates show the path of the tubes and channels and the relative depth of each well. (c) a close-up view of the two compartments showing the groove pattern on the bottom of the wells. (d) Effect of chip materials on absorption of nicotine. A solution of 10 mM nicotine in phosphate-buffered saline (PBS) was kept in the wells of the chips made from PEEK or PDMS for 8 hours before it was collected. The concentrations of nicotine were then measured using liquid chromatography coupled to high-resolution accurate mass spectrometry. Nicotine concentrations remaining in solution are expressed as % relative to the stock solution. Data are presented as mean ± SEM. N = 3. PEEK: Polyetheretherketone; PDMS: Polydimethylsiloxane [126].

resultant increased ocular pressure lead to a plethora of ocular disorders including diabetic retinopathy, macular degeneration etc. Nanoparticle mediated implants or delivery systems attempt to target pro-angiogenic markers in the eye. However, this needs to be carefully implemented with increased attention on not to induce any inflammation and reduction in angiogenesis.

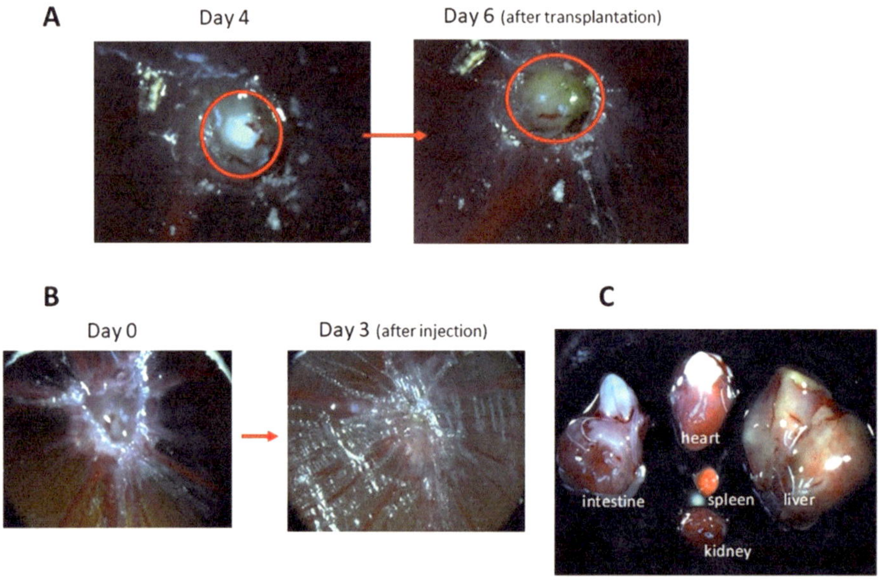

Figure 6.
Patient tumor sample transplanted on the CAM membrane. (A) Tumor formed by transplanting minced sample of ovarian cancer patient tumor. (B) Tumor is eliminated after intravenous injection of PMO-1 containing doxorubicin. (C) Chick embryo major organs look normal 3 days after injection [130].

Briefly fertilized eggs will be candled to spot the air sac and blood vessels. A small hole of 1×1cm will be made into the egg shell at air sac region. Similarly, a cut is made near the vascular region without disturbing the CAM. Sucking out air from the air sac will distant the CAM from the egg membrane at the vascular region. Once this is achieved, a small incision is made at the CAM of vascular region where implants are introduced through sterilized filters. These filters containing the particle are then incubated over desired time frame. At the end of experiment duration, CAM is fixed and imaged to count the no of blood vessels compared to untreated controls (**Figure 7** represents multiple ways of performing CAM assay for angiogenesis studies).

For a simplistic and cost-effective model like CAM assay, most often the only disadvantages lies in the frequent contamination of samples. This can be eliminated with limited exposure of the opened CAM to outside air and by following stringent sterilization practices [132].

4. In vivo models

Nanocarrier formulations are tested in in vivo models as pre-clinical studies to evaluate its feasibility to escalate to clinical trials and into commercial market at the end of pipeline. The maximum dosage of NPs which could be safely tolerated, the pharmacokinetics and elimination window of NPs from the tested organisms, the accumulation and effect of long-term exposure is analyzed during these experiments. Vertebrates and invertebrate groups of animals are used for these studies to scale up from simplistic biological networks to understanding the effects on complex

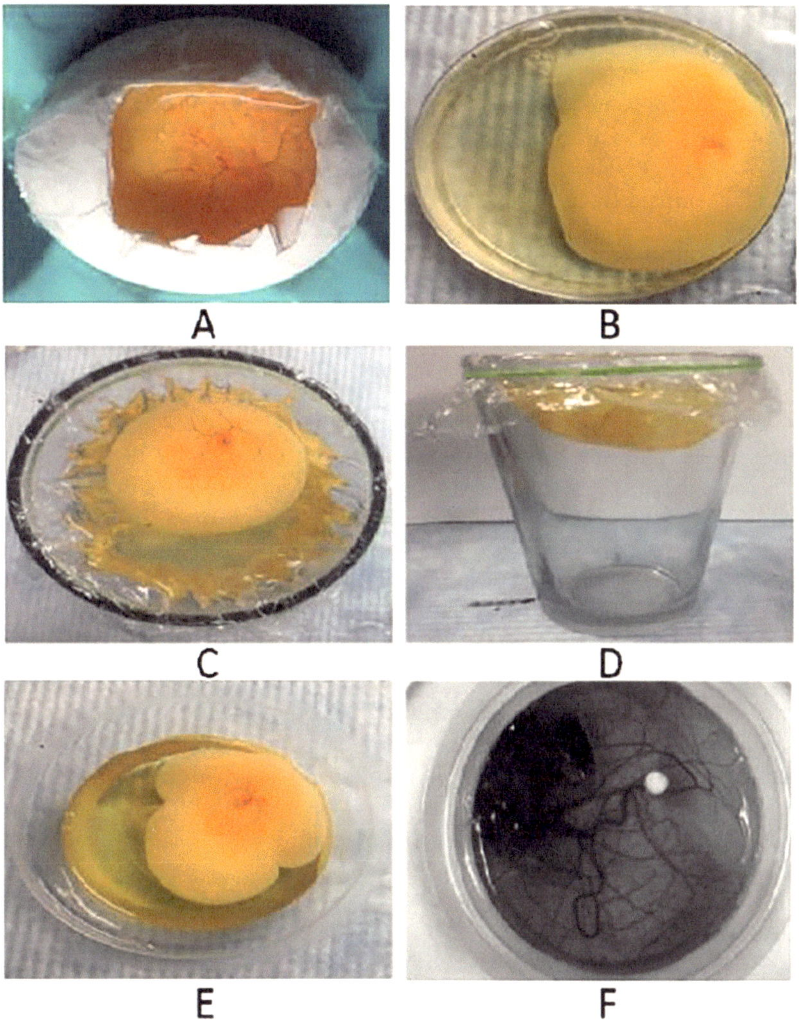

Figure 7.
Representative images of chorioallantoic membrane (CAM) variants. (A) in-ovo setup by windowing method on day of incubation; (B) ex-ovo setup in a petri plate; (C) ex-ovo setup in a glass-vertical view; (D) ex-ovo setup in a glass-horizontal view; (E) ex-ovo setup on plastic cups, image taken by a camera and; (F) ex-ovo setup in plastic cups, image taken by a Chemidoc (charge-coupled device (CCD) camera) [131].

organisms genetically closer to human beings. Good lab practices are stringently adhered to while conducting these studies with appropriate ethical standards.

4.1 Invertebrates

Invertebrate models often have a shorter lifespan which aids researchers in testing nanoparticle toxicities. Due to their shorter life cycle, it's feasible to understand and compare the effects of exposure of NPs in their developmental stages. It also benefits to conduct multiple rounds of testing within a shorter duration of time. The most established invertebrate model systems for nanoparticle toxicity include *Caenorhabditis elegans* and *D. melanogaster.*

4.1.1 C. elegans

C. elegans are nematodes which can grow up to 1 mm in size in its fully developed adult stage. They are often used for understanding nanotoxicity through oral uptake which is also the major form of nanotoxicity in human beings [133–138]. They pose around 70–80% of gene homology with humans and have around 70% of major signal transduction pathways conserved as compared to human beings. They are also transparent in nature allowing to visualize and track the accumulation of fluorescent labeled NPs. (**Figure 8**).

Major studies conducted on testing nanoparticle formulation and its effect on *C. elegans* has found that there have been significant changes in oxidative response, reproduction and lysosomal signaling after oral uptake of particles. For example, treatment with TiO_2 NPs have reported to alter the expression levels of glutathione-s-transferase gene. Exposure to NPs in *C. elegans* could also be through their vulval slit or opening. The nematodes being hermaphrodite, interaction of NPs at vulval site and spermathecae, where sperms are stored and oocyte fertilization happens, could provide preliminary results on how NPs affect reproduction in organisms. Interestingly, a recent study has reported that exposure of TiO_2 NPs in *C. elegans* leads to decreased expression levels of pod-2 a gene known to have role in reproduction in the nematodes [133, 136]. Another study has reported the exposure of silver NPs leading to altered expression levels of proteases involved in lysosomal pathway related genes.

Generation of mutant strains of *C. elegans* has helped in understanding the effects of common particles like graphene oxide (GO), Silver, cadmium quantum dots and TiO_2. Tracing of these NPs is often reliable and easy in *C. elegans* due to the transparency of its body. However, *C. elegans* lacks organs like a well-developed lung, kidney, heart etc. which makes it difficult to draw comparisons with higher order species. Also, it only has a 70% homology with human genome with some critical signaling pathways completely absent.

4.1.2 D. melanogaster

D. melanogaster or fruit fly is another model well used for oral toxicity of NPs. Like *C. elegans*, fruit fly poses different life development stages which makes it suitable to study the effect of NPs on different life stages. Effects of NPs in the gut cells, eye and wing development are often studied for testing the toxicity of NPs (**Figure 9**). Along with this, behavioral studies are also conducted where crawling speed and path is monitored [140, 141]. Immune pathway in fruit flies is well studied and they also display a similarity of autophagy related genes with human beings. This makes *D. melanogaster* an ideal model for studying immune response and rate of autophagy with respect to NPs like GO [142]. The effect of nanoparticle exposure to the organisms is also analyzed by studying the effects on reproduction. Offspring number, morphology, development life stages are analyzed to understand the same [140, 143–145].

The studies of nanotoxicity using fruit flies are limited to survival, developmental stages, eclosion rate, fertility and geotaxis performance analysis. Though it provides a significant addition in terms of understanding the toxicity as compared to *in vitro* studies, parallels cannot be drawn between human beings. One of the major challenges in using *D. melanogaster* as a model organism is the absence of an adaptive immunity in the organism. Other limitations include insufficient evidence of the cognitive capabilities of the organism especially affecting behavioral studies. Also,

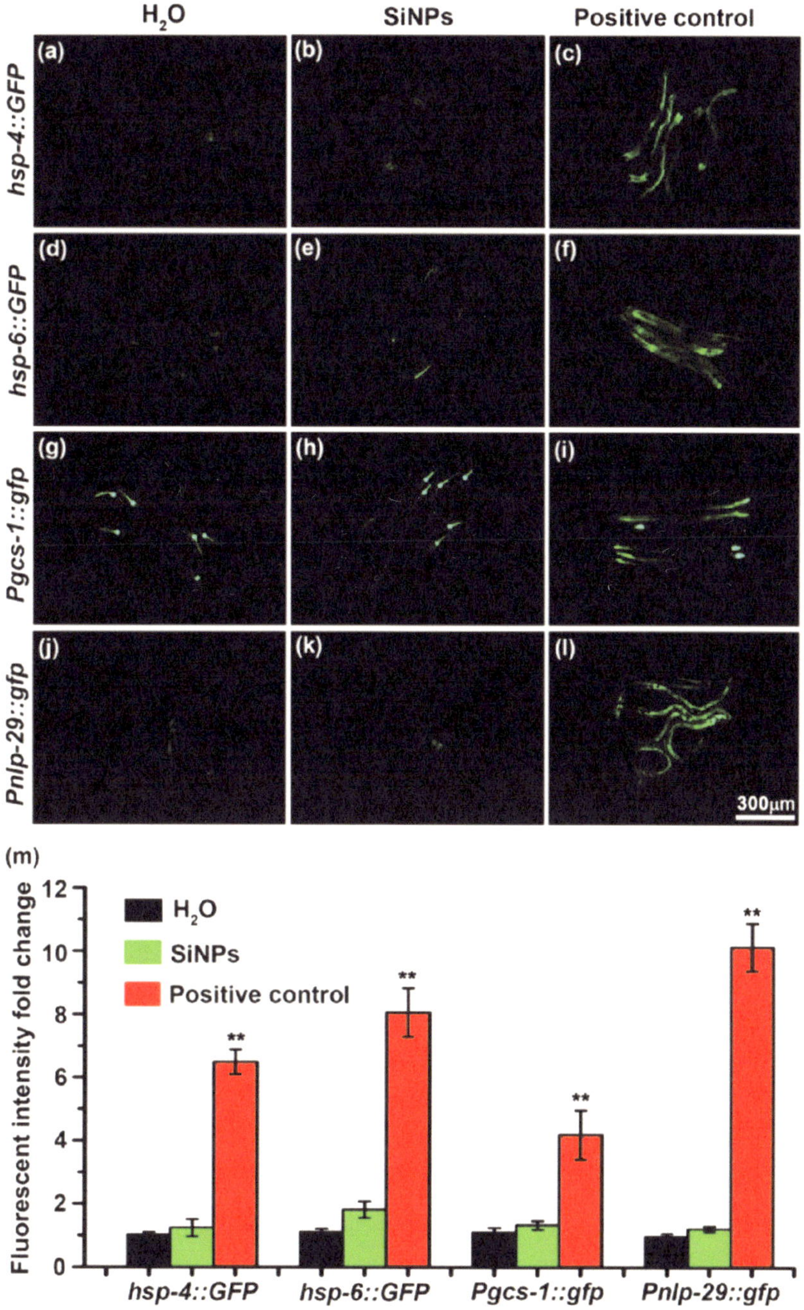

Figure 8.
*Induction of C. elegans major stress or host defense responses by SiNP treatment. Fluorescent images of worms treated with H_2O (a), SiNPs (b), and tunicamycin (c) with a GFP reporter for ER stress. Fluorescent images of worms treated with H_2O (d), SiNPs (e), and ethidium bromide (f) with a GFP reporter for mitochondrial stress. Fluorescent images of transgenic worms carrying the GFP reporter for oxidative stress subjected to H_2O (g), SiNPs (h), and H_2O_2 (i) treatment. Fluorescent images of transgenic worms carrying the GFP reporter for innate defense subjected to H_2O (j), SiNP treatment (k), and physical injury (l). The same magnification was used in all of the images (m). Quantitative analysis of fluorescent intensity fold change of worms treated with H_2O and SiNPs and a positive control group corresponding to (a–l). N ≥ 20. Error bars represent mean ± SEM; ** $p < 0.01$ [139].*

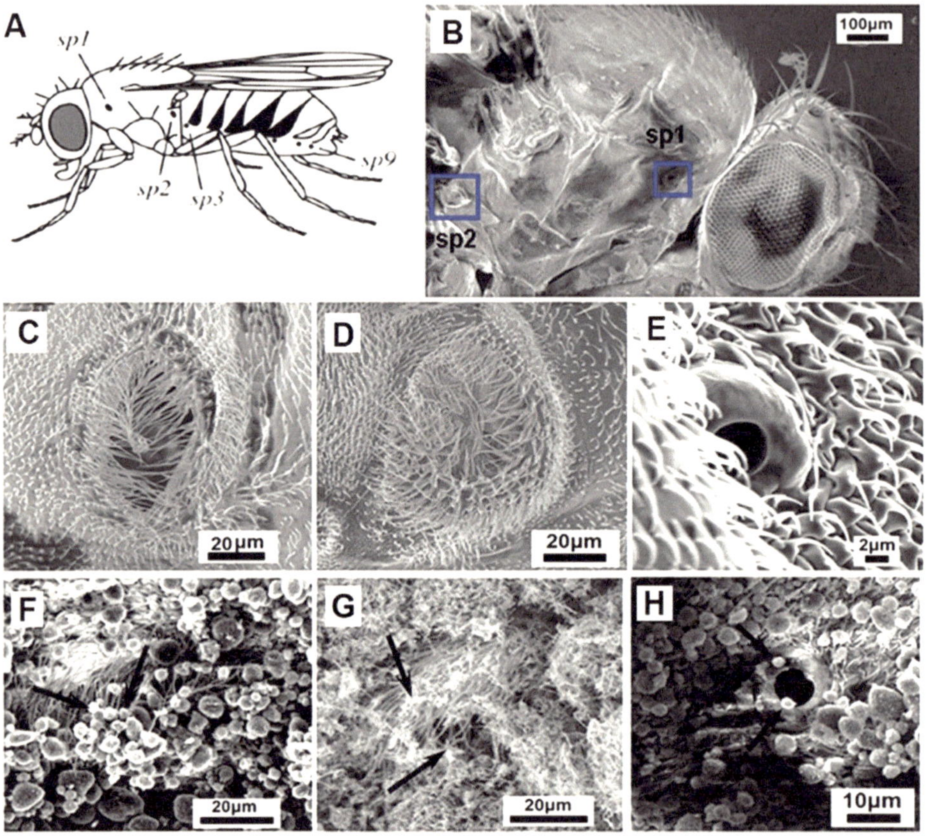

Figure 9.
Possible mechanism of nanoparticle-induced mortality in adult Drosophila. (A) Location of spiracles in drosophila: sp1, mesothoracic spiracle; sp2, metathoracic spiracle; sp3 to sp9, abdominal spiracles, image from Lehnmann et al. (B) SEM image shows mesothoracic and metathoracic spiracle of an adult drosophila (blue square) Center row (C – E): SEM images of spiracles in unexposed drosophila; sp1 (C), sp2 (D), both 20–50 μm, and an abdominal spiracle (E) at 5 μm. Bottom row (F – H): Spiracles are covered/decorated with nanomaterials (see arrows) after dry exposure of adults to CB (F); MWNTs (G); CB (H) differential toxicity of carbon nanomaterials in drosophila: Larval dietary uptake is benign, but adult exposure causes locomotor impairment and mortality [140].

vertebrate specific genetic disorder models cannot be developed in *D. melanogaster*. Other than this, unlike other models, amount of ingestion of NPs per flies cannot be accurately standardized as they are not gavage fed. Even after the above-mentioned limitations they remain one of the simplest models along with *C. elegans* to develop transgenic lines by breeding.

4.2 Vertebrate models

Vertebrate models for studying nanotoxicity include zebrafish, rabbit rodent models like mice, rat, hamsters etc. They share higher similarities with human beings in terms of the respiratory, circulatory, and nervous system. However, studying long term exposure of NPs in these organisms, especially rodents, becomes challenging due to their longer lifespan and gestation periods. The following section will briefly discuss the recent reports of using these vertebrate models for nanotoxicology studies and their limitations.

4.2.1 D. rerio

Zebrafish or *D. rerio* is one of the few vertebrate models with shorter breeding and offspring rearing period. This allows it to be a vertebrate model with ease to study developmental stages and the effect of NPs on life cycle. Also, oral and circulatory introduction of NPs is possible through zebrafish. The oral toxicity studies are conducted by introducing NPs in embryonic medium or fish water. The NPs are introduced in the blood stream directly using micro injections or intravitreal, intraperitoneal, intraventricular injections. Zebrafish models have been used to develop blood brain barrier models, tumor models and studied to see the effect and penetration properties of these nano formulations [146–150].

Toxicity of nanoparticle formulation is often assessed by survival analysis, cardiac rhythm studies, morphological changes in eye, spine and fin development, edema in cardiac sac etc. Behavioral changes like swimming pattern, response to tapping and light are also investigated. Like fruit flies, conservation of autophagy related genes to human beings allows using zebrafish models for studying the autophagy related gene expression with respect to nanoparticle treatment. Recently, treatment of ZnO nanoparticle in zebrafish model has displayed an increase in inflammation related gene over expression [151]. There are also reports of siler NPs affecting the gut microbiota of zebrafish [152]. Embryonic zebrafish studies have also reported the effect of silica based NPs in reducing the blood pressure in zebrafish and vascular endothelial cells (**Figure 10**) [153].

One of the major limitations of using zebrafish is the inability to use them for studying respiratory effects of aerosol based nano formulations. Other than this limitation they are also not ideal for breast cancer and prostrate cancer models as they lack the appropriate tissue of origin in their body architecture.

4.2.2 Rabbit and rodent models

Rabbits were classified as rodents till the early 20th century. They along with models with mice, rat and limitedly hamsters have been used for studying pharmacokinetics and pharmacodynamics of nanoparticle formulations. Their organs are also harvested and used for tissue distribution studies as they have striking similarity in tissue characteristics with human beings. Ocular and dermal toxicity of NPs are mostly studied in rabbits. Recently, silver NPs were tested on shaved skin regions on albino rabbits and the toxicity was analyzed using prefixed criteria. Dry skin, scaling in doses lower than 100 ppm and erythema in higher doses up to 4000 ppm was observed as part of this study [154]. Nano-hydroxyapatite was intravenously introduced to New Zealand white rabbits and it was noted that they does not affect liver function, and renal function in the animals [155]. In another study conducted to understand the toxicity of aflatoxin B1, treatment with curcumin and ZnO NPs prevented lipid and protein degradation via oxidation and showed better liver health as compared to aflatoxin B1 treated groups [156]. In corneal fibrosis (haze) model in rabbit using excimer laser performing -9D photorefractive keratectomy (PRK), nanoparticle formulation containing BMP7 topical application 5 minutes after PRK reduced the corneal haze by 50 percent with no toxicity [157]. Nephroprotective effects of NPs have also been studied using rabbit models. Studies using CaO NPs however have reported significant toxicity in liver and kidney after exposure [158].

Rodent models of nanotoxicity mainly includes rats and mice. Immune compromised, genetically altered rodents are very commonly used in nanotoxicology studies

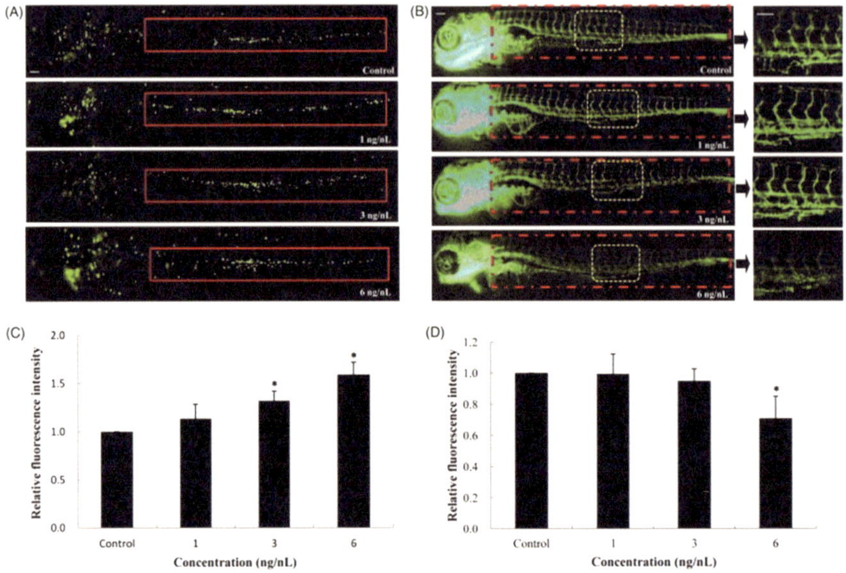

Figure 10.
*Inflammatory response and vascular endothelial cell dysfunction induced by SiNPs. (a,B) SiNPs increased the recruitment and chemotaxis of neutrophils in caudal vein of Tg(mpo:GFP) zebrafish. (C,D) SiNPs inhibited the expression of vascular endothelial cells in Tg(fli-1:EGFP) zebrafish. n = 30, data are expressed as mean ± standard deviation from three independent experiments (*p < 0.05). Scale bar: 100 μm [153].*

and have been reviewed better in various book chapters and reviews elaborately [159–162]. Briefly they are used for central nervous system disorder models, cancer models, hepatotoxicity models, aerosol treatment models. They provide relatively more homology with human beings; however, the laboratory maintenance of rodents and rabbits are often challenging. Also, historical trait analysis with respect to nanoparticle formulation is often limited by the number of offspring produced and longer gestational periods. Besides this, there is no accepted duration of days or guidelines recommending the number of days nanoparticle exposure in rodents to be monitored. This largely leads to inaccuracies and inconstancies in reporting toxicity of the same or similar nano formulation by different research groups.

4.2.3 Primate models

Primate models for nanotoxicity analysis rose to prominence after a study conducted on *Rhesus macaque* by intravenously injecting 25 mg/kg of phospholipid-micelle-encapsulated CdSe-Cds-ZnS QDs [163]. The study term was 90 days and the authors observed significant changes in behavioral patterns including loss of sleep, appetite, body weight, physical activity etc. By the end of study term, most of these effects were reversed to original state, however the particle displayed accumulation on liver and kidneys. Further studies on this were not carried out to understand the long term implications of hepatic accumulation. Quantum dots are generally considered to be safe to administer after conducting studies on *in vitro* and other *in vivo* models. However, this study indicated the need of behaviorally closer primate models to draw significant conclusion on nanoparticle toxicity before human trials.

Following this study, there were multiple reports of nanoparticle toxicity analysis using primates. Polylyisne conjugated DNA NPs for targeting retinal pigment epithelium was studied using baboons showing no inflammatory response in the eye [164]. Also in cynomolgus monkeys, the safety of gadolinium based NPs for imaging purpose was evaluated [165]. In another studies, PEG-bl-PPS polymerosomes were found to be nontoxic in non-human primates [166]. Also, cargo of siRNA in cyclodextrin with transferrin ligands as targeting moieties were also found to be relatively safe in cynomolgus monkeys [167]. However, there were elevated levels of creatinine and nitrogen along with certain inflammatory cytokines at higher dosage. In another study, mice and *Rhesus macaque* were compared on its effect of CdSe/CdS/ZnS semiconductor NPs in placental crossing and miscarriage rates. Interestingly, in the rodent models there was not toxicity recorded even though the particles were shown to cross the placenta. There was no miscarriage in rodents and fetus displayed no abnormalities. However, in primate models there was a 60 percentage rate of miscarriage establishing the fact that primate models are far superior in compared to rodent models for toxicity analysis [168].

Even though they are ideally the closest to understanding human body's response to nano formulations, testing these formulations on them require more human resources and expertise along with stricter ethical guidelines in handling them. Also, these magnificent creatures are often sacrificed at the end of the study to harvest organs and understand tissue damage. Most of them being social organisms, this would have larger implications on their family group and could even lead closer ones to depression. Infant carrying is one such response displayed by mothers losing infants in primates. Secondary responses of curiosity and stress to death of infants or members is often displayed by these primates [169–172]. A closer evaluation of the morality and scientific rationale should be evaluated before conducting such studies.

5. Conclusions

The past few decades witnessed the advent of nanoparticles and their potential use in multiple fields of biomedical sciences. From drug delivery to semiconductor devices, nanoparticles find applications around us. Informed use of nanomaterials, especially on its toxicity is highly relevant as more and more studies report the hazardous effects of these particles. The current chapter discussed *in vitro*, *in vivo*, *ex vivo* models for evaluating nanoparticle toxicity. As we analyze the plethora of assays conducted to study, in some cases, same particles in multiple model systems, we understand the varying toxicity reports. Such studies challenge the dangerous assumption of deeming a NP to be nontoxic by simply analyzing *in vitro* and in some cases rodent models. The need of non-human primate models closer to the genetic and physiological profile of human beings vs. the morale of sacrificing animal life for our benefit need to be carefully questioned. Alternate strategies like organ-on -chip models require further refinement and balance in incorporating parameters to better mimic study conditions.

Acknowledgements

Authors acknowledge the financial support provided by Blazer foundation and Medical Biotechnology Program at Department of Biomedical Sciences, UIC College of Medicine Rockford.

Conflict of interest

The authors express no conflict of interest.

Appendices and nomenclature

NP	Nanoparticles
ROS	Reactive Oxygen Species
MTT	3-[4,5-dimethylthiazole-2-yl]-2,5 diphenyltetrazolium bromide
XTT	2,3-bis[2-methoxy-4-nitro-5-sulfophenyl]-2H-tetrazolium-5-carboxanilide
CHO-K1	Chinese hamster ovary cells
Au-PEG-NPs	PEG coated gold nanoparticles
SWCNTs	Single Walled Carbon Nanotubes
GO	Graphene Oxide
FPG	formamidopyrimidine DNA glycosylase
TUNEL	terminal deoxynucleotidyl transferase biotin-dUTP nick end labeling
DCFH-DA	2′,7′-dichlorofluorescein diacetate
DNPH	2,4-dinitrophenylhydrazine
PUFA	Poly unsaturated fatty acids
TBA	Thio barbituric acid
CAM	Chick Chorioallantoic Membrane
PEG-bl-PPS	Polyethylene glycol-bI-polypropylene sulfide
AuNPs	Gold nanoparticles
PEEK	polyetheretherketone
PDMS	polydimethylsiloxane

Author details

Neeraja Revi, Oluwatosin D. Oladejo and Divya Bijukumar*
Department of Biomedical Sciences, Blazer Nanomedicine Lab, University of Illinois College of Medicine Rockford, Rockford, USA

*Address all correspondence to: drbiju2@uic.edu

References

[1] Nel A, Xia T, Mädler L, Li N. Toxic potential of materials at the nanolevel. Science. 2006;**311**:622-627. DOI: 10.1126/science.1114397

[2] Dreher KL. Health and environmental impact of nanotechnology: toxicological assessment of manufactured nanoparticles. Toxicological Sciences (n.p.). 2004;**77**:3-5. DOI: 10.1093/toxsci/kfh041

[3] Caruthers SD, Wickline SA, Lanza GM. Nanotechnological applications in medicine. Current Opinion in Biotechnology. 2007;**18**:26-30. DOI: 10.1016/j.copbio.2007.01.006

[4] Choi O, Hu Z. Size dependent and reactive oxygen species related Nanosilver toxicity to nitrifying bacteria. Environmental Science & Technology. 2008;**42**:4583-4588. DOI: 10.1021/es703238h

[5] Zoroddu MA, Medici S, Ledda A, Nurchi VM, Lachowicz JI, Peana M. Toxicity of nanoparticles. Current Medicinal Chemistry. 2014;**21**:3837-3853. DOI: 10.2174/092986732166614060116231 4

[6] Hagens WI, Oomen AG, de Jong WH, Cassee FR, Sips AJAM. What do we (need to) know about the kinetic properties of nanoparticles in the body? Regulatory Toxicology and Pharmacology. 2007;**49**:217-229. DOI: 10.1016/j.yrtph.2007.07.006

[7] Takenaka S, Karg E, Roth C, Schulz H, Ziesenis A, Heinzmann U, et al. Pulmonary and systemic distribution of inhaled ultrafine silver particles in rats. Environmental Health Perspectives. 2001;**109**:547-551. DOI: 10.1289/ehp.01109s4547

[8] Xia T, Kovochich M, Liong M, Mädler L, Gilbert B, Shi H, et al. Comparison of the mechanism of toxicity of zinc oxide and cerium oxide nanoparticles based on dissolution and oxidative stress properties. ACS Nano. 2008;**2**:2121-2134. DOI: 10.1021/nn800511k

[9] Vance ME, Kuiken T, Vejerano EP, McGinnis SP, Hochella MF, Rejeski D, et al. Nanotechnology in the real world: Redeveloping the nanomaterial consumer products inventory. Beilstein Journal of Nanotechnology. 2015;**6**:1769-1780. DOI: 10.3762/bjnano.6.181

[10] Chen H, Zhao R, Wang B, Cai C, Zheng L, Wang H, et al. The effects of orally administered Ag, TiO_2 and SiO_2 nanoparticles on gut microbiota composition and colitis induction in mice. NanoImpact. 2017;**8**:80-88. DOI: 10.1016/j.impact.2017.07.005

[11] Yang X, Liu J, He H, Zhou L, Gong C, Wang X, et al. SiO_2 nanoparticles induce cytotoxicity and protein expression alteration in HaCaT cells. Particle and Fibre Toxicology. 2010;**7**:1. DOI: 10.1186/1743-8977-7-1

[12] Pandey RP, Vidic J, Mukherjee R, Chang CM. Experimental methods for the biological evaluation of nanoparticle-based drug delivery risks. Pharmaceutics. 2023;**15**(2):612

[13] Mahajan SD, Law W-C, Aalinkeel R, Reynolds J, Nair BB, Yong K-T, et al. Chapter three - nanoparticle-mediated targeted delivery of Antiretrovirals to the brain (NY). In: Düzgüneş N, editor. Methods in Enzymology. Vol. 509. Academic Press; 2012. pp. 41-60. DOI: 10.1016/B978-0-12-391858-1.00003-4

[14] Patravale V, Dandekar P, Jain R. 4 - Nanotoxicology: Evaluating toxicity

potential of drug-nanoparticles. In: Patravale V, Dandekar P, Jain R, editors. Nanoparticulate Drug Delivery. Mumbai: Woodhead Publishing; 2012. pp. 123-155. DOI: 10.1533/9781908818195.123

[15] Marshall NJ, Goodwin CJ, Holt SJ. A critical assessment of the use of microculture tetrazolium assays to measure cell growth and function. Growth Regulation. 1995;**5**:69-84

[16] Kreft S, Kreft M. Quantification of dichromatism: A characteristic of color in transparent materials. Journal of the Optical Society of America. A, Optics, Image Science, and Vision. 2009;**26**:1576-1581. DOI: 10.1364/JOSAA.26.001576

[17] Rampersad SN. Multiple applications of Alamar blue as an indicator of metabolic function and cellular health in cell viability bioassays. Sensors (Basel). 2012;**12**:12347-12360. DOI: 10.3390/s120912347

[18] Messele T, Roos MTL, Hamann D, Koot M, Fontanet AL, Miedema F, et al. Nonradioactive techniques for measurement of In vitro T-cell proliferation: Alternatives to the [3H] thymidine incorporation assay. Clinical and Diagnostic Laboratory Immunology. 2000;7:687-692

[19] Cell Proliferation Assays Essential for Drug Discovery and Development. 2021. Available from: https://cytologicsbio.com/cell-proliferation-assays-essential-for-drug-discovery-and-development/ [Accessed April 8, 2023]

[20] Brown JM, Attardi LD. The role of apoptosis in cancer development and treatment response. Nature Reviews Cancer. 2005;**5**:231-237. DOI: 10.1038/nrc1560

[21] Ghasemi M, Turnbull T, Sebastian S, Kempson I. The MTT assay: Utility, limitations, pitfalls, and interpretation in bulk and single-cell analysis. International Journal of Molecular Sciences. 2021;**22**:12827. DOI: 10.3390/ijms222312827

[22] Johno H, Takahashi S, Kitamura M. Influences of acidic conditions on formazan assay: A cautionary note. Applied Biochemistry and Biotechnology. 2010;**162**:1529-1535. DOI: 10.1007/s12010-010-8934-z

[23] Han M, Li JF, Tan Q, Sun YY, Wang YY. Limitations of the use of MTT assay for screening in drug discovery. Journal of Chinese Pharmaceutical Sciences. 2010;**19**:195-200. DOI: 10.5246/jcps.2010.03.027

[24] Gormley AJ, Ghandehari H. Evaluation of Toxicity of Nanostructures in Biological Systems, Nanotoxicity. John Wiley & Sons, Ltd (n.p.); 2009. pp. 115-159. DOI: 10.1002/9780470747803.ch7

[25] Wang S, Yu H, Wickliffe JK. Limitation of the MTT and XTT assays for measuring cell viability due to superoxide formation induced by nano-scale TiO_2. Toxicology in Vitro. 2011;**25**:2147-2151. DOI: 10.1016/j.tiv.2011.07.007

[26] Herzog E, Casey A, Lyng FM, Chambers G, Byrne HJ, Davoren M. A new approach to the toxicity testing of carbon-based nanomaterials—The clonogenic assay. Toxicology Letters. 2007;**174**:49-60. DOI: 10.1016/j.toxlet.2007.08.009

[27] Casey A, Herzog E, Davoren M, Lyng FM, Byrne HJ, Chambers G. Spectroscopic analysis confirms the interactions between single walled carbon nanotubes and various dyes commonly used to assess cytotoxicity. Carbon. 2007;**45**:1425-1432. DOI: 10.1016/j.carbon.2007.03.033

[28] Monteiro-Riviere NA, Inman AO. Challenges for assessing carbon nanomaterial toxicity to the skin. Carbon. 2006;**44**:1070-1078. DOI: 10.1016/j.carbon.2005.11.004

[29] Ciofani G, Danti S, D'Alessandro D, Moscato S, Menciassi A. Assessing cytotoxicity of boron nitride nanotubes: Interference with the MTT assay. Biochemical and Biophysical Research Communications. 2010;**394**:405-411. DOI: 10.1016/j.bbrc.2010.03.035

[30] Al-Jamal KT, Kostarelos K. Assessment of cellular uptake and cytotoxicity of carbon nanotubes using flow cytometry. In: Balasubramanian K, Burghard M, editors. Carbon Nanotubes: Methods and Protocols. Totowa, NJ: Humana Press; 2010. pp. 123-134. DOI: 10.1007/978-1-60761-579-8_11

[31] Horáková K, Šovčíková A, Seemannová Z, Syrová D, Bušányová K, Drobná Z, et al. Detection of drug-induced, superoxide-mediated cell damage and its prevention by antioxidants. Free Radical Biology and Medicine. 2001;**30**:650-664. DOI: 10.1016/S0891-5849(00)00508-6

[32] Jones CF, Grainger DW. In vitro assessments of nanomaterial toxicity. Advanced Drug Delivery Reviews. 2009;**61**:438-456. DOI: 10.1016/j.addr.2009.03.005

[33] Monteiro-Riviere NA, Inman AO, Zhang LW. Limitations and relative utility of screening assays to assess engineered nanoparticle toxicity in a human cell line. Toxicology and Applied Pharmacology. 2009;**234**:222-235. DOI: 10.1016/j.taap.2008.09.030

[34] Ehmann UK, Williams JR, Nagle WA, Brown JA, Belli JA, Lett JT. Perturbations in cell cycle progression from radioactive DNA precursors. Nature. 1975;**258**:633-636. DOI: 10.1038/258633a0

[35] Hussain SM, Frazier JM. Cellular toxicity of hydrazine in primary rat hepatocytes. Toxicological Sciences. 2002;**69**:424-432. DOI: 10.1093/toxsci/69.2.424

[36] Solary E, Bertrand R, Jenkins J, Pommier Y. Radiolabeling of DNA can induce its fragmentation in HL-60 human promyelocytic leukemic cells. Experimental Cell Research. 1992;**203**:495-498. DOI: 10.1016/0014-4827(92)90027-6

[37] Cleaver JE, Thomas GH, Burki HJ. Biological damage from intranuclear tritium: DNA strand breaks and their repair. Science. 1972;**177**:996-998. DOI: 10.1126/science.177.4053.996

[38] Buch K, Peters T, Nawroth T, Sänger M, Schmidberger H, Langguth P. Determination of cell survival after irradiation via clonogenic assay versus multiple MTT assay - a comparative study. Radiation Oncology. 2012;**7**:1. DOI: 10.1186/1748-717X-7-1

[39] Ryter SW, Kim HP, Hoetzel A, Park JW, Nakahira K, Wang X, et al. Mechanisms of cell death in oxidative stress. Antioxidants & Redox Signaling. 2007;**9**:49-89. DOI: 10.1089/ars.2007.9.49

[40] Li G-Y, Osborne NN. Oxidative-induced apoptosis to an immortalized ganglion cell line is caspase independent but involves the activation of poly(ADP-ribose)polymerase and apoptosis-inducing factor. Brain Research. 2008;**1188**:35-43. DOI: 10.1016/j.brainres.2007.10.073

[41] Demchenko AP. Beyond annexin V: Fluorescence response of cellular membranes to apoptosis.

Cytotechnology. 2013;**65**:157-172. DOI: 10.1007/s10616-012-9481-y

[42] Vanden Berghe T, Grootjans S, Goossens V, Dondelinger Y, Krysko DV, Takahashi N, et al. Determination of apoptotic and necrotic cell death in vitro and in vivo. Methods. 2013;**61**:117-129. DOI: 10.1016/j.ymeth.2013.02.011

[43] Lu X, Qian J, Zhou H, Gan Q, Tang W, Lu J, et al. In vitro cytotoxicity and induction of apoptosis by silica nanoparticles in human HepG2 hepatoma cells. International Journal of Nanomedicine. 2011;**6**:1889-1901. DOI: 10.2147/IJN.S24005

[44] Baharara J, Ramezani T, Divsalar A, Mousavi M, Seyedarabi A. Induction of Apoptosis by green synthesized gold nanoparticles through activation of Caspase-3 and 9 in human cervical cancer cells, Avicenna J med. Biotechnol. 2016;**8**:75-83

[45] van Engeland M, Nieland LJW, Ramaekers FCS, Schutte B, Reutelingsperger CPM. Annexin V-affinity assay: A review on an apoptosis detection system based on phosphatidylserine exposure. Cytometry. 1998;**31**:1-9. DOI: 10.1002/(SICI)1097-0320(19980101)31:1<1::AID-CYTO1>3.0.CO;2-R

[46] Kisin ER, Murray AR, Keane MJ, Shi X-C, Schwegler-Berry D, Gorelik O, et al. Single-walled carbon nanotubes: Geno- and cytotoxic effects in lung fibroblast V79 cells. Journal of Toxicology and Environmental Health. Part A. 2007;**70**:2071-2079. DOI: 10.1080/15287390701601251

[47] Bajpayee M, Kumar A, Dhawan A. The comet assay: Assessment of in vitro and in vivo DNA damage. Methods in Molecular Biology. 2013;**1044**:325-345. DOI: 10.1007/978-1-62703-529-3_17

[48] Kumar A, Sharma V, Dhawan A. Methods for detection of oxidative stress and genotoxicity of engineered nanoparticles. Methods in Molecular Biology. 2013;**1028**:231-246. DOI: 10.1007/978-1-62703-475-3_15

[49] Kumar A, Dhawan A. Genotoxic and carcinogenic potential of engineered nanoparticles: An update. Archives of Toxicology. 2013;**87**:1883-1900. DOI: 10.1007/s00204-013-1128-z

[50] Bijukumar D, Segu A, Chastain P, Mathew MT. Implant-derived CoCrMo alloy nanoparticle disrupts DNA replication dynamics in neuronal cells. Cell Biology and Toxicology (IL). 2021;**37**:833-847. DOI: 10.1007/s10565-020-09577-7

[51] Karlsson HL, Nygren J, Möller L. Genotoxicity of airborne particulate matter: The role of cell–particle interaction and of substances with adduct-forming and oxidizing capacity. Mutation Research/Genetic Toxicology and Environmental Mutagenesis. 2004;**565**:1-10. DOI: 10.1016/j.mrgentox.2004.07.015

[52] Stone V, Johnston H, Schins RP. Development of in vitro systems for nanotoxicology: Methodological considerations. Critical Reviews in Toxicology. 2009;**39**:613-626. DOI: 10.1080/10408440903120975

[53] Shukla RK, Kumar A, Gurbani D, Pandey AK, Singh S, Dhawan A. TiO2 nanoparticles induce oxidative DNA damage and apoptosis in human liver cells. Nanotoxicology. 2013;**7**:48-60. DOI: 10.3109/17435390.2011.629747

[54] Schiavo S, Oliviero M, Miglietta M, Rametta G, Manzo S. Genotoxic and cytotoxic effects of ZnO nanoparticles for Dunaliella tertiolecta and

comparison with SiO_2 and TiO_2 effects at population growth inhibition levels. Sci Total Environ. 2016;**550**:619-627. DOI: 10.1016/j.scitotenv.2016.01.135

[55] Handy RD, van den Brink N, Chappell M, Mühling M, Behra R, Dušinská M, et al. Practical considerations for conducting ecotoxicity test methods with manufactured nanomaterials: What have we learnt so far? Ecotoxicology. 2012;**21**:933-972. DOI: 10.1007/s10646-012-0862-y

[56] David SS. Chemistry of glycosylases and endonucleases involved in base-excision repair. Chemical Reviews. 1998;**98**:1221-1261

[57] Cordelli E, Bignami M, Pacchierotti F. Comet assay: A versatile but complex tool in genotoxicity testing. Toxicology Research (Camb). 2021;**10**:68-78. DOI: 10.1093/toxres/tfaa093

[58] Sasaki YF, Sekihashi K, Izumiyama F, Nishidate E, Saga A, Ishida K, et al. The comet assay with multiple mouse organs: comparison of comet assay results and carcinogenicity with 208 chemicals selected from the IARC monographs and US NTP Carcinogenicity Database. Critical Reviews in Toxicology. 2000;**30**:629-799. DOI: 10.1080/10408440008951123

[59] Arends MJ, Morris RG, Wyllie AH, Apoptosis. The role of the endonuclease. The American Journal of Pathology. 1990;**136**:593-608

[60] Bortner CD, Oldenburg NBE, Cidlowski JA. The role of DNA fragmentation in apoptosis. Trends in Cell Biology. 1995;**5**:21-26. DOI: 10.1016/S0962-8924(00)88932-1

[61] Loo DT. TUNEL Assay. In: Didenko VV, editor. In Situ Detection of DNA Damage: Methods and Protocols. Totowa, NJ: Humana Press; 2002. pp. 21-30. DOI: 10.1385/1-59259-179-5:21

[62] Gavrieli Y, Sherman Y, Ben-Sasson SA. Identification of programmed cell death in situ via specific labeling of nuclear DNA fragmentation. Journal of Cell Biology. 1992;**119**:493-501. DOI: 10.1083/jcb.119.3.493

[63] Elmore S. Apoptosis: A review of programmed cell death. Toxicologic Pathology. 2007;**35**:495-516. DOI: 10.1080/01926230701320337

[64] Mirzayans R, Murray D. Do TUNEL and other Apoptosis assays detect cell death in preclinical studies? International Journal of Molecular Sciences. 2020;**21**:9090. DOI: 10.3390/ijms21239090

[65] Ledbetter JA, Rouse RV, Micklem HS, Herzenberg LA. T cell subsets defined by expression of Lyt-1,2,3 and Thy-1 antigens. Two-parameter immunofluorescence and cytotoxicity analysis with monoclonal antibodies modifies current views. Journal of Experimental Medicine. 1980;**152**:280-295. DOI: 10.1084/jem.152.2.280

[66] Ibuki Y, Toyooka T. Nanoparticle uptake measured by flow cytometry. In: Reineke J, editor. Nanotoxicity: Methods and Protocols. Totowa, NJ: Humana Press; 2012. pp. 157-166. DOI: 10.1007/978-1-62703-002-1_11

[67] Mourdjeva M, Kyurkchiev D, Mandinova A, Altankova I, Kehayov I, Kyurkchiev S. Dynamics of membrane translocation of phosphatidylserine during apoptosis detected by a monoclonal antibody. Apoptosis. 2005;**10**:209-217. DOI: 10.1007/s10495-005-6076-5

[68] Balasubramanian K, Bevers EM, Willems GM, Schroit AJ. Binding of

Annexin V to membrane products of lipid peroxidation. Biochemistry. 2001;**40**:8672-8676. DOI: 10.1021/bi010841y

[69] Kain J, Karlsson HL, Möller L. DNA damage induced by micro- and nanoparticles—Interaction with FPG influences the detection of DNA oxidation in the comet assay. Mutagenesis. 2012;**27**:491-500. DOI: 10.1093/mutage/ges010

[70] Loo DT. In situ detection of Apoptosis by the TUNEL assay: An overview of techniques. In: Didenko VV, editor. DNA Damage Detection In Situ, Ex Vivo, and In Vivo: Methods and Protocols. Totowa, NJ: Humana Press; 2011. pp. 3-13. DOI: 10.1007/978-1-60327-409-8_1

[71] LearnHaem, Limitations, LearnHaem | Haematology Made Simple. n.d. Available from: https://www.learnhaem.com/courses/flow-cytometry/lessons/limitations/ [accessed April 17, 2023]

[72] Klein S, Petersen S, Taylor U, Rath D, Barcikowski S. Quantitative visualization of colloidal and intracellular gold nanoparticles by confocal microscopy. Journal of Biomedical Optics. 2010;**15**:036015. DOI: 10.1117/1.3461170

[73] Attarilar S, Yang J, Ebrahimi M, Wang Q, Liu J, Tang Y, et al. The toxicity phenomenon and the related occurrence in metal and metal oxide nanoparticles: A brief review from the biomedical perspective. Frontiers in Bioengineering and Biotechnology. 2020;**8**:822. DOI: 10.3389/fbioe.2020.00822

[74] Winckler J. Vital staining of lysosomes and other cell organelles of the rat with neutral red (author's transl). Progress in Histochemistry and Cytochemistry. 1974;**6**:1-91

[75] Nemes Z, Dietz R, Lüth JB, Gomba S, Hackenthal E, Gross F. The pharmacological relevance of vital staining with neutral red. Experientia. 1979;**35**:1475-1476. DOI: 10.1007/BF01962793

[76] Repetto G, del Peso A, Zurita JL. Neutral red uptake assay for the estimation of cell viability/cytotoxicity. Nature Protocols. 2008;**3**:1125-1131. DOI: 10.1038/nprot.2008.75

[77] Borenfreund E, Babich H, Martin-Alguacil N. Comparisons of two in vitro cytotoxicity assays—The neutral red (NR) and tetrazolium MTT tests. Toxicology in Vitro. 1988;**2**:1-6. DOI: 10.1016/0887-2333(88)90030-6

[78] Arranz MJ, Festing MFW. Prior use of the neutral red assay and reduction of total protein determination in 96-well plate assays. Toxicology in Vitro. 1990;**4**:211-212. DOI: 10.1016/0887-2333(90)90024-N

[79] Mello DF, Trevisan R, Rivera N, Geitner NK, Di Giulio RT, Wiesner MR, et al. Caveats to the use of MTT, neutral red, Hoechst and Resazurin to measure silver nanoparticle cytotoxicity. Chemico-Biological Interactions. 2020;**315**:108868. DOI: 10.1016/j.cbi.2019.108868

[80] Yip DK, Auersperg N. The dye-exclusion test for cell viability: Persistence of differential staining following fixation. In Vitro. 1972;**7**:323-329. DOI: 10.1007/BF02661722

[81] Lovitt CJ, Shelper TB, Avery VM. Advanced cell culture techniques for cancer drug discovery. Biology. 2014;**3**:345-367. DOI: 10.3390/biology3020345

[82] Kim SI, Kim HJ, Lee H-J, Lee K, Hong D, Lim H, et al. Application of

a non-hazardous vital dye for cell counting with automated cell counters. Analytical Biochemistry. 2016;**492**:8-12. DOI: 10.1016/j.ab.2015.09.010

[83] Hamida RS, Albasher G, Bin-Meferij MM. Oxidative stress and apoptotic responses elicited by Nostoc-synthesized silver nanoparticles against different cancer cell lines. Cancers. 2020;**12**(8):2099

[84] Bass DA, Parce JW, Dechatelet LR, Szejda P, Seeds MC, Thomas M. Flow cytometric studies of oxidative product formation by neutrophils: A graded response to membrane stimulation. The Journal of Immunology. 1983;**130**:1910-1917. DOI: 10.4049/jimmunol.130.4.1910

[85] Levine RL, Garland D, Oliver CN, Amici A, Climent I, Lenz AG, et al. Determination of carbonyl content in oxidatively modified proteins. Methods in Enzymology. 1990;**186**:464-478. DOI: 10.1016/0076-6879(90)86141-h

[86] Levine RL, Williams JA, Stadtman ER, Shacter E. Carbonyl assays for determination of oxidatively modified proteins. Methods in Enzymology. 1994;**233**:346-357. DOI: 10.1016/s0076-6879(94)33040-9

[87] Ayala A, Muñoz MF, Argüelles S. Lipid peroxidation: Production, metabolism, and signaling mechanisms of malondialdehyde and 4-hydroxy-2-nonenal. Oxidative Medicine and Cellular Longevity. 2014;**2014**:360438. DOI: 10.1155/2014/360438

[88] Ohkawa H, Ohishi N, Yagi K. Assay for lipid peroxides in animal tissues by thiobarbituric acid reaction. Analytical Biochemistry. 1979;**95**:351-358. DOI: 10.1016/0003-2697(79)90738-3

[89] Chemistry and biochemistry of 4-hydroxynonenal, malonaldehyde and related aldehydes – ScienceDirect. n.d. Available from: https://www.sciencedirect.com/science/article/pii/0891584991901926?casa_token=-y-RVlMtBvgAAAAA:Bkl6SwWpdrCKLebPQxi52q66lV_TojjgBw5VG2ZcC-lGx-9g6vOB6bkoPp9EC7VDTVfHHm0BOw [Accessed April 14, 2023]

[90] Weber D, Milkovic L, Bennett SJ, Griffiths HR, Zarkovic N, Grune T. Measurement of HNE-protein adducts in human plasma and serum by ELISA—Comparison of two primary antibodies. Redox Biology. 2013;**1**:226-233. DOI: 10.1016/j.redox.2013.01.012

[91] Wang H, Joseph JA. Quantifying cellular oxidative stress by dichlorofluorescein assay using microplate reader. Free Radical Biology and Medicine. 1999;**27**:612-616. DOI: 10.1016/S0891-5849(99)00107-0

[92] Savage DT, Hilt JZ, Dziubla TD. In vitro methods for assessing nanoparticle toxicity. Methods in Molecular Biology;**2019**(1894):1-29. DOI: 10.1007/978-1-4939-8916-4_1

[93] Wohlsen A, Martin C, Vollmer E, Branscheid D, Magnussen H, Becker W-M, et al. The early allergic response in small airways of human precision-cut lung slices. The European Respiratory Journal. 2003;**21**:1024. DOI: 10.1183/09031936.03.00027502

[94] Henjakovic M, Martin C, Hoymann HG, Sewald K, Ressmeyer AR, Dassow C, et al. Ex vivo lung function measurements in precision-cut lung slices (PCLS) from chemical allergen–sensitized mice represent a suitable alternative to In vivo studies. Toxicological Sciences. 2008;**106**:444-453. DOI: 10.1093/toxsci/kfn178

[95] Liu Y, Wu P, Wang Y, Liu Y, Yang H, Zhou G, et al. Application of

precision-cut lung slices as an In vitro model for research of inflammatory respiratory diseases. Bioengineering (Basel). 2022;**9**:767. DOI: 10.3390/bioengineering9120767

[96] Morin J-P, Baste J-M, Gay A, Crochemore C, Corbière C, Monteil C. Precision cut lung slices as an efficient tool for in vitro lung physio-pharmacotoxicology studies. Xenobiotica. 2013;**43**:63-72. DOI: 10.3109/00498254.2012.727043

[97] Bartucci R, Åberg C, Melgert BN, Boersma YL, Olinga P, Salvati A. Time-resolved quantification of nanoparticle uptake, distribution, and impact in precision-cut liver slices. Small. 2020;**16**:1906523. DOI: 10.1002/smll.201906523

[98] Bartucci R, van der Meer AZ, Boersma YL, Olinga P, Salvati A. Nanoparticle-induced inflammation and fibrosis in ex vivo murine precision-cut liver slices and effects of nanoparticle exposure conditions. Archives of Toxicology. 2021;**95**:1267-1285. DOI: 10.1007/s00204-021-02992-7

[99] Palma E, Doornebal EJ, Chokshi S. Precision-cut liver slices: A versatile tool to advance liver research. Hepatology International. 2019;**13**:51-57. DOI: 10.1007/s12072-018-9913-7

[100] Piña-Olmos S, Dolores-Hernández M, Diaz-Torres R, Ramírez-Bribiesca JE, López-Arellano R, López Barrera LD, et al. Precision-cut liver slices as a model for assess hepatic cellular response of chitosan-glutathione nanoparticles on cultures treated with zilpaterol and clenbuterol. Toxicology Mechanisms and Methods. 2022;**32**:313-324. DOI: 10.1080/15376516.2021.2002992

[101] Dragoni S, Franco G, Regoli M, Bracciali M, Morandi V, Sgaragli G, et al. Gold nanoparticles uptake and cytotoxicity assessed on rat liver precision-cut slices. Toxicological Sciences. 2012;**128**:186-197. DOI: 10.1093/toxsci/kfs150

[102] Arık YB, Buijsman W, Loessberg-Zahl J, Cuartas-Vélez C, Veenstra C, Logtenberg S, et al. Microfluidic organ-on-a-chip model of the outer blood–retinal barrier with clinically relevant read-outs for tissue permeability and vascular structure. Lab on a Chip. 2021;**21**:272-283. DOI: 10.1039/D0LC00639D

[103] Jeon J, Hotaling N, Zamani M, Dejene R, Ingber D, Bharti K. Tissue engineered human blood-retinal barrier-on-a-Chip. Investigative Ophthalmology & Visual Science. 2016;**57**:5325

[104] Park TS, Nguyen E, Pakhchanian H, Ortolan D, Vettikatu N, Dejene R, et al. Modeling the outer blood-retina barrier using microfluidic chips and iPSC-derived RPE and endothelial cells. Investigative Ophthalmology & Visual Science. 2021;**62**:2192

[105] Ragelle H, Goncalves A, Kustermann S, Antonetti DA, Jayagopal A. Organ-on-A-Chip Technologies for advanced blood–retinal barrier models. Journal of Ocular Pharmacology and Therapeutics. 2020;**36**:30-41. DOI: 10.1089/jop.2019.0017

[106] Baptista D, Moreira Teixeira L, Barata D, Tahmasebi Birgani Z, King J, van Riet S, et al. 3D lung-on-Chip model based on biomimetically microcurved culture membranes. ACS Biomaterials Science & Engineering. 2022;**8**:2684-2699. DOI: 10.1021/acsbiomaterials.1c01463

[107] Huang D, Liu T, Liao J, Maharjan S, Xie X, Pérez M, et al. Reversed-engineered human alveolar

lung-on-a-chip model. Proceedings of the National Academy of Sciences. 2021;**118**:e2016146118. DOI: 10.1073/pnas.2016146118

[108] (Dan) Huh D. A human breathing lung-on-a-Chip. Annals of the American Thoracic Society. 2015;**12**:S42-S44. DOI: 10.1513/AnnalsATS.201410-442MG

[109] Huh D, Matthews BD, Mammoto A, Montoya-Zavala M, Hsin HY, Ingber DE. Reconstituting organ-level lung functions on a Chip. Science. 2010;**328**:1662-1668. DOI: 10.1126/science.1188302

[110] Zamprogno P, Wüthrich S, Achenbach S, Thoma G, Stucki JD, Hobi N, et al. Second-generation lung-on-a-chip with an array of stretchable alveoli made with a biological membrane. Commun Biol. 2021;**4**:1-10. DOI: 10.1038/s42003-021-01695-0

[111] Ahn SI, Sei YJ, Park H-J, Kim J, Ryu Y, Choi JJ, et al. Microengineered human blood–brain barrier platform for understanding nanoparticle transport mechanisms. Nature Communications. 2020;**11**:175. DOI: 10.1038/s41467-019-13896-7

[112] Cui B, Cho S-W. Blood-brain barrier-on-a-chip for brain disease modeling and drug testing. BMB Reports. 2022;**55**:213-219. DOI: 10.5483/BMBRep.2022.55.5.043

[113] Mir M, Palma-Florez S, Lagunas A, López-Martínez MJ, Samitier J. Biosensors integration in blood–brain barrier-on-a-Chip: Emerging platform for monitoring neurodegenerative diseases. ACS Sens. 2022;**7**:1237-1247. DOI: 10.1021/acssensors.2c00333

[114] Peng B, Hao S, Tong Z, Bai H, Pan S, Lim K-L, et al. Blood–brain barrier (BBB)-on-a-chip: A promising breakthrough in brain disease research. Lab on a Chip. 2022;**22**:3579-3602. DOI: 10.1039/D2LC00305H

[115] Wevers NR, Kasi DG, Gray T, Wilschut KJ, Smith B, van Vught R, et al. A perfused human blood–brain barrier on-a-chip for high-throughput assessment of barrier function and antibody transport. Fluids and Barriers of the CNS. 2018;**15**:23. DOI: 10.1186/s12987-018-0108-3

[116] Leung CM, de Haan P, Ronaldson-Bouchard K, Kim G-A, Ko J, Rho HS, et al. A guide to the organ-on-a-chip. Nat Rev Methods Primers. 2022;**2**:1-29. DOI: 10.1038/s43586-022-00118-6

[117] Kwak B, Ozcelikkale A, Shin CS, Park K, Han B. Simulation of complex transport of nanoparticles around a tumor using tumor-microenvironment-on-chip. Journal of Controlled Release. 2014;**194**:157-167. DOI: 10.1016/j.jconrel.2014.08.027

[118] Albanese A, Lam AK, Sykes EA, Rocheleau JV, Chan WCW. Tumour-on-a-chip provides an optical window into nanoparticle tissue transport. Nature Communications. 2013;**4**:2718. DOI: 10.1038/ncomms3718

[119] Toy R, Peiris PM, Ghaghada KB, Karathanasis E. Shaping cancer nanomedicine: the effect of particle shape on the in vivo journey of nanoparticles. Nanomedicine (Lond). Jan 2014;**9**(1):121-134. DOI: 10.2217/nnm.13.191. PMID: 24354814; PMCID: PMC4057606

[120] Papademetriou I, Vedula E, Charest J, Porter T. Effect of flow on targeting and penetration of angiopep-decorated nanoparticles in a microfluidic model blood-brain barrier. PLoS One. 2018;**13**:e0205158. DOI: 10.1371/journal.pone.0205158

[121] Nasir I, Lundqvist M, Cabaleiro-Lago C. Size and surface chemistry of nanoparticles lead to a variant behavior in the unfolding dynamics of human carbonic anhydrase. Nanoscale. 2015;**7**:17504-17515. DOI: 10.1039/C5NR05360A

[122] Zhang M, Xu C, Jiang L, Qin J. A 3D human lung-on-a-chip model for nanotoxicity testing. Toxicology Research. 2018;**7**:1048-1060. DOI: 10.1039/c8tx00156a

[123] Yin F, Zhu Y, Zhang M, Yu H, Chen W, Qin J. A 3D human placenta-on-a-chip model to probe nanoparticle exposure at the placental barrier. Toxicology in Vitro. 2019;**54**:105-113. DOI: 10.1016/j.tiv.2018.08.014

[124] Fede C, Fortunati I, Weber V, Rossetto N, Bertasi F, Petrelli L, et al. Evaluation of gold nanoparticles toxicity towards human endothelial cells under static and flow conditions. Microvascular Research. 2015;**97**:147-155. DOI: 10.1016/j.mvr.2014.10.010

[125] van Meer BJ, de Vries H, Firth KSA, van Weerd J, Tertoolen LGJ, Karperien HBJ, et al. Small molecule absorption by PDMS in the context of drug response bioassays. Biochemical and Biophysical Research Communications. 2017;**482**:323-328. DOI: 10.1016/j.bbrc.2016.11.062

[126] Bovard D, Sandoz A, Luettich K, Frentzel S, Iskandar A, Marescotti D, et al. A lung/liver-on-a-chip platform for acute and chronic toxicity studies. Lab on a Chip. 2018;**18**:3814-3829. DOI: 10.1039/C8LC01029C

[127] Grace Intasa-Ard S, Birault A. Nanoparticles characterization using the CAM assay. Enzyme. 2019;**46**:129-160. DOI: 10.1016/bs.enz.2019.09.001

[128] Harmankaya OF, Hamsho A, Goztepe M, Baghirova S, Cirik S, Sirin H, et al. The effect of chitosan capped copper nanoparticles on angiogenesis in cam assay. Biomedical Journal of Scientific & Technical Research. 2021;**35**:27971-27975. DOI: 10.26717/BJSTR.2021.35.005755

[129] Pedrosa P, Heuer-Jungemann A, Kanaras AG, Fernandes AR, Baptista PV. Potentiating angiogenesis arrest in vivo via laser irradiation of peptide functionalised gold nanoparticles. Journal of Nanobiotechnology. 2017;**15**:85. DOI: 10.1186/s12951-017-0321-2

[130] Vu BT, Shahin SA, Croissant J, Fatieiev Y, Matsumoto K, Le-Hoang Doan T, et al. Chick chorioallantoic membrane assay as an in vivo model to study the effect of nanoparticle-based anticancer drugs in ovarian cancer. Scientific Reports. 2018;**8**:8524. DOI: 10.1038/s41598-018-25573-8

[131] Naik M, Brahma P, Dixit M. A cost-effective and efficient Chick ex-Ovo CAM assay protocol to assess angiogenesis. Methods Protoc. 2018;**1**:19. DOI: 10.3390/mps1020019

[132] Ribatti D. Advantages and limitations of Chorioallantoic membrane in comparison with other classical in vivo angiogenesis assays. In: Ribatti D, editor. The Chick Embryo Chorioallantoic Membrane in the Study of Angiogenesis and Metastasis: The CAM Assay in the Study of Angiogenesis and Metastasis. Netherlands, Dordrecht: Springer; 2010. pp. 75-85. DOI: 10.1007/978-90-481-3845-6_7

[133] Charão MF, Souto C, Brucker N, Barth A, Jornada DS, Fagundez D, et al. *Caenorhabditis elegans* as an alternative in vivo model to determine oral uptake, nanotoxicity, and efficacy of melatonin-loaded lipid-core nanocapsules on paraquat damage. International Journal of Nanomedicine. 2015;**10**:5093-5106. DOI: 10.2147/IJN.S84909

[134] Gubert P, Gubert G, de Oliveira RC, Fernandes ICO, Bezerra IC, de Ramos B, et al. *Caenorhabditis elegans* as a prediction platform for nanotechnology-based strategies: Insights on analytical challenges. Toxics. 2023;**11**:239. DOI: 10.3390/toxics11030239

[135] Hu C-C, Wu G-H, Lai S-F, Muthaiyan Shanmugam M, Hwu Y, Wagner OI, et al. Yen, toxic effects of size-tunable gold nanoparticles on *Caenorhabditis elegans* development and gene regulation. Scientific Reports. 2018;**8**:15245. DOI: 10.1038/s41598-018-33585-7

[136] Viau C, Haçariz O, Karimian F, Xia J. Comprehensive phenotyping and transcriptome profiling to study nanotoxicity in *C. elegans*. PeerJ. 2020;**8**:e8684. DOI: 10.7717/peerj.8684

[137] Wu T, Xu H, Liang X, Tang M. *Caenorhabditis elegans* as a complete model organism for biosafety assessments of nanoparticles. Chemosphere. 2019;**221**:708-726. DOI: 10.1016/j.chemosphere.2019.01.021

[138] Zhao X, Wan Q, Fu X, Meng X, Ou X, Zhong R, et al. Toxicity evaluation of one-dimensional nanoparticles using *Caenorhabditis elegans*: A comparative study of Halloysite nanotubes and chitin nanocrystals. ACS Sustainable Chemistry & Engineering. 2019;**7**:18965-18975. DOI: 10.1021/acssuschemeng.9b04365

[139] Wang Q, Zhu Y, Song B, Fu R, Zhou Y. The In vivo toxicity assessments of water-dispersed fluorescent silicon nanoparticles in *Caenorhabditis elegans*. International Journal of Environmental Research and Public Health. 2022;**19**:4101. DOI: 10.3390/ijerph19074101

[140] Liu X, Vinson D, Abt D, Hurt RH, Rand DM. Differential toxicity of carbon nanomaterials in drosophila: Larval dietary uptake is benign, but adult exposure causes locomotor impairment and mortality. Environmental Science & Technology. 2009;**43**:6357-6363

[141] Pappus SA, Mishra M. A drosophila model to decipher the toxicity of nanoparticles taken through Oral routes. Advances in Experimental Medicine and Biology. 2018;**1048**:311-322. DOI: 10.1007/978-3-319-72041-8_18

[142] Priyadarsini S, Sahoo SK, Sahu S, Mukherjee S, Hota G, Mishra M. Oral administration of graphene oxide nano-sheets induces oxidative stress, genotoxicity, and behavioral teratogenicity in *Drosophila melanogaster*. Environmental Science and Pollution Research. 2019;**26**:19560-19574. DOI: 10.1007/s11356-019-05357-x

[143] Ng CT, Yu LE, Ong CN, Bay BH, Baeg GH. The use of *Drosophila melanogaster* as a model organism to study immune-nanotoxicity. Nanotoxicology. 2019;**13**:429-446. DOI: 10.1080/17435390.2018.1546413

[144] Yan S, Li N, Guo Y, Chen Y, Ji C, Yin M, et al. Chronic exposure to the star polycation (SPc) nanocarrier in the larval stage adversely impairs life history traits in *Drosophila melanogaster*. Journal of Nanobiotechnology. 2022;**20**:515. DOI: 10.1186/s12951-022-01705-1

[145] Alaraby M, Abass D, Domenech J, Hernández A, Marcos R. Hazard assessment of ingested polystyrene nanoplastics in drosophila larvae, environmental science. Nano. 2022;**9**:1845-1857. DOI: 10.1039/D1EN01199E

[146] Cassar S, Adatto I, Freeman JL, Gamse JT, Iturria I, Lawrence C, et al. Use of zebrafish in drug discovery

toxicology. Chemical Research in Toxicology. 2020;**33**:95-118. DOI: 10.1021/acs.chemrestox.9b00335

[147] Lee KY, Jang GH, Byun CH, Jeun M, Searson PC, Lee KH. Zebrafish models for functional and toxicological screening of nanoscale drug delivery systems: Promoting preclinical applications. Bioscience Reports. 2017;**37**:BSR20170199. DOI: 10.1042/BSR20170199

[148] Paatero I, Casals E, Niemi R, Özliseli E, Rosenholm JM, Sahlgren C. Analyses in zebrafish embryos reveal that nanotoxicity profiles are dependent on surface-functionalization controlled penetrance of biological membranes. Scientific Reports. 2017;**7**:8423. DOI: 10.1038/s41598-017-09312-z

[149] Pereira AC, Gomes T, Ferreira Machado MR, Rocha TL. The zebrafish embryotoxicity test (ZET) for nanotoxicity assessment: From morphological to molecular approach. Environmental Pollution. 2019;**252**:1841-1853. DOI: 10.1016/j.envpol.2019.06.100

[150] Saleem S, Kannan RR. Zebrafish: A promising real-time model system for nanotechnology-mediated Neurospecific drug delivery. Nanoscale Research Letters. 2021;**16**:135. DOI: 10.1186/s11671-021-03592-1

[151] Choi JS, Kim R-O, Yoon S, Kim W-K. Developmental toxicity of zinc oxide nanoparticles to zebrafish (*Danio rerio*): A transcriptomic analysis. PLoS One. 2016;**11**:e0160763. DOI: 10.1371/journal.pone.0160763

[152] Chen P, Huang J, Rao L, Zhu W, Yu Y, Xiao F, et al. Resistance and resilience of fish gut microbiota to silver nanoparticles. MSystems. 2021;**6**:e00630-e00621. DOI: 10.1128/mSystems.00630-21

[153] Duan J, Liang S, Yu Y, Li Y, Wang L, Wu Z, et al. Inflammation–coagulation response and thrombotic effects induced by silica nanoparticles in zebrafish embryos. Nanotoxicology. 2018;**12**:470-484. DOI: 10.1080/17435390.2018.1461267

[154] Raesian A, Arbabi Bidgoli S, Rezayat Sorkhabadi SM. Dermal toxicity of colloidal Nanosilver in albino rabbit: A new approach to physicochemical properties. Nanomedicine Research Journal. 2017;**2**:142-149. DOI: 10.22034/nmrj.2017.03.001

[155] Liu L-P, Xiao Y-B, Xiao Z-W, Wang Z-B, Li C, Gong X. Toxicity of hydroxyapatite nanoparticles on rabbits. Wei Sheng Yan Jiu. 2005;**34**:474-476

[156] Atef HA, Mansour MK, Ibrahim EM, El-Ahl RMHS, Al-Kalamawey NM, Kattan YAE, et al. Efficacy of zinc oxide nanoparticles and curcumin in amelioration the toxic effects in Aflatoxicated rabbits. International Journal of Current Microbiology and Applied Sciences. 2016;**5**:795-818. DOI: 10.20546/ijcmas.2016.512.090

[157] Tandon A, Sharma A, Rodier JT, Klibanov AM, Rieger FG, Mohan RR. BMP7 gene transfer via gold nanoparticles into stroma inhibits corneal fibrosis In vivo. PLoS One. 2013;**8**:e66434. DOI: 10.1371/journal.pone.0066434

[158] Khaleel RI, Mohammed AH. Toxicity of CaO nanoparticles effect on haematology and histopathology of male rabbits. Journal of Kufa-Physics. 2021;**13**:9-16. DOI: 10.31257/2018/JKP/2021/130202

[159] Brohi RD, Wang L, Talpur HS, Wu D, Khan FA, Bhattarai D, Rehman ZU, Farmanullah F, Huo LJ. Toxicity of Nanoparticles on

the Reproductive System in Animal Models: A Review. Front Pharmacol. 5 Sep 2017;**8**:606. DOI: 10.3389/fphar.2017.00606. PMID: 28928662; PMCID: PMC5591883

[160] Chrishtop VV, Prilepskii AY, Nikonorova VG, Mironov VA. Nanosafety vs. nanotoxicology: Adequate animal models for testing in vivo toxicity of nanoparticles. Toxicology. 2021;**462**:152952. DOI: 10.1016/j.tox.2021.152952

[161] He X. In vivo Nanotoxicity assays in animal models. In: Toxicology of Nanomaterials. John Wiley & Sons, Ltd (n.p.); 2016. pp. 151-198. DOI: 10.1002/9783527689125.ch7

[162] Sharma V, Aneja B, Yata VK, Malakar D, Mohanty AK. Systemic Nanotoxicity and its assessment in animal models. In: Yata VK, Ranjan S, Dasgupta N, Lichtfouse E, editors. Nanopharmaceuticals: Principles and Applications. Vol. 3. Cham: Springer International Publishing; 2020. pp. 201-243. DOI: 10.1007/978-3-030-47120-0_7

[163] Peng L, He M, Chen B, Wu Q, Zhang Z, Pang D, et al. Cellular uptake, elimination and toxicity of CdSe/ZnS quantum dots in HepG2 cells. Biomaterials. 2013;**34**:9545-9558. DOI: 10.1016/j.biomaterials.2013.08.038

[164] Kelley RA, Conley SM, Makkia R, Watson JN, Han Z, Cooper MJ, et al. DNA nanoparticles are safe and nontoxic in non-human primate eyes. International Journal of Nanomedicine. 2018;**13**:1361-1379. DOI: 10.2147/IJN.S157000

[165] Kotb S, Piraquive J, Lamberton F, Lux F, Verset M, Di Cataldo V, et al. Safety evaluation and imaging properties of gadolinium-based nanoparticles in nonhuman primates. Scientific Reports. 2016;**6**:35053. DOI: 10.1038/srep35053

[166] Allen SD, Liu Y-G, Bobbala S, Cai L, Hecker PI, Temel R, et al. Polymersomes scalably fabricated via flash nanoprecipitation are non-toxic in non-human primates and associate with leukocytes in the spleen and kidney following intravenous administration. Nano Research. 2018;**11**:5689-5703. DOI: 10.1007/s12274-018-2069-x

[167] Heidel JD, Yu Z, Liu JY-C, Rele SM, Liang Y, Zeidan RK, et al. Administration in non-human primates of escalating intravenous doses of targeted nanoparticles containing ribonucleotide reductase subunit M2 siRNA. Proceedings of the National Academy of Sciences. 2007;**104**:5715-5721. DOI: 10.1073/pnas.0701458104

[168] Ye L, Hu R, Liu L, Liu J, Liu J, Chen H, et al. Comparing semiconductor nanocrystal toxicity in pregnant mice and non-human primates. Nano. 2019;**3**:54-65. DOI: 10.7150/ntno.27452

[169] Soldati A, Fedurek P, Crockford C, Adue S, Akankwasa JW, Asiimwe C, et al. Dead-infant carrying by chimpanzee mothers in the Budongo Forest. Primates. 2022;**63**:497-508. DOI: 10.1007/s10329-022-00999-x

[170] Do Monkeys Grieve for Fallen Mates?. n.d. Available from: https://www.science.org/content/article/do-monkeys-grieve-fallen-mates [Accessed April 27, 2023]

[171] Anderson JR. Responses to death and dying: Primates and other mammals. Primates. 2020;**61**:1-7. DOI: 10.1007/s10329-019-00786-1

[172] Baboons in Mourning Seek Comfort Among Friends, ScienceDaily. n.d. Available from: https://www.sciencedaily.com/releases/2006/01/060130154735.htm [Accessed April 27, 2023]

Chapter 9

Biological Agents for the Synthesis of Silver Nanoparticles and Their Applications

Krishna Gudikandula, Trimothi Dasari and Reeja Sundaram

Abstract

The field of nanotechnology is experiencing rapid growth owing to its distinctive functionality and diverse range of applications. Nanomedicine is a field of study that investigates the potential applications of nanotechnology in the areas of disease prevention, treatment, diagnosis, and control. The significance of silver nanoparticles lies in their distinct characteristics, capacity to generate varied nanostructures, extensive range of bactericidal and anticancer properties, wound healing and other therapeutic capabilities, and cost-effectiveness in manufacturing. These nanoparticles are particularly noteworthy due to their size, which can range from 1 to 100 nm. This paper provides an overview of diverse physical, chemical, and biological techniques employed for the synthesis of silver nanoparticles. The text delineates various methodologies utilizing silver nanoparticles as agents for combating microbial and biofilm infections, as well as for their potential as antitumorigenic agents. Additionally, the text explores the applications of silver nanoparticles in the field of dentistry and dental implants, their role in promoting bone regeneration, their use in cardiovascular implants, and their potential as promoters of wound healing. The present study investigates the mechanism of action, synthesis techniques, and morphological characterization of silver nanoparticles in order to evaluate their potential applications in medical therapies and disease control.

Keywords: silver nanoparticles, biological synthesis, characterization, antimicrobial agent, antibiofilm, health management activity

1. Introduction

The distinctive characteristics of size, shape, and morphology of nanoparticles facilitate their interaction with bacteria, plants, and animals [1–4]. Silver nanoparticles (Ag NPs) have demonstrated remarkable bactericidal efficacy against a diverse array of microorganisms [5–7]. These entities are formulated from diverse viewpoints, frequently for the purpose of examining their morphology or physical attributes. Certain authors have employed a chemical approach [8] and have erroneously conflated it with green synthesis, albeit unintentionally. The utilization of Ag NPs in various fields such as electronics, catalysis, pharmaceuticals, and biomedicine for regulating microorganism proliferation in biological systems has rendered them environmentally sustainable [6, 9]. The process of synthesizing Ag NPs through biogenic means entails the utilization of microorganisms such as bacteria, fungi, yeast, actinomycetes, as well as plant extracts [9, 10]. In contemporary times, various components of plants, including but not limited to flowers, leaves, and fruits, as well as enzymes, have been employed in the production of gold and silver nanoparticles. The physical characteristics of nanoparticles, including their dimensions, shape, and durability, are contingent upon various factors such as the preparation technique employed, the solvent utilized, the concentration of the solution, the potency of the reducing agent, and the temperature conditions [9, 10].

Among the various nanoparticles that have been developed and characterized, silver nanoparticles hold a prominent position due to their innate ability to function as an antimicrobial agent, even in their solid state. Despite being acknowledged for its importance at an earlier time, its potential was not fully utilized, with the exception of its application in traditional medicine and numismatics. Approximately 320 tons of silver nanoparticles (Ag NPs) are produced annually for utilization in various applications such as nanomedical imaging, biosensing, and food products [11, 12].

The prevalence of multidrug-resistant bacterial and viral strains is persistently rising, attributed to genetic mutations, environmental pollution, and alterations in ecological circumstances. In order to overcome this dilemma, researchers are endeavoring to create pharmaceuticals for the management of said microbial infections. Several metal salts and metal nanoparticles have demonstrated efficacy in impeding the proliferation of various pathogenic bacteria. Silver and silver nanoparticles (Ag NPs) hold a significant position in the category of metals utilized as antimicrobial agents since ancient times [13, 14]. Silver salts are employed as a means of impeding the proliferation of diverse bacterial strains within the human body. Antimicrobial agents are employed in medical applications such as catheterization, wound care, and burn treatment to safeguard against potential infection [15, 16]. According to Das et al. [17], the growth of certain bacteria can be effectively inhibited by small-sized silver nanoparticles (Ag NPs). Silver nanoparticles (Ag NPs) that are produced using silk sericin (SS), a protein that is soluble in water and extracted from silkworms at a pH of 11, are composed of hydrophilic proteins that possess polar groups such as hydroxyl, carboxyl, and amino functional groups. Functional groups present in the aforementioned molecules exhibit reducing properties towards AgNO3, resulting in the formation of metallic silver [18]. It has been proposed that the hydroxyl groups present in SS are capable of forming a complex with silver ions, thereby impeding their aggregation or precipitation [19, 20]. The elemental state of Ag NPs may experience segregation as a result of the presence of large molecules within the solvent. However, it is unlikely that they will form complexes since both entities are neutral. The screening of the antibacterial efficacy of silver nanoparticles (Ag NPs) capped with SS has been conducted against both gram-positive and gram-negative bacterial strains. The study revealed that the minimum inhibitory concentration (MIC) ranges from 0.001 to 0.008 mM for various microorganisms, including *Bacillus subtilis*, *Staphylococcus aureus*, *Escherichia coli*, *Pseudomonas aeruginosa*, and *Acinetobacter baumannii*.

While there have been numerous publications on the biosynthesis and characterization of silver nanoparticles, there is a dearth of information regarding their green synthesis, biological properties, and mechanism of action [18]. This review aims to provide a comprehensive overview of the biosynthesis process of silver nanoparticles (Ag NPs) using various sources such as plant extracts, bacteria, fungi, viruses, and actinomycetes. The potential of these agents as biological agents and their mechanisms of action have been the subject of discussion [21].

2. AgNP synthesis methods

Various techniques are utilized in the production of silver nanoparticles, encompassing physical, chemical, and biological methods. It is noteworthy that every approach possesses its own set of merits and demerits. In the process of biogenic synthesis of silver nanoparticles, the biological entity functions as a capping, reducing, or stabilizing agent, thereby facilitating the reduction of Ag^{+} to Ag^{0} [22]. In recent years, there has been a surge in the popularity of biological methods that rely on natural products derived from microorganisms and plants. This can be attributed to their low cost, high yields, and minimal toxicity to both the environment and human health [23]. The subsequent sections outline various techniques employed in the synthesis of silver nanoparticles.

2.1 Biological techniques

The manufacture of silver nanoparticles through physical and chemical means is a costly, protracted, and environmentally unfriendly process. Therefore, it is of utmost significance to devise a technique that is both ecologically and financially sustainable; the production method in question is devoid of hazardous chemicals [24] and circumvents the complications that arise from chemical and physical manufacturing techniques. Biological techniques serve to address these lacunae and possess diverse utilities in healthcare administration by means of controlling diverse biological processes. Biological production techniques encompass the utilization of microorganisms such as fungi, bacteria, and yeasts, as well as botanical sources. The popularity of utilizing nanoparticles in medical applications is largely attributed to the sources cited.

According to reports, the utilization of microorganisms and plants for nanoparticle production is a safe and cost-effective method that poses relatively lower environmental risks compared to chemical synthesis [25, 26]. Furthermore, it has been observed that microorganisms and plants possess the capability to assimilate and amass inorganic metallic ions from their ambient milieu [27]. The production of silver nanoparticles through biological means primarily entails the utilization of microorganisms and plant-based sources (**Figure 1**) [28].

2.2 Production in bacteria

A recent investigation was conducted to synthesize silver nanoparticles by means of reducing aqueous Ag + ions with the aid of culture supernatants derived from diverse bacterial strains. The expeditiousness of this methodology was exhibited, as the amalgamation of silver ions with the cellular filtrate resulted in the production of silver nanoparticles in a mere 5-minute timeframe. Additionally, the present study documented that piperine exhibited partial inhibition of the process of reducing Ag^{+} to metallic silver nanoparticles [29]. It is noteworthy that the nitro reduction activity exhibited by *Enterobacteriaceae* is impeded by the naturally occurring compound piperine. It is postulated that the process of bioreduction, which involves the conversion of silver ions to silver nanoparticles, may experience partial hindrance due to the presence of various strains of Enterobacteriaceae, including *Klebsiella pneumonia*.

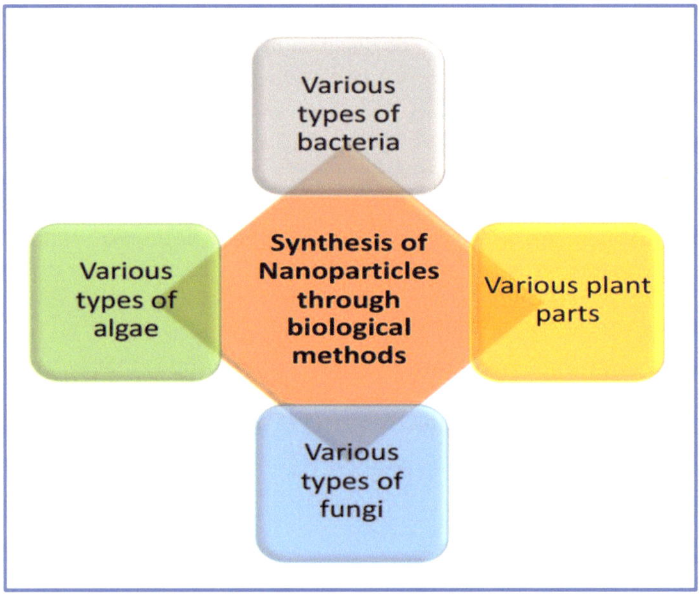

Figure 1.
Various biological approaches for the production of silver nanoparticles.

The optimization of silver nanoparticle production by *Lactobacillus casei* subspecies casei was investigated by Korbekandi et al. The study confirmed the bio-reductive synthesis of silver nanoparticles [30]. The study conducted by Liu and colleagues demonstrated the production of nanoparticles through the desiccation of *Bacillus megaterium* cells [31]. According to Das et al. a particular bacterial strain is used in the extracellular manufacture of silver nanoparticles. According to the study, the application of AgNO3 to Bacillus strain CS 11 led to the extracellular production of silver nanoparticles [32]. It was also reported that piperine exhibited partial inhibition of the reduction process of Ag + to form metallic silver nanoparticles [29]. It is interesting that the nitro reduction activity exhibited by Enterobacteriaceae is impeded by the naturally occurring compound piperitone. It is postulated that the process of bio-reduction, which involves the conversion of silver ions to silver nanoparticles, may experience partial inhibition due to the presence of various strains of Enterobacteriaceae, including *Klebsiella pneumoniae*. The optimization of silver nanoparticle production by *Lactobacillus casei* subspecies casei was investigated by Korbekandi et al. who confirmed the bio-reductive synthesis of silver nanoparticles [30]. The study conducted by Liu and colleagues demonstrated the generation of nanoparticles through the desiccation of *Bacillus megaterium* cells [31]. The extracellular production of silver nanoparticles by a bacterial strain has been described by Das et al. The research findings indicate that the application of $AgNO_3$ to Bacillus strain CS 11 led to the extracellular synthesis of silver nanoparticles [32].

2.3 Synthesis/production based on fungi

Diverse fungal species have been documented to participate in the synthesis of silver nanoparticles [33]. The rapidity of silver nanoparticle synthesis by fungi has been observed to be significant. The biosynthesis of silver nanoparticles by fungi has been a subject of study for numerous researchers [34]. A study has demonstrated the extracellular synthesis of spherical silver nanoparticles through the interaction between Fusarium solani and silver nitrate [35]. The biosynthesis of silver nanoparticles by the Humicola sp. has been reported by Syed et al. The study demonstrated that the reduction of a precursor solution occurred through the interaction between Humicola sp. and Ag + ions, resulting in the production of extracellular nanoparticles [36]. The production of silver nanoparticles through bio-reduction of silver nitrate, induced by the extract of *Pleurotus cornucopia*, has been reported by Owaid et al. [37]. The biosynthesis of silver nanoparticles with antifungal properties was investigated by Xue et al. through the utilization of Arthroderma fulvum in an experimental setting [38]. According to Vigneshwaran et al.'s findings, the fungus *Aspergillus flavus* exhibited the accumulation of silver nanoparticles on its cell wall surface as a result of its interaction with a silver nitrate solution [39]. Additionally, Bhainsa and D'Souza conducted a study on the extracellular synthesis of silver nanoparticles utilizing *Aspergillus fumigatus*. The findings suggest that the amalgamation of silver ions and the cell filtrate resulted in the production of silver nanoparticles within a brief duration [40]. The utilization of *Fusarium oxysporum* leads to the extracellular synthesis of silver nanoparticles, which exhibit a size range of 5–50 nm [41]. Furthermore, the introduction of a silver nitrate solution to *Phanerochaete chrysosporium* mycelium resulted in the production of silver nanoparticles [42]. The bio-reductive synthesis of silver nanoparticles was demonstrated by Korbekandi et al. through the utilization of *Fusarium oxysporum* [43].

2.4 Synthesis/production in algae

The aforementioned methodology presents a viable alternative to conventional physical and chemical techniques for synthesizing nanoparticles, owing to its cost-effectiveness and environmentally sustainable nature [44]. In addition, it has been observed that algae exhibit a significant ability to absorb metals. Observations have indicated that biological entities, such as marine algae, possess the ability to facilitate particular chemical reactions. The ability to perform this function is crucial for contemporary and pragmatic biosynthetic strategies [45]. According to a recent study utilizing algae extract, the alteration of hue from yellow to brown may serve as an indicator of the reduction of silver ions to silver nanoparticles. Furthermore, Rajesh Kumar et al. observed a significant deepening of the brown hue of silver nanoparticles after 32 hours of incubation. This finding suggests a positive correlation between the duration of incubation and the intensity of the observed coloration [46]. The present study reports the synthesis of silver nanoparticles via reduction of aqueous solutions of silver nitrate, utilizing powder and solvent extracts of *Padina pavonia*. Moreover, the nanoparticles obtained exhibited notable stability, rapid formation kinetics, and diminutive dimensions [47]. The production of silver nanoparticles was reported by Salari et al. via bio-reduction of silver ions, which was induced by Spirogyra variants [48].

2.5 Synthesis/production in virus

The utilization of viruses for the production of artificial nanocrystals, including but not limited to cadmium sulfide, silicon dioxide, ferrous oxide, and zinc sulfide, represents a distinctive methodology. The investigation of methods for generating semiconductor nanoparticles, including zinc sulfide and cadmium sulfide, is currently a topic of great interest in the electronics industry and the field of green chemistry. The employment of intact viruses in the synthesis of nanomaterials has been a topic of investigation for numerous years [49]. The extrinsic capsid protein of the virus serves a beneficial function in the creation of nanoparticles through the production of a metal ion-binding interface with notable reactivity [50]. At the exterior of the tobacco-mosaic virus [TMV], the total number of capsid proteins is 2130. Peptides possess the potential to serve as interlocking components for the construction or production of three-dimensional conduits intended for diverse medical applications [51]. The reduction in size of the synthesized nanoparticles was observed upon the addition of Au and Ag salts in moderate quantities to TMV, prior to the introduction of plant extracts from *Nicotiana benthamiana* or Hordeum vulgare. The augmented quantities of free nanoparticles observed at higher concentrations of tobacco mosaic virus (TMV) in comparison to the control group indicate a relatively modest production of the nanoparticles. The utilization of TMV as a biological template for the metallization of nanowires was also observed [52]. In contrast to a scenario without the virus, the existence of a pathogen not only resulted in a decrease in the length of biosynthesized nanoparticles but also significantly augmented their synthesis.

2.6 Synthesis/production in Actinomycetes

Actinomycetes exhibit the ability to produce both intracellular and extracellular nanoparticles through their metabolic processes. The process of intracellular synthesis takes place at the mycelia surface as a result of electrostatic attraction between Ag + ions and the negatively charged carboxylate groups found in the enzyme located on the mycelia cell wall. Subsequently, the enzymes present in the cellular wall facilitate the reduction of Ag + ions, leading to the formation of silver nuclei. The aggregation of silver nuclei results in the generation of silver nanoparticles at the nanoscale level [53]. The process of synthesizing gold nanoparticles was carried out using Rhodococcus sp., which is a type of alkalotolerant Actinomycetes. The presence of nanoparticles on the Actinomycetes' cell walls was confirmed by transmission electron microscopy (TEM) images, indicating that the nanoparticles were synthesized intracellularly [54]. It has been revealed that Rhodococcus NCIM 2891 can be used for the intracellular manufacture of silver nanoparticles [55]. After 72 hours of contact with HAuCl4, the previously yellow biomass of *Streptomyces hygroscopicus* changed pink, indicating the production of extracellular gold nanoparticles [56]. The production of silver, manganese, and zinc nanoparticles by Streptomyces sp. HBUM171191 was observed upon exposure of the wet biomass to the respective metal solutions. The alteration in hue of the biomass, transitioning from a light-yellow shade to brown, dark yellow, and dark yellow, respectively, is indicative of the synthesis of silver, manganese, and zinc nanoparticles [57]. The enzymatic processes associated with the nitrogen cycle have been identified as a potential mechanism for the extracellular biosynthesis of nanoparticles. It is possible that they could assume responsibility for the enzymatic reduction of metals through electron shuttle [58]. Thea-NADH-dependent nitrate reductase is a significant contributor to the process of reducing Ag + ions to silver nuclei. Actinomycetes were responsible for mediating the extracellular creation of nanoparticles by the use of Streptomyces glaucus71MD [59]. Within 12 hours, a pure culture of Streptomyces sp. ERI-3 was able to convert a colorless solution of silver nitrate to a color that was more reddish-brown [60], indicating the creation of silver nanoparticles.

3. Biological applications of AgNPs

Because of their one-of-a-kind qualities, silver nanoparticles (AgNPs) have found widespread use in a variety of fields, including the healthcare sector, the food storage business, as well as environmental and medicinal applications. There have been a number of reviews and book chapters written on the subject of the application of AgNPs in a variety of different fields. Antibacterial, antifungal, antiviral, antiinflammatory, anticancer, and antiangiogenic treatments are only a few of the biological and medicinal applications that we would like to highlight here using silver nanoparticles (AgNPs). In this article, we focused specifically on previously published seminal works, and we concluded with more current revisions. **Figure 2** provides a diagrammatic representation of a number of different applications utilizing AgNPs.

3.1 Anti-bacterial activity of AgNPs

Silver nanoparticles (AgNPs) have emerged as a potential substitute for antibiotics in combating bacterial infections, owing to their capacity to surmount the bacterial resistance that has developed against conventional antibiotics (**Figure 3**). Consequently, the development of silver nanoparticles (AgNPs) as antibacterial agents is deemed imperative. AgNPs exhibit potential as antibacterial agents owing to their crystallographic surface structure and large surface-to-volume ratios, among other promising nanomaterials. Sondi and Salopek-Sondi's seminal study, as documented in [61], showcased the efficacy of AgNPs in combating *Escherichia coli*. The study revealed that the treatment of *E. coli* cells with AgNPs resulted in the accumulation of AgNPs in the cell wall and the formation of "pits" in the bacterial cell walls, ultimately leading to cell death. Additionally, the study demonstrated the antimicrobial properties of AgNPs. The antibacterial activity of smaller particles with a greater surface-to-volume ratio was found to be more efficient than that of larger particles in the identical *E. coli* strain [62]. Moreover, it should be noted that the antibacterial efficacy of silver nanoparticles (AgNPs) is contingent not only on their size but also on their shape [63]. Silver nanoparticles (AgNPs) were produced using four distinct saccharides, resulting in an average size of 25 nm. The AgNPs exhibited notable antimicrobial and bactericidal properties against both Gram-positive and Gram-negative bacteria, including methicillin-resistant *Staphylococcus aureus* and other highly resistant strains. As

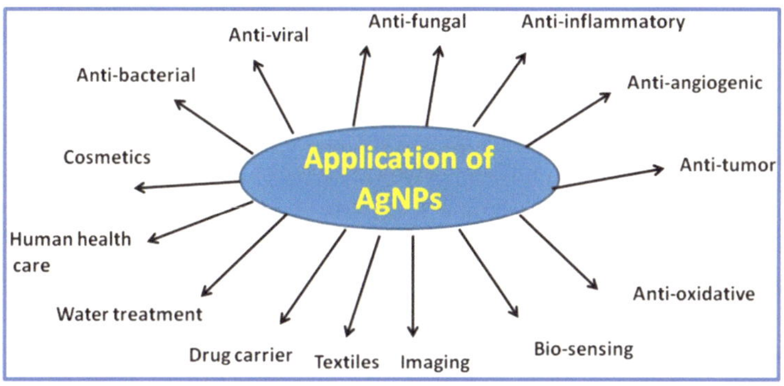

Figure 2.
AgNPs have a wide range of potential applications.

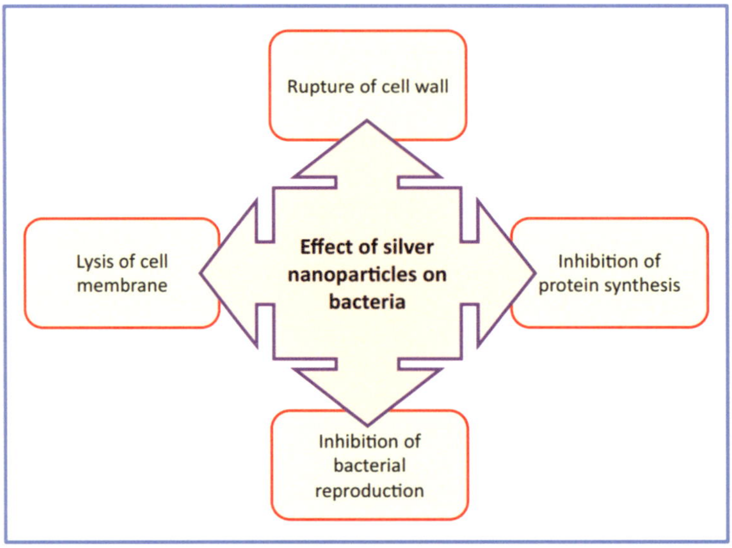

Figure 3.
A schematic representation of the mechanisms underlying the antibacterial activity of silver nanoparticles.

previously noted, the efficiency of AgNPs is determined not only by their size but also by their shape, as they exhibit a shape-dependent interaction with the Gram-negative organism *E. coli* [64]. Additionally, a comprehensive investigation was conducted to assess the efficacy of silver nanoparticles (AgNPs) in combating yeast, *E. coli*, and *Staphylococcus aureus* antimicrobial activity. The findings indicate that yeast and *E. coli* exhibited complete growth inhibition at low concentrations of AgNPs, whereas a minor impact was observed in *S. aureus* [65]. The present study assessed AgNPs that were biologically synthesized from the culture supernatants of *Klebsiella pneumoniae*. The impact of Ag-NPs on the efficacy of different antibiotics, including penicillin G, amoxicillin, erythromycin, clindamycin, and vancomycin, was evaluated against *Staphylococcus aureus* and *E. coli*. The results indicated that the presence of Ag-NPs led to an increase in the effectiveness of the aforementioned antibiotics [66]. In comparison to silver nanoparticles, hydrogel-silver nanocomposites exhibited remarkable antibacterial efficacy against *Escherichia coli*. The composite of chitosan-Ag-nanoparticle, synthesized in a single reaction vessel, exhibited superior antimicrobial properties compared to its constituent parts at equivalent concentrations. This can be attributed to the preferential formation of small AgNPs bound to the polymer during the one-pot synthesis process, which allows for dispersion in media with a pH of 6.3 or lower [67]. The utilization of culture supernatants of *Staphylococcus aureus* for the biological synthesis of AgNPs resulted in noteworthy antimicrobial activity against methicillin-resistant *S. aureus*, methicillin-resistant *Staphylococcus epidermidis*, and *Streptococcus pyogenes*. However, the antimicrobial activity against *Salmonella typhi* and *Klebsiella pneumoniae* was only moderate [68]. The present study investigated the cellular mechanisms underlying the induction of cell death by AgNPs in *E. coli*. Specifically, the study examined the leakage of reducing sugars and proteins as indicators of cell death. Moreover, it has been observed that AgNPs possess the ability to disrupt the permeability of bacterial membranes by creating numerous pits and gaps. This suggests that AgNPs have the potential to impair the structural integrity of bacterial cell membranes [69]. The AgCHX complex consisting of silver nanocrystalline and chlorhexidine, exhibited potent antibacterial efficacy against a range of Gram-positive/negative bacterial strains and methicillin-resistant *Staphylococcus aureus* (MRSA) strains. The results indicate that the nanocrystalline Ag (III)CHX exhibited significantly lower minimal inhibitory concentrations (MICs) compared to the ligand (CHX), AgNO3, and the established benchmark. Silver sulfadiazine is a topical antimicrobial agent commonly used in the treatment of burns and other skin injuries [70].

3.2 Antiviral activity of AgNPs

Nanoparticles present a viable substitute to pharmaceuticals in the management and regulation of viral pathogen proliferation. The process of biosynthesizing silver nanoparticles has the potential to yield highly effective antiviral agents that can effectively impede the functions of viruses. The bio-silver nanoparticles were investigated by Suriyakalaa et al. for their anti-HIV properties during the initial phase of the reverse transcription process, yielding compelling results [71]. Metallic nanoparticles synthesized through biosynthesis possess numerous binding sites for gp120 present in the viral membrane, thereby regulating the virus's functionality. According to another study, bio-based nanoparticles have been found to be efficient virucidal agents against both free HIV and cell-associated virus [72]. Studies have shown that silver nanoparticles exhibit antiviral properties against HIV-1 at concentrations that do not cause harm to cells. The present study aimed to assess the mechanism of antiviral activity of silver nanoparticles against HIV-1 through a range of in vitro assays [73]. According to a different study, the monkeypox virus was resistant to the antiviral effects of silver nanoparticles with or without a polysaccharide covering. The present investigation has revealed that silver nanoparticles exhibit significant inhibitory effects against monkeypox virus infection *in vitro* [74].

Pre-exposure of Tacaribe virus to silver nanoparticles was found to enhance the virus's cellular uptake in host cells. The observation was made that the introduction of silver treatment resulted in a noteworthy decrease in the production of viral RNA. This discovery serves as evidence that silver nanoparticles possess the ability to impede the infection of arenavirus *in vitro* [75]. According to a recent study, it was found that out of the three types of silver nanoparticle-MHCs that were tested, Ag30-MHCs exhibited the most effective viral inactivation [76].

3.3 Antifungal activity of AgNPs

The antifungal properties of silver nanoparticles have been demonstrated against various fungal species [77, 78]; however, the underlying mechanism remains incompletely comprehended. The presence of silver nanoparticles has been observed to disrupt the integrity of the cellular membrane structure. The suggested mechanism for the antifungal activity of silver nanoparticles against *Candida albicans* species involves the inhibition of the budding process and damage to the membrane integrity [79]. The present investigation employed nano-Ag sepiolite fibers that contained monodispersed silver nanoparticles as the silver source to examine their antibacterial and antifungal properties. Soda with a low melting point. The incorporation of nanoparticles into lime glass powder resulted in favorable antibacterial and antifungal properties [80]. According to a study, the combination of fluconazole and silver nanoparticles exhibited the most significant inhibition against *Candida albicans*. The present investigation employed *Alternaria alternata* fungus for the purpose of extracellular biosynthesis of silver nanoparticles [81]. The study determined that the growth of fungi was significantly reduced by the presence of silver nanoparticles at concentrations ranging from 30 to 200 mg/L [82]. Additionally, the supernatant of the GP-23 strain was utilized in the production of silver nanoparticles, which exhibited potent antifungal properties [83]. The utilization of *Trichoderma harzianum* cell filtrate was employed in the synthesis of silver nanoparticles, yielding their production in a mere 3 hours. Subsequent analysis via TEM revealed the presence of both ellipsoid and spherical nanoparticles, with a size range spanning from 19 to 63 nm and an average size of 34.77 nm [84]. According to Jalal et al.'s findings through transmission electron microscopy analysis, the application of silver nanoparticles on Candida cells led to a significant distortion of the cellular structure. Moreover, the augmentation of cell contraction was observed as a result of the interaction between nanoparticles and the fungal cell wall and membrane. The observed outcome was the disruption of the cellular membrane structure, which impeded the typical budding process as a result of compromised membrane integrity and damage [85]. In their study, Jalal et al. demonstrated the antimicrobial properties of silver nanoparticles derived from Syzygium cumini against Candida species. The authors concluded that these nanoparticles possess the ability to inhibit the proliferation, germ tube, and biofilm formation, as well as the secretion of hydrolytic enzymes by Candida species [86].

3.4 Antiparasitic action of AgNPs

The larvicidal properties of silver nanoparticles against Aedes aegypti [87] and Culex quinquefasciatus, which are dengue vectors, have been identified. A study was conducted by Allahverdiyev et al. to assess the impact of silver nanoparticles on the biological parameters of Leishmania tropica. The findings of this investigation have substantiated the antileishmanial properties of silver nanoparticles, which can be attributed to their ability to impede the proliferation activity of promastigotes. Additionally, it was observed that silver nanoparticles exhibited the ability to impede the viability of amastigotes within host cells, and this phenomenon was augmented by the existence of ultraviolet radiation [88]. The antiparasitic activity of silver and copper nanoparticles synthesized by Saad et al. was investigated. The results indicated that the viability of *Cryptosporidium parvum* oocysts was significantly reduced by silver nanoparticles. The results indicate that silver nanoparticles exhibited notable efficacy and safety in combatting parasitic infections caused by *Entamoeba histolytica* and *Cryptosporidium parvum* [89].

3.5 Antifouling action of AgNPs

Biofouling represents a significant obstacle encountered by the water industry and public health. The efficacy of silver nanoparticles derived from the *Rhizopus oryzae* fungal species has been evaluated for the remediation of water contaminated with pollutants. The utilization of *Lactobacillus fermentum* cells in the production of silver nanoparticles has been observed to effectively regulate biofilm formation. Additionally, the antifouling characteristics of these nanoparticles have been verified. In addition, silver nanoparticles have been utilized in various environmental applications, including but not limited to air, water, and surface disinfection [90]. A recent investigation has indicated that the direct application of silver nanoparticle coatings onto ecologically sound surfaces can lead to a proficient control of biofouling [91].

3.6 Antibiofilm activity

In contemporary times, silver nanoparticles have emerged as a potential agent for impeding biofilm formation. However, the precise mechanism underlying the inhibitory effect of silver nanoparticles remains elusive. The classification of antibiofilm strategies was conducted by Chen et al. who identified two distinct categories: (i) interventions that specifically impede the formation of biofilms and (ii) the utilization of modified biomaterials in biomedical devices to prevent and resist biofilm formation [92]. Prior studies have corroborated novel methodologies for surface modification of biomedical apparatuses with the aim of impeding microbial attachment, adhesion, and proliferation [93]. The present investigation examined the antibiofilm efficacy of silver nanoparticles against multidrug-resistant Gram-negative bacterial isolates, and it was found that they successfully inhibited biofilm formation [94]. Martinez-Gutierrez et al. drew the conclusion that silver nanoparticles effectively hindered the formation of biofilms and exhibited bactericidal properties against established biofilms based on their research findings [95].

A study was conducted by Palanisamy et al. to investigate the impact of silver nanoparticles on biofilm formation. The study exhibited that the silver nanoparticles effectively hindered the development of biofilms in antibiotic-resistant strains [96]. A recent study was conducted to assess the interaction between silver nanoparticles and Pseudomonas putida biofilms. The study demonstrated that the utilization of silver nanoparticles effectively inhibited the formation of biofilms [97]. The antibiofilm activity of silver nanoparticles against biofilms created by Pseudomonas aeruginosa and Staphylococcus epidermidis was examined by Kalishwaralal et al. The application of silver nanoparticles on these organisms resulted in the suppression of biofilm formation [98]. Mohanty et al. conducted a study to assess the antibacterial efficacy of silver nanoparticles against a range of human pathogens.

4. Conclusion

The broad spectrum of applications of silver nanoparticles, including their use as antimicrobial and antitumor agents, as well as in food packaging, agriculture, and healthcare, positions them as a crucial component in health management. Moreover, it is widely recognized that the majority of practical applications of antibiotics demonstrate resistance, resulting in a lack of efficacy. Therefore, the presence of bacteria that form biofilms poses a significant issue. The issue of antibiotic resistance has garnered heightened global attention, leading to a focus on alternative treatment approaches. The potential employment of silver nanoparticles and the surface coating or impregnation of nanomaterials are among the alternative strategies that can be utilized as antibiofilm agents. Furthermore, silver nanoparticles have been extensively researched and employed in the treatment of diverse ailments, encompassing cancer, wound healing, dental implants, and other therapeutic interventions that involve the modulation of biological activities. Through increased comprehension and advanced technological capabilities, the utilization of these innovative particles within the medical field will establish a standardized framework for preventing and treating multidrug resistance and biofilm pathogens.

Conflict of interest

The authors state no conflict of interest.

Author details

Krishna Gudikandula[1*], Trimothi Dasari[1] and Reeja Sundaram[2]

1 Department of Microbiology, Kakatiya University, Warangal, Telangana, India

2 Department of Tree Breeding and Improvement, Forest College and Research Institute (FCRI), Hyderabad at Mulugu, Telangana, India

*Address all correspondence to: krishnagudikandula@gmail.com

References

[1] Husen A, Siddiqi KS. Phytosynthesis of nanoparticles: Concept, controversy and application. Nanoscale Research Letters. 2014;**9**:1-24

[2] Husen A, Siddiqi KS. Plants and microbes assisted selenium nanoparticles: Characterization and application. Journal of Nanobiotechnology. 2014;**12**(1):1

[3] Siddiqi KS, Husen A. Green synthesis, characterization and uses of palladium/platinum nanoparticles. Nanoscale Research Letters. 2016;**11**(1):1-3

[4] Siddiqi KS, Rahman A, Husen A. Biogenic fabrication of iron/iron oxide nanoparticles and their application. Nanoscale Research Letters. 2016;**11**:1-3

[5] Wei L, Lu J, Xu H, Patel A, Chen ZS, Chen G. Silver nanoparticles: Synthesis, properties, and therapeutic applications. Drug Discovery Today. 2015;**20**(5):595-601

[6] Lara HH, Garza-Treviño EN, Ixtepan-Turrent L, Singh DK. Silver nanoparticles are broad-spectrum bactericidal and virucidal compounds. Journal of Nanobiotechnology. 2011;**9**:1-8

[7] Siddiqi KS, Husen A. Fabrication of metal nanoparticles from fungi and metal salts: Scope and application. Nanoscale Research Letters. 2015;**11**:1-5

[8] Zaheer Z. Silver nanoparticles to self-assembled films: Green synthesis and characterization. Colloids and Surfaces B: Biointerfaces. 2012;**90**:48-52

[9] Lokina S, Stephen A, Kaviyarasan V, Arulvasu C, Narayanan V. Cytotoxicity and antimicrobial activities of green synthesized silver nanoparticles. European Journal of Medicinal Chemistry. 2014;**76**:256-263

[10] Siddiqi KS, Husen A. Fabrication of metal and metal oxide nanoparticles by algae and their toxic efects. Nanoscale Research Letters. 2016;**11**:363

[11] Ahamed M, AlSalhi MS, Siddiqui MK. Silver nanoparticle applications and human health. Clinica Chimica Acta. 2010;**411**(23-24):1841-1848

[12] Chen X, Schluesener HJ. Nanosilver: A nanoproduct in medical application. Toxicology Letters. 2008;**176**(1):1-2

[13] Jones SA, Bowler PG, Walker M, Parsons D. Controlling wound bioburden with a novel silver-containing Hydrofiber® dressing. Wound Repair and Regeneration. 2004;**12**(3):288-294

[14] Silver S, Phung LT. Bacterial heavy metal resistance: New surprises. Annual Review of Microbiology. 1996;**50**(1):753-789

[15] Catauro MI, Raucci MG, De Gaetano F, Marotta A. Antibacterial and bioactive silver-containing Na2O·CaO 2SiO2 glass prepared by sol-gel method. Journal of Materials Science: Materials in Medicine. 2004;**15**(7):831

[16] Crabtree JH, Burchette RJ, Siddiqi RA, Huen IT, Hadnott LL, Fishman A. The efficacy of silver-ion implanted catheters in reducing peritoneal dialysis-related infections. Peritoneal Dialysis International. 2003;**23**(4):368-374

[17] Nakhjavani M, Nikkhah V, Sarafraz MM, Shoja S, Sarafraz M. Green synthesis of silver nanoparticles using green tea leaves: Experimental study

on the morphological, rheological and antibacterial behaviour. Heat and Mass Transfer. 2017;**53**:3201-3209

[18] Aramwit P, Bang N, Ratanavaraporn J, Ekgasit S. Green synthesis of silk sericin-capped silver nanoparticles and their potent anti-bacterial activity. Nanoscale Research Letters. 2014;**9**(1):1-7

[19] Vigneshwaran N, Nachane RP, Balasubramanya RH, Varadarajan PV. A novel one-pot 'green'synthesis of stable silver nanoparticles using soluble starch. Carbohydrate Research. 2006;**341**(12):2012-2018

[20] Shin Y, Bae IT, Exarhos GJ. "Green" approach for self-assembly of platinum nanoparticles into nanowires in aqueous glucose solutions. Colloids and Surfaces A: Physicochemical and Engineering Aspects. 2009;**348**(1-3):191-195

[21] Naikoo GA, Mustaqeem M, Hassan IU, Awan T, Arshad F, Salim H, et al. Bioinspired and green synthesis of nanoparticles from plant extracts with antiviral and antimicrobial properties: A critical review. Journal of Saudi Chemical Society. 2021;**25**(9):101304

[22] Zewde B, Ambaye A, Stubbs Iii J, Raghavan D. A review of stabilized silver nanoparticles–synthesis, biological properties, characterization, and potential areas of applications. Nanomed. 2016;**4**(1043):1-4

[23] Shanmuganathan R, Karuppusamy I, SaravananM,MuthukumarH,PonnuchamyK, Ramkumar VS, et al. Synthesis of silver nanoparticles and their biomedical applications-a comprehensive review. Current Pharmaceutical Design. 2019;**25**(24):2650-2660

[24] Iravani S. Bacteria in nanoparticle synthesis: Current status and future prospects. International Scholarly Research Notices. 29 Oct 2014;**2014**:359316

[25] Makarov VV, Love AJ, Sinitsyna OV, Makarova SS, Yaminsky I, Taliansky ME, et al. Green nanotechnologies: Synthesis of metal nanoparticles using plants. Acta Nature. 2014;**6**(1):35-44

[26] Gowramma B, Keerthi U, Rafi M, Rao DM. Biogenic silver nanoparticles production and characterization from native stain of Corynebacterium species and its antimicrobial activity. Apr 2015;**5**(2):195-201. DOI: 10.1007/s13205-014-0210-4. Epub 2014 Apr 8. PMID: 28324578; PMCID: PMC4362738

[27] Shah M, Fawcett D, Sharma S, Tripathy SK, Eddy G, Poinern J. Green synthesis of metallic nanoparticles via biological entities. Materials (Basel). 2015, 2015;**8**(11):7278-7308

[28] Tran QH, Le AT. Silver nanoparticles: Synthesis, properties, toxicology, applications and perspectives. Advances in Natural Sciences: Nanoscience and Nanotechnology. 2013;**4**(3):033001

[29] Shahverdi R, Minaeian S, Shahverdi HR, Jamalifar H, Nohi AA. Biosynthesis and application of silver and gold nanoparticles. Process Biochemistry. 2007;**42**:919-923

[30] Korbekandi H, Iravani S, Abbasi S. Optimization of biological synthesis of silver nanoparticles using lactobacillus casei subsp. casei. Journal of Chemical Technology & Biotechnology. 2012;**87**(7):932-937

[31] Liu Y, Fu J, Chen P, Yu X, Yang P. Studies on biosorption of Au3+ by bacillus megaterium. Wei Sheng wu xue bao=Acta Microbiologica Sinica. 2000;**40**(4):425-429

[32] Das VL, Thomas R, Varghese RT, Soniya EV, Mathew J, Radhakrishnan EK. Extracellular synthesis of silver nanoparticles by the bacillus strain CS 11 isolated from industrialized area. 3 Biotech. 2014;**4**:121-126

[33] Casagrande MG, De Lima R. Synthesis of Silver nanoparticles mediated by fungi: A review. Frontiers of Bioengineering & Biotechnology. 2019;**7**:287

[34] Iravani S. Bacteria in nanoparticle synthesis: Current status and future prospects. International scholarly research notices. Journal of Nanomaterials. 2014;**2014**:359316

[35] Ingle A, Rai M, Gade A, Bawaskar M. Fusarium solani: A novel biological agent for the extracellular synthesis of silver nanoparticles. Journal of Nanoparticle Research. 2009;**11**:2079-2085

[36] Syed A, Saraswati S, Kundu GC, Ahmad A. Biological synthesis of silver nanoparticles using the fungus Humicola sp. and evaluation of their cytoxicity using normal and cancer cell lines. Spectrochimica Acta Part A: Molecular and Biomolecular Spectroscopy. 2013;**114**:144-147

[37] Owaid MN, Raman J, Lakshmanan H, Al-Saeedi SS, Sabaratnam V, Abed IA. Mycosynthesis of silver nanoparticles by Pleurotus cornucopiae var. citrinopileatus and its inhibitory effects against Candida sp. Materials Letters. 2015;**153**:186-190

[38] Xue B, He D, Gao S, Wang D, Yokoyama K, Wang L. Biosynthesis of silver nanoparticles by the fungus Arthroderma fulvum and its antifungal activity against genera of Candida, aspergillus and Fusarium. International Journal of Nanomedicine. 2016;**11**:1899

[39] Vigneshwaran N, Ashtaputre NM, Varadarajan PV, Nachane RP, Paralikar KM, Balasubramanya RH. Biological synthesis of silver nanoparticles using the fungus aspergillus flavus. Materials Letters. 2007;**61**(6):1413-1418

[40] KC B. D'Souza SF. Extracellular biosynthesis of silver nanoparticles using the fungus aspergillus fumigates. Colloids and Surfaces. B, Biointerfaces 2006;47(2):160-164.

[41] Ahmad A, Mukherjee P, Senapati S, Mandal D, Khan MI, Kumar R, et al. Extracellular biosynthesis of silver nanoparticles using the fungus Fusarium oxysporum. Colloids and Surfaces B: Biointerfaces. 2003;**28**(4):313-318

[42] Vigneshwaran N, Kathe AA, Varadarajan PV, Nachane RP, Balasubramanya RH. Biomimetics of silver nanoparticles by white rot fungus, Phaenerochaete chrysosporium. Colloids and Surfaces B: Biointerfaces. 2006;**53**(1):55-59

[43] Korbekandi H, Ashari Z, Iravani S, Abbasi S. Optimization of biological synthesis of silver nanoparticles using Fusarium oxysporum. Iranian Journal of Pharmaceutical Research: IJPR. 2013;**12**(3):289

[44] Hamouda RA, Hussein MH, Abo-Elmagd RA, Bawazir SS. Synthesis and biological characterization of silver nanoparticles derived from the cyanobacterium Oscillatoria limnetica. Scientific Reports. 2019;**9**(1):1-7

[45] Govindaraju K, Kiruthiga V, Kumar VG, Singaravelu G. Extracellular synthesis of silver nanoparticles by a marine alga, Sargassum wightii Grevilli and their antibacterial effects. Journal of Nanoscience and Nanotechnology. 2009;**9**(9):5497-5501

[46] Diamond DM, Campbell AM, Park CR, Halonen J, Zoladz PR. The temporal dynamics model of emotional

memory processing: A synthesis on the neurobiological basis of stress-induced amnesia, flashbulb and traumatic memories, and the Yerkes-Dodson law. Neural Plasticity. 2007;**2007**:60803. DOI: 10.1155/2007/60803. PMID: 17641736; PMCID: PMC1906714

[47] Obaid AY, Al-Thabaiti SA, El-Mossalamy EH, Al-Harbi LM, Khan Z. Extracellular bio-synthesis of silver nanoparticles. Arabian Journal of Chemistry. 2017;**10**(2):226-231

[48] Salari Z, Danafar F, Dabaghi S, Ataei SA. Sustainable synthesis of silver nanoparticles using macroalgae Spirogyra varians and analysis of their antibacterial activity. Journal of Saudi Chemical Society. 2016;**20**(4):459-464

[49] Zeng Q, Wen H, Wen Q, Chen X, Wang Y, Xuan W, et al. Cucumber mosaic virus as drug delivery vehicle for doxorubicin. Biomaterials. 2013;**34**(19):4632-4642

[50] Pokorski JK, Steinmetz NF. The art of engineering viral nanoparticles. Molecular Pharmaceutics. 2011;**8**(1):29-43

[51] Royston E, Ghosh A, Kofinas P, Harris MT, Culver JN. Self-assembly of virus-structured high surface area nanomaterials and their application as battery electrodes. Langmuir. 2008;**24**(3):906-912

[52] Love AJ, Makarov V, Yaminsky I, Kalinina NO, Taliansky ME. The use of tobacco mosaic virus and cowpea mosaic virus for the production of novel metal nanomaterials. Virology. 2014;**449**:133-139

[53] Abdeen S, Isaac RR, Geo S, Sornalekshmi S, Rose A, Praseetha PK. Evaluation of antimicrobial activity of biosynthesized iron and Silver nanoparticles using the fungi Fusarium Oxysporum and Actinomycetes sp. on human pathogens. Nano Biomedicine & Engineering. 2013;**5**(1):39-45

[54] Ahmad A, Senapati S, Khan MI, Kumar R, Ramani R, Srinivas V, et al. Intracellular synthesis of gold nanoparticles by a novel alkalotolerant actinomycete, Rhodococcus species. Nanotechnology. 2003;**14**(7):824

[55] Otari SV, Patil RM, Nadaf NH, Ghosh SJ, Pawar SH. Green biosynthesis of silver nanoparticles from an actinobacteria Rhodococcus sp. Materials Letters. 2012;**72**:92-94

[56] Waghmare SS, Deshmukh AM, Sadowski Z. Biosynthesis, optimization, purification and characterization of gold nanoparticles. African Journal of Microbiology Research. 2014;**8**(2):138-146

[57] Waghmare SS, Deshmukh AM, Kulkarni SW, Oswaldo LA. Biosynthesis and characterization of manganese and zinc nanoparticles. Universal Journal of Environmental Research & Technology. 2011;**1**(1):6303-6310

[58] Karthik L, Kumar G, Kirthi AV, Rahuman AA, Bhaskara Rao KV. Streptomyces sp. LK3 mediated synthesis of silver nanoparticles and its biomedical application. Bioprocess and Biosystems Engineering. 2014;**37**:261-267

[59] Tsibakhashvili NY, Kirkesali EI, Pataraya DT, Gurielidze MA, Kalabegishvili TL, Gvarjaladze DN, et al. Microbial synthesis of silver nanoparticles by Streptomyces glaucus and Spirulina platensis. Advanced Science Letters. 2011;**4**(11-12):3408-3417

[60] Sondi I, Salopek-Sondi B. Silver nanoparticles as antimicrobial agent: A case study on E. coli as a model

for gram-negative bacteria. Journal of Colloid and Interface Science. 2004;**275**(1):177-182

[61] Baker C, Pradhan A, Pakstis L, Pochan DJ, Shah SI. Synthesis and antibacterial properties of silver nanoparticles. Journal of Nanoscience and Nanotechnology. 2005;**5**(2):244-249

[62] Morones JR, Elechiguerra JL, Camacho A, Holt K, Kouri JB, Ramírez JT, et al. The bactericidal effect of silver nanoparticles. Nanotechnology. 2005;**16**(10):2346

[63] Tak YK, Pal S, Naoghare PK, Rangasamy S, Song JM. Shape-dependent skin penetration of silver nanoparticles: Does it really matter? Scientific Reports. 2015;**5**(1):1

[64] Kim JS, Kuk E, Yu KN, Kim JH, Park SJ, Lee HJ, et al. Antimicrobial effects of silver nanoparticles. Nanomedicine: Nanotechnology, Biology and Medicine. 2007;**3**(1):95-101

[65] Shahverdi AR, Fakhimi A, Shahverdi HR, Minaian S. Synthesis and effect of silver nanoparticles on the antibacterial activity of different antibiotics against Staphylococcus aureus and Escherichia coli. Nanomedicine: Nanotechnology, Biology and Medicine. 2007;**3**(2):168-171

[66] Sanpui P, Murugadoss A, Prasad PD, Ghosh SS, Chattopadhyay A. The antibacterial properties of a novel chitosan–Ag-nanoparticle composite. International Journal of Food Microbiology. 2008;**124**(2):142-146

[67] Nanda A, Saravanan M. Biosynthesis of silver nanoparticles from Staphylococcus aureus and its antimicrobial activity against MRSA and MRSE. Nanomedicine: Nanotechnology, Biology and Medicine. 2009;**5**(4):452-456

[68] Li WR, Xie XB, Shi QS, Zeng HY, Ou-Yang YS, Chen YB. Antibacterial activity and mechanism of silver nanoparticles on Escherichia coli. Applied Microbiology and Biotechnology. 2010;**85**:1115-1122

[69] Pal S, Yoon EJ, Tak YK, Choi EC, Song JM. Synthesis of highly antibacterial nanocrystalline trivalent silver polydiguanide. Journal of the American Chemical Society. 2009;**131**(44):16147-16155

[70] Alam MA, Al RK. Shear strengthening of reinforced concrete beam using natural fibre reinforced polymer laminates. Construction and Building Materials. 2018;**162**:683-696

[71] Sun RW, Chen R, Chung NP, Ho CM, Lin CL, Che CM. Silver nanoparticles fabricated in Hepes buffer exhibit cytoprotective activities toward HIV-1 infected cells. Chemical Communications. 2005;**40**:5059-5061

[72] Walter JG, Petersen S, Stahl F, Scheper T, Barcikowski S. Laser ablation-based one-step generation and bio-functionalization of gold nanoparticles conjugated with aptamers. Journal of Nanobiotechnology. 2010;**8**(1):1

[73] Rogers JV, Parkinson CV, Choi YW, Speshock JL, Hussain SM. A preliminary assessment of silver nanoparticle inhibition of monkeypox virus plaque formation. Nanoscale Research Letters. 2008;**3**(4):129-133

[74] Speshock JL, Murdock RC, Braydich-Stolle LK, Schrand AM, Hussain SM. Interaction of silver nanoparticles with Tacaribe virus. Journal of Nanobiotechnology. 2010;**8**(1):1-9

[75] Kim JS, Kuk E, Yu KN, Kim JH, Park SJ, Lee HJ, et al. Antimicrobial

effects of silver nanoparticles. Nanomedicine. Mar 2007;**3**(1):95-101. DOI: 10.1016/j.nano.2006.12.001. Erratum in: Nanomedicine. 2014 Jul;10(5):e1119. PMID: 17379174

[76] Bankar A, Joshi B, Kumar AR, Zinjarde S. Banana peel extract mediated novel route for the synthesis of silver nanoparticles. Colloids and Surfaces A: Physicochemical and Engineering Aspects. 2010;**368**(1-3):58-63

[77] Raut RW, Mendhulkar VD, Kashid SB. Photosensitized synthesis of silver nanoparticles using Withania somnifera leaf powder and silver nitrate. Journal of Photochemistry and Photobiology B: Biology. 2014;**132**:45-55

[78] Kim KJ, Sung WS, Suh BK, Moon SK, Choi JS, Kim JG, et al. Antifungal activity and mode of action of silver nano-particles on Candida albicans. Biometals. 2009;**22**:235-242

[79] Esteban-Tejeda L, Malpartida F, Esteban-Cubillo A, Pecharromán C, Moya JS. The antibacterial and antifungal activity of a soda-lime glass containing silver nanoparticles. Nanotechnology. 2009;**20**(8):085103

[80] Gajbhiye M, Kesharwani J, Ingle A, Gade A, Rai M. Fungus-mediated synthesis of silver nanoparticles and their activity against pathogenic fungi in combination with fluconazole. Nanomedicine: Nanotechnology, Biology and Medicine. 2009;**5**(4):382-386

[81] Ogar A, Tylko G, Turnau K. Antifungal properties of silver nanoparticles against indoor mould growth. Science of the Total Environment. 2015;**521**:305-314

[82] El-Boraey HA, El-Gammal OA. New 15-membered tetraaza (N4) macrocyclic ligand and its transition metal complexes: Spectral, magnetic, thermal and anticancer activity. Spectrochimica Acta Part A: Molecular and Biomolecular Spectroscopy. 2015;**138**:553-562

[83] Abdelghany TM, Al-Rajhi AM, Al Abboud MA, Alawlaqi MM, Ganash Magdah A, Helmy EA, et al. Recent advances in green synthesis of silver nanoparticles and their applications: About future directions. A review. BioNanoScience. 2018;**8**:5-16

[84] Jalal M, Ansari MA, Alzohairy MA, Ali SG, Khan HM, Almatroudi A, et al. Biosynthesis of silver nanoparticles from oropharyngeal Candida glabrata isolates and their antimicrobial activity against clinical strains of bacteria and fungi. Nanomaterials. 2018;**8**(8):586

[85] Jalal M, Ansari MA, Alzohairy MA, Ali SG, Khan HM, Almatroudi A, et al. Anticandidal activity of biosynthesized silver nanoparticles: Effect on growth, cell morphology, and key virulence attributes of Candida species. International Journal of Nanomedicine. 2019;**28**:4667-4679

[86] Suresh G, Gunasekar PH, Kokila D, Prabhu D, Dinesh D, Ravichandran N, et al. Green synthesis of silver nanoparticles using Delphinium denudatum root extract exhibits antibacterial and mosquito larvicidal activities. Spectrochimica Acta Part A: Molecular and Biomolecular Spectroscopy. 2014;**127**:61-66

[87] Allahverdiyev AM, Abamor ES, Bagirova M, Ustundag CB, Kaya C, Kaya F, et al. Antileishmanial effect of silver nanoparticles and their enhanced antiparasitic activity under ultraviolet light. International Journal of Nanomedicine. 2011;**6**:2705-2714. DOI: 10.2147/IJN.S23883. Epub 2011 Nov 3. PMID: 22114501; PMCID: PMC3218584

[88] Saad AH, Soliman MI, Azzam AM, Mostafa AB. Antiparasitic activity of silver and copper oxide nanoparticles against Entamoeba histolytica and Cryptosporidium parvum cysts. Journal of the Egyptian Society of Parasitology. 2015;**45**(3):593-602

[89] Minaeian S, Shahverdi AR, Nouhi AA, Shahverdi HR. Extracellular biosynthesis of silver nanoparticles by somebacteria. J. Sci. I. A. U. 2008;**17**:1-4

[90] Tran QH, Le AT. Silver nanoparticles: Synthesis, properties, toxicology, applications and perspectives. Advances in Natural Sciences: Nanoscience and Nanotechnology. 2013;**4**(3):033001

[91] Ren J, Han P, Wei H, Jia L. Fouling-resistant behavior of silver nanoparticle-modified surfaces against the bioadhesion of microalgae. ACS Applied Materials & Interfaces. 2014;**6**(6):3829-3838

[92] Chen M, Yu Q, Sun H. Novel strategies for the prevention and treatment of biofilm related infections. International Journal of Molecular Sciences. 2013;**14**(9):18488-18501

[93] Desrousseaux C, Sautou V, Descamps S, Traoré O. Modification of the surfaces of medical devices to prevent microbial adhesion and biofilm formation. Journal of Hospital Infection. 2013;**85**(2):87-93

[94] Ramachandran R, Sangeetha D. Antibiofilm efficacy of silver nanoparticles against biofilm forming multidrug resistant clinical isolates. The Pharma Innovation. 2017;**6** (11, Part A):36

[95] Martinez-Gutierrez F, Boegli L, Agostinho A, Sánchez EM, Bach H, Ruiz F, et al. Anti-biofilm activity of silver nanoparticles against different microorganisms. Biofouling. 2013;**29**(6):651-660

[96] Palanisamy NK, Ferina N, Amirulhusni AN, Mohd-Zain Z, Hussaini J, Ping LJ, et al. Antibiofilm properties of chemically synthesized silver nanoparticles found against Pseudomonas aeruginosa. Journal of Nanobiotechnology. 2014;**12**:1-7

[97] Fabrega J, Renshaw JC, Lead JR. Interactions of silver nanoparticles with pseudomonas putida biofilms. Environmental Science & Technology. 2009;**43**(23):9004-9009

[98] Kalishwaralal K, BarathManiKanth S, Pandian SR, Deepak V, Gurunathan S. Silver nanoparticles impede the biofilm formation by Pseudomonas aeruginosa and staphylococcus epidermidis. Colloids and Surfaces B: Biointerfaces. 2010;**79**(2):340-344

Chapter 10

Bioengineered Nanoparticle and Environmental Particulate Matter Toxicity: Mechanisms, Regulations and Applications

Hemant Sarin

Abstract

Bioengineered nanoparticles, and the inorganic fume agglomerates and detritus mineral ores include soft and hard particulates that differ in size distribution, surface properties and metabolites, and in dissolution kinetics. The subtypes of detritus-class microparticulates include the polyhedrally-bonded and ionic mineral- containing, inaddition to the other transition metal -oxide or -silicon oxide forms. Exposure to particle cumuli and any effect modifiers will result in the particulate matter-related disease. The initial observations on exposure-related effects of incompletely combusted products, while the remainder of earlier evidence on the association stems from epidemiologic studies. Both native and combustion composition particulates are associated with pathology, chemically synthesized nanoparticles have been designed for capillary type interstitium-pore selective passive theranostic applicability and high-affinity targeted binding to cell surface proteins with the aim of exterior bio-compatibility. In this chapter, the existing knowledge on methodologies for *in vitro* characterization of particulate matter, systemic biodistribution modeling of pharmacodynamic toxicokinetics and assessment of small molecule chemoxenobiotics efficacy, determination of environmental particulate matter exposure-related causation, standards for air sampling and exposure limits, surveillance monitoring and implementation of bioengineering controls, is covered.

Keywords: transition metals, mixtures, dissolution, genomics, energetics, bond structure, hard nanoparticle, soft nanoparticle, mutagen, epidemiology, surveillance, hierarchy of controls, imaging, targeting, pharmacokinetic modeling, In silico, permeability

1. Introduction

Particulates span bioengineered nanoparticles (NP) to geologic detri and fume agglomerates, and either soft nanoparticles or hard particulates with distinct dissolution barrier energetics [1, 2] and forms of toxicity [3]. These types of nanoparticles include colloidal such as carbon black and soot [4]. Beryllium and transition metal oxide agglomerates or aggregates [5], or unionized gold or silver colloidal nanoparticles with the oxidized electrophilic surface [6], shell-coated, aminosilane ionic coat on shell surface modification [7], and PEGylated with covalently-bound etheroylated isophilic repeating unit chain to coat [8]. Size distribution is between nanometers [9] to microns; composition, size, aspect and surface properties [10] are associated with particulate matter exposure-related toxicity [8, 11], which includes the nuisance dusts [12]. The initial observations on exposure-related effects of incompletely combusted products begin in 1775 with those of Percivall Pott on the soot composition being carcinogenic in chimney sweeps [13] as the initial cross-sectional study. The remainder of the evidence on association and causality subsequent to is by the application of non-parametric statistical methods, and is from the epidemiologic studies, both case-control with comparative groups and retrospective or prospective cohort [14].

Particulate toxicity can be studied at the single cell level, in experimental small animal subjects and by time-weighted average (TWA) air sampling-coupled to functional assessments of exposed persons, which is by air sampling with filter-threshold devices or flow density separation elutriation [15, 16], and study of nanoparticulate matter particle size distributions adsorbed on grids or of the sub-cellular morphology with electron diffraction imaging (TEM) [17], or by detection of size differences in solution with dynamic-light scattering (DLS) [18], and enhanced dark field microscopy (EDFM) with hyperspectral imaging (HSI) for accurate detection of sampled less dense NPs on filters (i.e. MCE) [19]. Isolation can be also be by specimen digestion and particle fractionation with detection at 10^{-12} M concentration resolution [20], and nanoparticle properties characterization is coupled to toxicity assessments at single cell molecular scale resolution; and in this day combined to high-sensitivity study of gene expression by quantitative PCR (qPCR), RNA sequencing, and epigenetic changes by bisulfite genomics imaging [21].

The *d*-orbital block detritus minerals between Group 3 – Group 12 are the transition metals with polyhedral bonding configuration to Group 16 nonmetals, and the subtypes of dichotomous earth particulates include the inosilicates such as Silicon dioxide (SiO_2) or partial oxidation state ideal ores such as Copper ore ($Cu_{12}As_4S_{13}$) with polyhedral bond configuration crystal lattice structures determinable by X-ray diffraction [22], inaddition to the asbestos classes of ionic minerals-containing rock with silica-based ionic composition; and there are also the other non-crystalline structure oxides such as the synthetic amorphous and biogenic silicates. There are several compositions of ore lithificates, and regional contamination secondary to industrial processes.

Inaddition to the naturally-occurring detritus particulates, there are the combustion-generated particulate oxides that agglomerate over time [23] with increased particle size with lower degradability and increased toxicity risk, and the chemically-synthesized monodisperse Zinc (II)- or Cadmium (II)- based transition metal-nonmetal (Se^{2-}, S^{2-}) semi-conductor materials that are valence-conduction band gap size-tunable for variable wavelength emission properties with applicability to electronic systems [24]. The hard NPs, ferrous or ferric iron oxides (FeO, Fe_2O_3) have been utilized for supraparamagnetic MRI (T_2W) [25] and cell tracking by transfection loading [26]; whereas the others, soft nanoparticles, with the classic four-electron C-atom bonding arrangement and exterior biocompatibility are within liposomal phospholipid encapsulation and in dendritic forms with diaminobutane cores that are utilized for biomedical application small molecule chemoxenobiotic enhanced permeation and retention (EPR). Toxicities include immediate systemic inflammatory response and hypersensitivity with earlier formulations, immunogenic sensitization with repeated administrations that also applies to nervous tissue treatments [27], while biodistribution to reticuloendothelial cell-containing tissues also limits efficacy [28].

Both hard and soft matter are in the inhalable size range (1–100 μm) [29], and result in risk of direct toxicity through air, water or food, and waste bioaccumulates to environment including plastics. In this chapter, the current day principles on bioengineered NP and environmental particulate matter exposure-related pathology causation, particulates molecular structure and cell biomolecular pathways activation, regulatory standards for exposure limits, industrial hygiene and cell responses to the potential for either immediate or delayed cellular toxicity are presented.

2. Particulates, aerosols and droplets and respiratory tree deposition

Nano-sized particulates (NPs), agglomerated nano- or micro-nanoparticles, can be defined as nanometer (nm) to micron size agglomerates with nanoparticle size in one dimension (1-D) being ≤100 nm, while particulate or agglomerate structural irregularity necessitates characterization by aerodynamic size with adjustment for air flow effects inhomogeneity in real conditions. The inhalable size range of particles, agglomerates and large aerosols includes Zika virus (45 nm, single), smoke (400–700 nm), bacterium (1–5 μm), dust particle (2.5 μm), cells (RBC, 8 μm), pollen (15 μm) and extends into droplets (> 100 μm), **Table 1**. Detritus minerals and particulates. Fine particles are the 2 nm to 2 μm size range of nanoparticulates including atmospheric aerosols with several different molecules that are found in association include sulfate (SO_4^{2-}), carbon (soot), lead, ammonium (NH_4^+), As, Se and protons (H^+); and coarse particles constitute the 2 μm to 100 μm interval of nanoparticulates with iron, calcium, titanium, magnesium, potassium, phosphate (PO_4^{2-}), silicon, aluminum and organic (i.e. pollen and plant matter). The formation of droplets in trimodal distributions results from volatile gas to hot vapor and condensation upon cooling to primary particles and chain aggregates (~5 nm – 100 nm) [51], or gas chemical conversion to low volatility vapor, nucleation and condensation growth of aggregated nuclei into the larger droplet forms with coagulation (50 nm – 8 μm), while the mechanically-generated aerosol range (1–90 μm) is for combination aerosol particulates with emissions, dust, volcano ash and plant matter; with formation, the weighted increase in size results in rainout or sedimentation depending on size range, and the 90 nm to 2 micron range is known as the accumulation range.

The human respiratory tree accommodates a certain size range of particulate agglomerates in the breathing zone [29], and the particle aerodynamic equivalent diameter (D_{ae}) is the diameter of a sphere with same falling velocity (D_a) corrected for particle density (ρ). Based on the initial studies on human exposure to Amosite (A), Crocidolite (CR) or Chrysotile (CH) asbestos or glass (G) fibers, sampled and measured by the aerosol spectrometer [52], i) the fiber length/fiber diameter (aspect ratio) becomes independent of the aerodynamic equivalent diameter (D_{ae}, D_e)/fiber diameter ratio at aspect ratios greater than around 10: 1 and suggests that the width becomes the primary determinant of deposition, where D_e/D_f follows a fractional base-variable power function as particle diameters do not increase much for lengthier fibers; and ii) the equivalent diameter to fiber diameter relationship for three of the four fiber types is non-linear and weighted towards the equivalent diameter (A, CR, G) over the actual diameter (3.5 μm, 3 μm, 2.5 μm) but less weighted to the same in

Particulate (Size&1)	Example	Formula / Composition&2	Gene expression (Inc, no δ)&3	Gene expression (dec)	Regulation / IARC classification&4 {IARC, 2012 [30]}	Ref. {Hygienists, 2021 [15]}
Inosilicate (> 5 μm (l.))	Crocidolite	$[(Na^{+},Fe^{2+}{}_{3}\ Fe^{3+}{}_{2})\ Si_8O_{22}(OH)_2]_n$	Ceacam1, Espl1, Sult1b1, Sult2b1; Sema3a (A)	Abcc10, Rbck1	TLV-TWA, PEL-TWA, REL-TWA, 0.1 f/mL!	{Boulanger, 2014; Selikoff, 1978; Pascolo, 2013; Perkins, 2015} [31–34]
	Chrysotile!	$[(Mg^{+}{}_{3})\ Si_2O_5(OH)_4]_n$	—			—
	Amosite (Cummingtonite-Grunerite)%	$[(Mg^{2+}, Fe^{2+})_7Si_8O_{22}(OH)_2]_n$	Cxcl6 (Il-8), Cxcl6, Ptgs2, Tnf (B)	n/a		{Pascolo, 2013; Duncan, 2014} [33, 35]
	Anthophyllite	$[(Mg^{2+}, Fe^{2+})_7Si_8O_{22}(OH)_2]_n$	—			—
	Ferroactinolite	$[Ca^{2+}{}_2(Mg^{2+}, Fe^{2+})_5Si_8O_{22}(OH)_2]_n$				
	Tremolite	$[Ca^{2+}{}_2(Mg^{2+}{}_5,Fe^{2+})\ Si_8O_{22}(OH)_2]_n$	—			{Webber, 2008; Duncan, 2014} [35, 36]
Silicate (2.16 μm, A - 3.55 μm, C)	Amorphous (Monodisperse)	$(SiO_{2,4})_n$	Txbip (A)	Fst, Rrm2	TLV-TWA, 10 mg/m^3; PEL-TWA, 80 mg/m^3/ (%SiO_2)	{Dove, 2008; Albers, 2015; Meijerink, 2019; Perkins, 2012} [1, 2, 37, 38]
	Crystalline – Cristobalite (Disperse)*		Trfc, Cxcr4, Mx2, Col4a6, Tlr1, Cxcl6 (Il-6), Stat5a, Fas, Jun, Mx2, Mmp1, Il-8, Bcl2a1, Tnfaip3 (A)	Ppargc1a, Tp53, Vegfa, Wnt5a, Stat5a, P13k, CD14, Cebpa, Bcl2, Sulf1	TLV-TWA, 0.05 mg/m^3; PEL-TWA$_{Quartz}$ = 10 mg/ m^3/ (%SiO_2 + 2), Quartz; PEL-TWA$_{Cristobalite}$, 0.5 x PEL-TWA$_{Quartz}$	{Dove, 2008; Vallières, 2016; Wan, 2017; Perkins, 2015; Perkins, 2012; Jiang, 2008; Uboldi, 2016} [2, 34, 38–42]
Nesosilicates (–)	Spessartine	$[3\ Mn^{2+}, 2\ Al^{3+}(SiO_4)_3]_n$*	—	—	—	—
Sorosilicates (–)	Axinite	$[3\ (Ca^{2+}, Fe^{2+}, Mn^{2+}), 2\ Al^{3+}\ (BO_3)(Si_4O_{12})(OH)]_n$				
Cyclosilicates (–)	Tourmaline	$[(Na^{+},Ca^{2+}), 3\ (Al^{3+}, Li^{+}, Mg^{2+}), 6\ (Al^{3+}, Fe^{2+}, Mn^{2+})\ (Si_6O_{18})(BO_3)_3(OH)_4]_n$				
	Beryl	$[(3\ Be^{+}, 2\ Al^{3+})\ Si_6O_{16}]_n$	—		—	—

Particulate (Size[&1])	Example	Formula / Composition[&2]	Gene expression (Inc, no δ)[&3]	Gene expression (dec)	Regulation / IARC classification[&4] {IARC, 2012 [30]}	Ref. {Hygienists, 2021 [15]}
Copper ore / Copper NP (1–5, 500 nm)	Cupric, ionic (n/a)	$Cu^{2+}(C_2H_3O_2)_2$, $CuSO_4$	Hmox1, Nfkb1, Il-8, Thbs1, Relb, Egf, Icam1 (C)	Angpt2, Igfbp3, RelA	TLV-TWA, PEL-TWA, 1 mg/m^3 (dust)	{Meijerink, 2019; McElwee, 2009} [37, 43]
	Tetrahedrite	12 (Cu^{2+}, $Fe^{3+/2+}$) Sb_4S_{13}				
	Chrysocolia	2 (Cu^{2+}, Al^{3+}) $H_2Si_2O_5$ $(OH)_4$ $\cdot n$ (H_2O)				
	Tenorite (or smelter oxide)	CuO	—			
	Cuprite	Cu_2O				
	Chalcopyrite	$CuFeS_2$				
	Digenite	Cu_9S_5				
	Enargite	Cu_3AsS_4	Nfe2l2 (Nrf-2), Hif1a, Cox2^	Mir199a^		{He, 2014; Smith, 1998} [44, 45]
	Tennantite	12 (Cu^+, Ag^+, Zn^{2+}, $Fe^{2+/3+}$) As_4S_{13}				
Nano-Cobalt (20 nm, μ; 260 nm, agglom.)	Cobalt (Colloidal, Oxide)	CoO, $Co(OH)_3$	Cxcl1, Mki67, Pcna (D)	—	TLV-TWA, 0.02 mg/m^3; PEL-TWA, 0.1 mg/m^3	{Wan, 2017; Meijerink, 2019} [37, 40]
Nano-Silver (114–159 nm, 1.48 μm)	Silver (Colloidal, Oxide)	AgCl, Ag_2O, $AgClO_4$	Cxcl8 (Il-8), CCL3/−4 (Mip-1α/−1β) (E)	—	PEL-TWA, 0.01 mg/m^3	{Vallières, 2016} [39]
Nano-Titania (12–95 nm, 28 nm (μ); 280 nm (μ, agglom.), 2.48 μm)	Titanium (Oxide)	TiO_2	n/c	n/c	—	{Vallières, 2016; Meijerink, 2019; Uboldi, 2016} [37, 39, 42]

Particulate (Size[&1])	Example	Formula / Composition[&2]	Gene expression (Inc, no δ)[&3]	Gene expression (dec)	Regulation / IARC classification[&4] {IARC, 2012 [30]}	Ref. {Hygienists, 2021 [15]}
Metal oxides (i.e. Lead fume, 12–95 nm, 2.9 μm)	Lead oxide	PbO, $Pb_{3,4}O_4$	Cxcl6, IL1b, Tnfα, Ccl3; Tek, Vegfa, Flt1, Fgf2 (E)	—	PEL-TWA, 50 μg/m^3 =	{Landrigan, 2022; Steenland, 1992; Machoń-Grecka, 2017} [5, 46, 47]
	Zinc oxide	ZnO_2	Ccl2 (Mcp-1), Rantes, Tnfα, Cxcl8; Vimentin (protein) (D)	—	PEL-TWA, 5 mg/m^3 (fume), 15 mg/m^3 (total dust); STEL-TWA$_{15\ min}$, 10 mg/m^3; REL, 2 mg/m^3	{Vallières, 2016} [39]
	Oxides mixture (5 nm – 20 μm, 10–30% ≥ 1 μm)	Fe_mO_n, Cr_mO_n	—	—	TLV$_{mix}$▲	{Park, 2014} [23]
Metal halogen (weld fume, 3–180 nm)*	Zinc (smelt, oxide; ionic)	$ZnCl_2$	—	—	PEL-TWA 1 (STEL) – 2 mg/m^3	{Vallières, 2016} [39]
	Manganese (weld, μ 11–47 nm)	MnF_2	Dmt1#	Slc30a10, Slc39a14, Rbfox1	TLV-TWA, 0.02 mg/m^3 (resp), 0.1 mg/m^3 (inh); REL-TWA 1 mg/m^3; PEL, STEL-TWA, 5 mg/m^3 (dust), 3 mg/m^3	{Bozack, 2021; Lindner, 2022; Balachandran, 2020} [48, 49]
Metal acetate (−)	Lead, ionic	$Pb^{2+}(C_2H_3O_2)_2$	Flt1, Kdr, Vegfa, Vegfb (F); Bmp6 (no δ), Fos/Jun (no δ) (G)	NF-κB, Col10a	=	{Machoń-Grecka, 2017; Zuscik, 2002} [47, 50]

**, Zn or Mn (2^+, ox), soluble ion form as most common at pH < 8 (Pourbaix), or 3^+ oxidation in presence of ROS; #MPP+ responsive; ^Arsenite (AsO_3) exposure-applicable differential gene expression; n/c, no change; A, amorphous; C, Cristobalite; &1, a) Size, Phase contrast (40-100x, objective), TEM, STEM, XRD, DSL; b) %, Amosite (RTI or UICC) or LA2000 (mixture); * TWA$_{Tridymite\ (Mohs,\ 7)}$, 0.5 x PEL-TWA$_{Cristobalite\ (Mohs,\ 6.5)}$; &2, a) Composition, MS, ICP-MS or oxygen electrochemical cell (Pb^{2+}, non-oxide); b) x, another metal; l, length; aspect ratio, ≥ 3 (l): 1; m, min; c) n, x, number of repeating molecules; d) Inhalable dust (total), ∑ (Alveolar, Ultra-fine, < 3.5 μm$_{respirable}$; Bronchiolar, 3.5–10 μm; Bronchi (Thoracic) - Nasopharynx, ≥ 10–100 μm); and e) Copper-containing oxides, pure, Tenorite, Cupric (2^+) oxide; Cuprite, Cuprous (1^+) oxide, and Copper (II) acetate; &3, a) Detection: Gene, cDNA micro-array fluorescence, RNA-seq, gRNA Cas9-seq with qT-PCR amplification fluorescence; b) A) normal human bronchiolar epithelium, B) human airway epithelium, C) hematoma cell (HepG2), D) pulmonary, bronchiolar epithelium, pneumocytes (Type II); E) eosinophil; F) blood serum, human; G) chondrocyte; and &4, a) TLV, ACGIH exposure limit-TWA$_{8\ hr}$; PEL, OSHA exposure limit-TWA$_{8\ hr}$; REL, STEL, NIOSH exposure limit-TWA$_{8\ or\ 10\ hr}$ (400 L air sample, 100 min), short-term exposure limit (15 min); and b) sampling (collection), gravimetric cassette method or cyclone-cassette assembly, 0.8 μm metal fume/asbestos (mixed cellulose ester filter), and 5 μm dusts/crystalline silica (polyvinyl chloride filter), or vertical density elutriator, ≤ 1 μm; ▲, Appendix I, Sect. III; and Inhalable particle sizes: Aerosol particulates, Smoke soot (100 nm – 1 μm), Combusted matter fume (1 nm - 1 μm); Mist (0.1–10 μm), Dust PNOC/R (ambient inhalable, 1–100 μm; coarse PM, 2.5–10 μm, fine PM/respirable, < 3.5–2.5 μm).*

Table 1.
Detritus minerals and particulates.

case of the Chrysotile fiber type. Thus, i) the falling velocity of aspected fiber particle agglomerates is predictable mostly by the particulate equivalent fiber diameter (D_e) that is inclusive of internal voids present in fiber aggregates; ii) particulates that possess minimal diameters are not subject to sedimentation or inertial impaction in the upper airway, and due to width dimension-weighting result in deep deposition within the respiratory tree [52]; and iii) particulate interception occurs within the respiratory tree for compact particles of different equivalent diameters than aspected particles of different lengths with earlier stage penetration within the respiratory tree for compact large size particulates, i.e. > 10 microns.

Based on review of studies 1969–1974, an inhalable particulate (IP) is defined as being ≤15 microns, and at between 2 and 3.5 microns is considered the limit by the ACGIH based on aerodynamic equivalent diameter [51] with the diameter being at 2.5 to 3.5 microns in maximal alveolar tissue accretion, and at 50% human airway penetration efficiency (P_{TB}, P_{ET}), agglomerate particle size is at 15 microns at normal flow during light exercise [29]. In the same study, the deposition of larger particles by impaction at either the laryngeal or tracheobronchial region is best-fitted by a non-linear model with co-variables, aerodynamic diameter (d_a), inspiratory flow rate (Q_{total}), tidal volume (V_T), and a gender- and age- category-specific scaling factor (SF_t), which models the deposition efficiency for a range of particle sizes and shows a higher deposition efficiency in the tracheobronchial (TB) region for a particle of the same size consistent with the findings of, 70%, $P_{TB\ (tracheobronchial)}$ versus 50%, $P_{ET\ (laryngeal)}$.

3. Particulates characterization by shape, irregularity and charge in flow conditions

The Knudsen (K_n) variable is the mean free path length (λ, distance) to particle physical diameter (d) ratio, $K_n > 0.01 < 10$; and the slip correction for frictional drag velocity is a reciprocal function of the Knudsen variable [53]. The aerodynamic diameter, D_{ae}, to K_n flow regime transition relationship applies to non-homogenous air viscosity in flow conditions [16]; and the direct particle density (ρ_p) to dynamic shape factor (X) relationship (ρ_p/X) is applied to normalize the volume equivalent diameter (d_{ve}). Inaddition to particle morphology and density, the other independent variables for elutriator filtering are particle flow and charge, and include the Millikan apparatus [16].

Certain relationships are known: i) the volume equivalent diameter (d_{ve}) is the flow voids-adjusted mass equivalent diameter ($d_{me} \cdot \delta$) to shape factor adjusted for standard density; ii) aerodynamic diameter (d_{ae}, d_a) is the shape increases with increasing particle density and the diameter with electrical charge (d_m) is larger than the d_a; iii) particle density (ρ_p) increases semi-exponentially with the dynamic shape factor, X, which is for irregular particles with no flow voids [16]; and iv) the ratio of the d_a to d_{ve} (d_a/d_{ve}) increases between continuum and free molecule air flow types (transition regime) for particulates with shape factor-normalized particulate density > 1 ($\rho_p/X \cdot \rho_0$; ρ_0, 1.0 g/cm^3) ratios, and it decreases in transition flow for particulates with <1 density. The electrical mobility diameter (d_m) is greater than the volume equivalent diameter (d_{ve}) for internal void-containing particles, irregular non-sphere and aggregate particulates with voids in-between.

As per the above discussion, i) for less irregular particulates ($X < 1$) there is a smaller aerodynamic diameter as per a decrease in the d_a/d_{ve} ratio over the flow regime Knudsen-Weber (K_n) path length (λ), and there is an increase in the aerodynamic diameter with more shape irregularity ($X > 1$). In Ref. to the effective particle density (${\rho_{eff}}^{II}$) defined as the *measured* particle mass (m_p) divided by (i.e. normalized to) 0.125 π-adjusted spherical volume (${d_m}^3$), the relationship between the internal flow voids-adjusted particle density (ρ_p) and the dynamic shape factor is ρ_p/X, in which case an inverse relationship holds between the particle dynamic shape factor and effective particle density [15], where particle density is considered constant (k). There is also a relationship between the electrical mobility diameter (d_m) and the particle shape with the d_m increased with more particulate matter irregularity (X, dynamic shape factor > 1–2.5) [15], which is consistent with aggregation or agglomeration via ionic or non-ionic colloidal interactions, and in which case, there is an infra-*log* linear (saturable) increase in particles diameter (d_p) with increasing particle number (N_{pp}) consistent with overlapping of particles in aggregates (in solution) and agglomerates (upon condensation).

As there are lesser than expected boundary effects in the gaseous medium, there is an altered frictional drag (F_d) relation, the Navier-Stokes function with variables, coefficient of viscosity (η) and velocity of spherical particle (R_v) is modified [53], where a correction factor is applied for air velocity slip due to a lower F_d in real air flow. For imperfectly spherical nanoparticulates, the determination of particle aerodynamic equivalent diameter (D_{ae}) is based on the baseline normalized measure, the particle density to standard density ratio (P_p/P_o; g/cm^3 fract) [16]; the effective dynamic shape factor (X'; F_p/F_{me}) for particle is the ratio of the resistance (drag) force of the actual particle to that of its mass equivalent diameter (d_{me}), which is the diameter for a non-spherical irregular particle un-adjusted for internal flow voids as compared to the volume equivalent diameter (d_{VE}); and the fraction of internal voids correction factor (δ) is for particulates that possess internal voids and/or exterior irregularity with voids. Furthermore, the determined effective agglomerate diameter of irregularly shaped nanoparticles follows a non-linear relationship for its impaction along the respiratory tree during normal inspiratory flow of 30 liters per minute.

Inaddition to the direct relationships between particle size and settling velocity (V_s) or sedimentation time (t_s), relative humidity and hydration-based exponential growth, several other aspects have been further characterized by study of virus aerosols and droplets with particle number and volume size concentration and emission small and large particle distribution (B_1, B_2; Q_1, Q_2 > 5 μm) comparison modeling of pore size threshold limits of facial PPE (fitted masks) to larger diameter particles [54], which build on earlier works on determining of flow void- adjusted aerodynamic equivalent diameters (D_{ae}) from mass equivalent (D_{me}) of a 0.125x volume spherical particle as reference.

4. Particulate matter dissolution properties and solution percolation through detritus

The atom arrangement of the dichotomous earth metal oxide morphs is polyhedral, defined by the number faces of the metal (Titanium, ^{22}Ti) or metalloid (Silicon, ^{14}Si) and oxygen (^{8}O) bonding, and the configuration forms for TiO_2 are anatase metastable octahedral (6 Ti, 9 O, 1, Ti, 6, 9) or rutile stable tetrahedral (8 Ti, 6 O) with differences in atom bonding ratios. The crystalline forms of oxides possess some shape asymmetry-related two-dimensional aspect (2-D, x, width, y, length), an example of which includes rutile titania and α-quartz silica (crystalline SiO_2). The solubility of the transition metal oxides varies over a range, and is measured as absolute dissolution in moles of solute per liter (M) or fractional solubility (Log *S*). For amorphous Silica ($aSiO_2$), Magnesium Oxide (MgO) and Ceria (III; Ce_2O_3), the solubility amount is 1.6e-03 - 1.6e-06 M (high solubility); the solubility of that of TiO_2, CuO and Fe_2O_3 is 1.0e-09 - 4.0e-12 M (low solubility); and solubility of Al_2O_3 and Nb_2O_3 is intermediate at around 1.6e-08 – 2.5e-08 molar [37] as compared to sodium chloride (NaCl) as an example of alkali metal halogen molecule with equivalents with a solubility of around 6.1 molar.

There are certain relationships in the solubility differences between polymorphs, crystalline and amorphous [2] based on dissolution properties of the crystalline (Rutile) and amorphous forms (synthetic) as in Titania particulates; it is a two phase dissolution, early with a steep decrease in rate of dissolution, and less of a decrease in second phase of the dissolving process. The dissolution kinetics of hard particulates are dependent on free energetics ($-G_{critical}$, α) with a decrease the critical *G* threshold (less negative) that favors the process with a decrease in interfacial energy (α^2), specific volume (ω), and/or increase in temperature (T), in addition to solution electrolyte saturation ($1 - \sigma$). As the degree of electrolyte undersaturation (σ) increases, there is lower solution saturation, and the rate of dissolution increases, but the dissolution of all forms of hard nanoparticles, as compared to ionic particulates, is slow, with the rate constant on the order of 6 nmol per m^2 per sec due to the presence of the solute-solvent interaction energy barrier, a.k.a. surface tension, as the primary variable (Appendix IV. End of chapter educational objectives exercise – Particulate matter toxicity). Particle solubility can also be related by its partition coefficient (+Log *P*, −Log *P*), and its entropy of defusion as the temperature (T) · product parameter in the Gibb's law relationship in which fractional solubility (Log *S*) is favored for solutes with a lower melting point (*MP*) [55] as per Gibbs free energetics. Therefore, there is experimentally-determined indifference in solvation rates between amorphous and para-crystalline forms related to solution desaturation, in which case electrolyte saturation favors regrowth and supports the bio-persistence of both synthetic and harder polyhedrally-bonded particulate matter.

There is the Dreiding potential for the van der Waals (vdW) forces between molecules (E_{vdw}), intramolecular electrostatic attractions and repulsions between atoms of a molecule (E_Q) and also intramolecular hydrogen bond energies (E_{HB}) in between molecule atoms, which is represented in summation form, although the three functions themselves are non-linear. In the example of $aSiO_2$ and $cTiO_2$, as to the latter two energies (E_Q, E_{HB}), the inter-molecular ionic bond breakage energy is $0.9\ J \cdot m^{-2}$ [1] for the crystalline particulate and is $1\ J \cdot m^{-2}$ for the amorphous particulate, and is similar for both as it is the aggregate-to-aggregate interaction energy. The intramolecular (internal) oxide bond cleavage energy difference for the crystalline and amorphous particulates is $10\ J \cdot m^{-2}$ for the crystalline particulate but is $4\ J \cdot m^{-2}$ for the amorphous particulate, and is 2.5x for the polyhedral bond configuration of the crystalline rutile [1], which has closer transition element *d*-orbital lattice spacing than the anatase form. The surface adsorbate-dipalmitoylphosphatidylcholine (DPPC) interaction vWV energy is $0.2\ J \cdot m^{-2}$ for amorphous silica and $0.05\ J \cdot m^{-2}$; and is 1/20th of that of the bulk cleavage energy for amorphous SiO_2 and 1/200th of the cleavage energy for crystalline TiO_2, also as per the decreased bond length polyhedral bond structure.

Variables apply to water diffusion and also to the diffusion of water-dissolved solutes through a matrix, also known as percolation (Appendix IV). The diffusivity of the test substance though the matrix that can be normalized to its free diffusivity

(D_0), and the experimentally-determined relationship for diffusion is modelable as an exponential decay function with a respective decrease and increase in solution diffusability with effective molecule size (a) and matrix pore size (r_m) [56]. In the modeling of the relationship percolate diffusability through porous media or matrix, there are differences in relationship between percolate (p) permeability or diffusability ($D(p)/D_0$) and matrix porosity (φ)-normalized diffusability of water content through the matrix (D_s (θ) / D_s (φ, phi)). There are both lesser and greater slope, linear and non-linear approximation, fits between solution diffusability and matrix volumetric water content (θ range, 0.0–0.5 a.u.) with cross-overs (θ_X), 0.0–0.08 (linear fit model), 0.08–0.145 (non-linear fit), 0.145–0.155 (linear fit) and 0.155–0.195 (non-linear fit) [57]. Thus, the intervals of the quasi-linear and -exponential relationship are modelable by the power law with different base (x-axis) and exponent (slope) variables in which percolation (p) is related by the fraction of occupied bonds during filtering flow and the irregular or regular lattice pore filter threshold at critical bond occupation probability (p - p_c, Δp) to the power law rule variable (t or q) [58] in which a decrease or increase in diffusion is modelable with $q < 1$ (infra-linear) or with a $q > 1$ (supra-linear), while an indirect (inverse) linear relationship between pore ionic interaction coordination number (Z) and critical percolation threshold limit (p_c) exists in three-dimensional systems.

5. Bioengineered particulates

The particulates include the nano-sized small molecule amphiphile-coated iron oxides (8.9–16 nm), the size tunable wavelength-emission quantum dots (5–18 nm), i.e. CdSe interior, ZnS shell ± RGD-Lys, and proteolytically-degradable protein-aggregate suspension xenobiotic forms such as Abraxane (130 nm), or liposomal formulations of the same [59], size range 100–300 nm to 580 nm [60], with sustained drug release kinetics for biomedical applications and in *in vivo* circulation or as lower distribution size fractions (**Table 2**. Bioengineered nanoparticles). Engineered particles have increased surface area with capacity of exterior surface polyvalency by covalent bound formation to free exterior terminal groups such as amine- or carboxyl- with 2^x exterior terminal groups per n^{th} dendrimer generation and potential for conjugation by covalent linkages; and in addition to the Neocarzinostatin poly(-styrene-co-maleic acid) conjugate (SMANCS) [78], the soft nanoparticles include dendrimers and poly-(lactic-co-glycolic acid) (PLGA) in monodisperse size distributions within the NP size range [71], and also include the emulsion- and liposome-based that are approved for human use [70]. By covalent linkage of high-MW poly-ethylene glycol (PEG) for stealth properties, biocompatibility and blood plasma half-life are improved with a shift to either hepatic and/or splenic RES sequestration [8, 9, 70, 91]; and while soft nanoparticle size and exterior percent conjugation can be tuned by divergent synthesis in comparison the non-covalent type by affinity matching for functionalizing of emulsion- or inclusion-type nanoparticle phospholipid head groups, and emulsion polymerization in case of reactive end group types (i.e. cyano-acrylate monomer).

Generation of uniform distributions of hard nanoparticles is by either bottom-up methods such as by chemical reduction and molecular condensation of multi-atom nuclei [6], or by top-down bulk mechanical milling for example [92]. In addition to the growth of polymorph-bonded atoms, further coating of the surface layer (shell) is required for exterior biocompatibility such as for quantum dots [8, 9], iron oxide [7],

Bioengineered particle	Example(s)[&1]	Formula[&2]	Overall or effective particle size ($x10^{-9}$ m), Mean or range	Physicochemical properties – Exterior/Interior	Application(s)	Ref.
Soft						
Dendrimer	PAMAM, Poly-L-Lysine, Polyester bowtie	2^n	1.5 (Core), 2–14 ($G_1 - G_8$)	Cationic (terminal amine)/ DAB, EDA*; Anionic (chelate, Succinate)/*, Ionic Neutral (Benzene disulfonate)/*	(T_1^{-1}, per msec)-low relaxivity (r_1^{-1}) contrast MRI (passive EPR / active)	{Jackson, 1998; Tomalia, 2007; Rupp, 2007; Wang, 2022; Sarin, 2021; Kaminskas, 2011; Olson, 2022; Sousa, 2009; Kobayashi, 2003; Banerjee, 2011} [17, 61–69]
Liposome	Doxil, Myocet	X-$(PEG_{5\ kDa})_n$	100–580 nm	Phospholipid bilayer (±PEG), polar periphery, apolar core	Stealth imaging, EPR	{Prabhakar, 2013; Pillai, 2013} [59, 70]
PLGA	Disulfiram Folate nanocapsule	$PLGA_n$ -N(H)OC $(PEG_{5\ kDa})_n$- N (H)OC-sm. mol. · sm. mol.	1.5–10 $[(PLGA_n)_1$-drug$]_1$; 150–210, 0.5 µm (polymer)	D, L-lactic acid, glycolic acid – Folate (c.), Disulfiram (n.c.)	Polydrug targeting	{Danhier, 2012; Fasehee, 2016; Makadia, 2011} [71–73]
Carbon nanotube (SWCNT)	SWNT-*l*-, –*br*-PEG	$(Cyclohexane)_n$ · $[C(O)O]_2$-$(CH_2)_2(PO_4)$-NC (O)-$(PEG_{ethyl\ ether})_n$	5–10 (w.)	Poly-cyclohexane · PEG phospholipid (n.c.)	Raman imaging, passive EPR*	{Liu, 2008; Singh, 2006; Kotagiri, 2014} [74–76]
Uni-micelle	PEG-IR780-C_{13}	$(PEG)_n$-IR780-$(C)_{13}$	≤ 120	Phospholipid layer, apolar periphery, polar core	Imaging, passive EPR	{Åslund, 2022} [77]
Suspension	Abraxane	$(Taxol, f.f.)_n$ · $(Alb_{6.9\ nm})_m$	130	Taxol tetra-ester (Paclitaxel) · Albumin (n.c.) · (Y solvent)	Nano-sized suspension EPR	{Prabhakar, 2013; Pillai, 2013} [59, 70]
Inclusion emulsion	SMANCs	NCS-$(SMA)_2$	3.75–5	Neocarzinostatin-SMA_2 · Lipiodol (n.c.)	EPR	{Tsuchiya, 2000; Deschamps, 2017; Ahnfelt, 2019; He, 2022} [78–81]
	PACA	$(Taxol, f.f.)_n$ · $(ACA)_m$	50–130	Cabazitaxel ·PEBCA (n.c.)	EPR, fluoro-imaging	{Åslund, 2022} [77]

Bioengineered particle	Example(s)[&1]	Formula[&2]	Overall or effective particle size ($x10^{-9}$ m), Mean or range	Physicochemical properties – Exterior/Interior	Application(s)	Ref.
Colloidal	Gold	$(Au)_n$	10–50	Non-polar inorganic atom (n. c.)	X-ray imaging, Gold radio-therapy	{Dykman, 2019; Arnida, 2011} [6, 10]
Shape polymer	MPE	$(PEG_n)_m(PE)_o$-Ph_p	2–12 μm	Isophilic / non-polar organic atom (Polyethylene $glycol_4)_2$-(erythritol $phenyl)_2$ (c.): PVA (n.c.)	Matrix fabrication / Bioapplication	{Kinbara, 2018} [82]
Hard						
Iron oxide	VSIOP, SPION	$(Fe_2O_3)_n$^	8–30	Fe_2O_3 core: polymer coat (n.c.): PEG (c. with stabilizer)	(T_2^{-1})-high relaxivity (r_2^{-1}) contrast MRI	{Nabavinia, 2020; Bulte, 2002; Wu, 2015; Jeon, 2021} [25, 26, 83, 84]
Quantum dot	ZnS, CdS, ZnSe shell QD	$(CdSe)_n$	2–10	Valance band core: shell (n.c.):: PEG (c.); Inverse band gap size-dependence	Semi-conductive transmission (non-band gap); Narrowband fluorescence emission (Red, 7 nm; Light blue, 2 nm)	{Schipper, 2009; Choi, 2007; Dhenadhayalan, 2018; Singh, 2019} [8, 9, 85, 86]
Aminosilane silica-coated iron oxide	SIONP, CLIO	$(C_2O_2)_n)_{5\ kDa}$ - H_3SiNH_2 [SiO_2 (am) [$(Fe_2O_3)_n$]	111–202 (whole); 9.7 (core); 8.5 (shell)	Isophilic PEG 5 kDa – amphiphilic aminosilane coat (n.c.): Silica shell (n.c.): iron oxide	Stealth; High r_1/r_2 (T_1W MRI); High r_2/r_1 ratio (T_2W MRI)	{Bruce, 2005; Bumb, 2008; Mathieu, 2019; Iqbal, 2015; Wunderbaldinger, 2002} [7, 87–90]

[&1] *Polyamidoamine (PAMAM) DAB core, diaminobutane ($C_4H_{14}Cl_2N_2$); EDA core, ethanediamine ($C_2H_8O_2$); MPE, monodisperse polyethylene; SPION, supra-paramagnetic iron oxide nanoparticle; PLGA, poly(lactic-co-glycolic) acid; PEBCA, poly(2-ethylbutyl cyanoacrylate); CLIO, cross-linked iron oxide* [&2] where x = n + 2, *and* n = 0, *dendrimer G_0 with 4 terminal amines,* n = 1, *dendrimer G_1, 2^3 = 8 terminal groups; and* ^*Iron-containing oxides, Hematite, α-Fe_2O_3 (rhombohedral); Magnetite, Fe_3O_4 (cubic), Maghemite, γ-Fe_2O_3 (tetragonal); PVA, polyvinyl alcohol; EPR, passive enhanced permeation and retention, or targeted EPR at *some CM receptor or channel; n.c., non-covalent; c., covalent;* l, *linear;* br, *branched; w, width; and X, head group (zwitterion, carboxyl or amine, i.e. NH_2-PLGA-NH_2; sm. mol., small molecule*

Table 2.
Bioengineered nanoparticles.

silicon dioxide [87, 88], emulsion and colloid type [88]. The size and shape of hard nanoparticles can be further modified by differing reaction times, concentration, and via thermal decomposition [83] followed by encapsulation of dispersed monomeric particles by surfactant addition and sonication with resultant droplet formation containing magnetite particles (Fe_3O_4) as an example of methods applicable for high MW PEG-stealthing of simple phospholipid bilayer liposomes with interior hydrophilic contents [70]. Functionalization of nanoparticles is by surface layer and condensation reaction such in synthesis of silane coat-bonded PEG on SiO_2 encapsulated Magnetite cores (Fe_3O_4) [83], or by the ethero-isophilicity of the exterior surface PEG portion of the dialkene (i.e. 15-C) diphosphorylo- group covalently bound to PEG for shift from renal clearance [74, 75] and stealth property to limit opsonization [76]; the interaction with the SWCNT itself is non-covalent with the alkane part molecule with similar apolarity affinity for the SWCNT wall with dimensional aspect and a + Log P lipophilicity. The degradation of the covalent linkage is hydrolytic time constant dependent for the dendritic conjugates, it could be that earlier disassociation of the two portions of the similar partition coefficient (P) results in SWCNT nanoparticle opsonization and splenic accumulation, and the free alkyl chain phosphorylo-PEG portion could result in delayed sensitivity by B-lymphocyte IgM, IgG or IgA immunoglobulin response. The less biocompatible polymers include the more hydrophobic, and large diameter hexagonal polymers such as 6·(PEG_4-erythritol) phenyl group type ([82], **Table 2**) that instead have applications in matrix fabrication, and the polyvinyl chloride (PVC) PEG proportioned crystalline-amorphous composites that have lower transition, melting temperatures (T_m, DSC) [93] and are more deformable than PVC with uses in plastics.

5.1 Dendrimers

The sub-classes of dendrimers include amido-amino dendrimers with core ethylene diamine, (EDA) and diaminobutane (DAB) with amine, carboxyl, hydroxyl or polyethylene glycol (PEG) terminal groups, **Table 2**). Naked heavy metal dye-stained PAMAM dendrimers range in between 1.9 (G_1) – 9.8 (G_8) nanometers, and with DPTA-/DOTA-chelate functionalization range in between 12.7 ± 0.7 and 13 ± 1.4 nm (Gd-G8; Rh-, Gd-G8) with the number of Gd at 350 atoms per dendrimer (Gd-G8). Inaddition to the monodispersity and narrow size distribution as evident with (Na_n) phospho-Tungstate staining by catanionic affinity [17], terminal PAMAM group synthesis is an applicable property as it results in polar molecular anisotropy for DNA van der Waals (vdW) ionic affinity; as result, it is neutral surface nucleic acid transfection by electropolation [61]. The naphtha-sulfonate functionalized Lysine amino acid-based G4 dendrimer [62], BHA.$Lys_{15}Lys_{16}(NHCOCH_2O)$1-(3,6-naphth$(SO_3Na)_{32}$ (BHA = benzhydrylamin), MW 16.6 kDa, is an applied gel sol barrier cream with overall neutral surface and low risk for nanotoxicity; and most recent, there is a polyfunctional group inseries dendrimer hydrogel with internal PEG linker (oxyethane$_n$) and hydrolysis-sensitive ester linkage for improved degradability and less toxicity [63].

The surface anionic PAMAM dendrimers are the half-generations ($-COO^-$: G1.5, n = 16; G3.5, 12.9 kDa, 5.2 nm; G5.5, 52 kDa, 7.9 nm), and by small angle scattering (i.e. SANS) are spherical shape in deionized solution with low polydispersity (PDI); and there are certain surface physical interaction properties that are associated with anionic macromolecule tissue deposition: i) Caveolin-1-associated protein (AP)-mediated endocytic uptake component exists at higher concentration (0.13 μM) since there is lung tissue deposition at lower concentration (0.020 μM) for example in the case of AT1-like alveolar cell/macrophage; and ii) inter-epithelial cellular permeability decreases with increasing half-dendrimer generation between 7.9 nm and 9 nm (-FITC-anionic dextran-70 kDa) due to the junctional complex pore size threshold. A cationic or anionic exterior results in interactions with cell surface receptors or channels with varying affinity, and also results in uptake via internalization mechanisms by non-linear rate kinetics at binding, which is evident by the less than expected blood serum half-live ($t_{1/2}$) to plasma/cell surface interaction as in the case of certain dendrimer types. Since an earlier blood $t_{1/2}$ than expected for the mass density size product of a nanoparticle is consistent with rapid non-selective internalization, this pharmacokinetic parameter can be considered an indicator of toxicity potential [64]. In contrast, the benzene disulphonic acid (BDS) dendrimer ionically-neutralized anionic exterior dendrimer (near neutral) has extended blood half-life [65] with enhanced passive permeation accumulation in tumor tissue. Mixed-surface charge - type dendrimers with less effective cationic charge per area for higher affinity binding of cell- and pathogen-released nucleic acid and molecule byproducts, which decreases inflammation pathway activation and toxicity by D- and P-associated molecular pathways (AMPs) with maximum potency and efficacy of deactivation effect for G4 50:50 dendrimers with equivalent terminal amine and hydroxyl groups [66].

High-resolution characterization of Gadolinium (Gd^{3+})-chelated mass-dense dendrimers is by transmission electron (TEM) with a combination of low- and high-dose diffraction techniques, annular dark field (ADF) STEM and energy-filtering (EFTEM) [67], for contrasted detection at high resolution of macromolecule mass and the number of heavy atoms as in the example of Gadolinium (Gd^{3+})-chelated ($DTPA^{5-}$) monodisperse nanoparticles without the need for heavy metal, i.e. Osmium (190 Da) tetraoxide (OsO_4) staining for visualization for a distribution with a median frequency of between 25 and 30 individual nanoparticles within a size range of 12.7 ± 0.7 nanometers (nm) for the Gd-generation 8 dendrimer (600 kDa). With neutral surface nanoparticles there is limited toxic potential, and the mechanism for improved efficacy is by the Maeda effect of passively-enhanced accumulation and retention ('EPR') for tumor treatment is based on blood capillary pore size selectivity [78], and further improvements by conjugated exterior ligand targeting of cell surface receptor overexpression or receptor-specific monoclonal recombinant IgG (12.6 nm)-based therapeutics with non-renal, hepatic and/or splenic clearance-dependent circulatory half-lives. As far as macromolecular imaging agents of this class for macromolecular imaging by dendritic conjugates are concerned [68], cyclic chelate Gd-DOTA poly-Lysine dendritic architecture conjugate, Gadomer-17 kDa, is a clinical use due to the concern that acyclic chelates such as diethylenetriamene pentaacetate (DTPA) have the potential for toxicity *in vivo* due to free Gd^{3+} ion release from chelate. As paramagnetic atom (i.e. Gd^{3+}) chelation affinity increases, water relaxation rate decreases per M*msec, which results in lower contrast enhancement intensity.

5.2 Micro-emulsion and inclusion emulsion

Microparticles with insoluble components dispersed in or out of a mixture that scatter light are either: i) colloidal are defined as homogenous non-crystalline matter dispersed into the other but non-sedimentable, and also consist of elements in their unionized element core form in self-association for example in a medium or in deionized water; ii) an emulsion as a colloid with single layer exterior phospholipid coat (amphiphile); or iii) a suspension with sedimentation possible of one of the substances.

Lipiodol (ethiodized oil) is an ethyl ester of iodized fatty acids with the other lipophilic component a single layered amphiphile stabilizer with 1% of poppy seed oil linoleic and oleic acids, a simple form of emulsion. The toxicity profile of Lipiodol use as radiopaque contrast agent for angiography or lymphography is related to: its rate of infusion and risk for fat emulsion embolism [79], and to it lipophilic constituent composition that results in delayed type hypersensitivity response (Type IV) with repeated administrations. Emulsion or emulsion inclusion type particle size distributions depend on water: oil ratio and are within the millimeter (mm) droplet size range (**Table 2**). The ethiodized oil-in-water (immiscible), or water-in-oil (miscible), emulsion is the two liquid emulsion form that is made a more complex drug-inclusion type emulsion containing less hydrophilic drugs such as Doxorubicin (Daunorubicin, Log D/vdWD: $-0.74\ nm^{-1}$) to improve the drug half-life and effectiveness; it is the water-in-oil emulsion that is stable and the improved systemic toxicity profile is by slow continuous infusion of the 62.5% water-in-oil emulsion showing statistical significance by categorical comparison [79]. Doxorubicin (Dox) has greater solubility in Iohexol (75 mg/mL) than in saline (50 mg/mL) than in ethiodized oil; the miscible 1:4 aqueous-to-lipid phase ratio has low stability and disperse distribution of particle sizes; and the release constant (K_{rel}; per hr) is most favorable for the 1:4 saline-in-lipiodol phase ratio form [80]. The formulation can be made more stable by inclusion of nano-sized hydrophilic NPs such as the PLGA monomers [81] that is tested for its *in vivo* properties such as in the rabbit HCC model in which it is shown that the homogenous intermixed emulsion inclusion Dox form (SSIF) with initial polymer droplet size 100–150 nm (non-covalent, n.c.; **Table 2**) has improved stability by which the Dox ($C_{27}H_{29}NO_{11}$, MW 544 Da) in PLGA monomer association is more effective than the traditional iodinated formulation (TIF) through blood-tumor barrier (BTB) capillary endothelium fenestrae-interstitium with pore size upper limit 12 nm (un-hydrated particle size/TEM-based; [64]) .

Other emulsion inclusion-based nanostructured lipid carriers also include the PACA sub-class PEBCA (**Table 2**) with more and less lipophilic phases and Cabazitaxel inclusion [77] with low-loading dose renal clearance and high-loading dose (higher MW size) splenic clearance over hepatic due to improved plasma half-life ($t_{1/2}$); and PEG-IR780-C_{13} uni-micelle (LipImage) has temporal biodistribution kinetics in comparison, first into liver and then to spleen, both tissues being RES cell tissues. The remainder of such soft nanoparticles include the targeted with drug delivery capability based on the poly-lactic glycolic acid (PLGA) repeating unit hydrophile structure and hydrophilicity-based non-covalent polymerization affinity (150–210 nm) [72] with the potential for site bioactivity *in vivo* in monomeric form (1.5–10 nm).

5.3 Silver and gold colloids

Colloidal particle synthesis is by mixture of an ion solution and hydrogel with applied heat, adsorption upon reduction of oxidation potential, and can be of defined sizes [6]; if uncoated, then surface oxidation results. The colloidal Silver nanoparticles, $AgNP_{20\ nm}$ and $AgNP_{70\ nm}$, has been studied in the human eosinophil cell culture model [39], and of the two groups, the smaller NPs with greater surface area-to-volume ratio ($AgNP_{20\ nm}$) result in a wide range of sizes in media solution, i.e. 158.9 nm (μ_1) – 1482 nm (μ_2) secondary to aggregation by DLS, while the $AgNP_{70\ nm}$

have a lower left-side distribution size (114 nm) and do not aggregate to the same extent due to less oxidizable surface area available for assoc. to the anionic composition of the media. Of these two size distributions of colloidal NPs, only the AgNP-20 nm show an increase in cell apoptosis by Annexin-V staining confirmed by pro-Caspase-3 and -7 decrease, inaddition to Laminin B1 filament protein cleavage; thus, the difference in the potential for cell apoptosis of colloidal elements is related to colloid element oxidation state, and available aggregated colloid anionic surface area for cell surface interaction.

Concerning the size difference between TEM measurement and DLS in the case of the AgNPs, for example the AgNP (NM300K) it can be noted that these are 7.75 ± 2.48 nm ($\mu \pm \sigma$) in diameter, while size range in solution even after dispersion for this set of colloidal NPs with a much narrower dry distribution ($AgNP_{8\ nm}$) than the $AgNP_{20\ nm}$ that aggregate is 28.71 nm or 38.46 nm (μ_1, μ_2), and 81.37 nm or 97.23 nm (μ_3, μ_4) in media [39], which varies with concentration (μ_1, μ_2) and time (μ_3, μ_4) in solution. For this particle distribution, the aggregation is less skewed the effect remains cell viability decreases with concentration dependence with more of a decrease in viability for the Beas-2B immortalized cell line over the A549 bronchial adenocarcinoma cell line; and due to the poor solubility of AgNP with a less electronegative electrical potential to ionize as per the Pourbaix-pH relationship. Cell viability percent (%) assay, and Comet DNA tail sign assay for chromatin DNA genotoxicity, and cellular oxidative stress-determining assays such as the FPG enzyme assay for oxidized base carbon detection (i.e. oxo-Guanine) or SOD activity and GSH level reduction assays are utilized for determination of such effects [94, 95].

The toxicity of PEGylated colloidal Gold NPs, in the form of spheres or rods, has also been characterized with size and shape determined by TEM, which is similar in length dimension (50 nm, 45 x 10 nm) [10]. The nanospheres are monodisperse with a PDI of 0.02 and with a negative zeta (ζ)-potential (-27.1 mV) and agglomerate in solution as per a size of 89 nm by DLS and this *in vivo* would be by association of Citrate impurity, or could be due to association with the R-groups of native albumin (pI = 4.5, anionic), whereas rod-like NPs with aspect will remain near neutral ζ (+1.13 mV). It appears that the presence of surface charge is the primary determinant to particle clearance *in vivo*, which limits tissue interstitial toxicity since the effective exterior anionic spherical particles are hepatic > splenic RES substrates, while the rod-like particles with slight exterior positive charge and 1-D aspect smaller than the tumor tissue pore size accumulate passively in tumor tissue. Based on a recent study of the biodistribution of AgNP or AgNP and AuNP of similar sizes (*log*-normal, $\mu = 10.82$, 10.86, GSM) after co-inhalation exposure to rodents over 28 days [96], at day 1, the AgNP aggregates distribute to the liver and olfactory bulb over the spleen; whereas, with AuNP co-inhalation, the biodistribution of the Silver (Ag, ng/g) is to the spleen over the liver, and is due to the opsonization of the particles, larger effective sizes and retained permeability to the splenic arterial side capillary beds of the red pulp; this is consistent with the more lipophilic and less soluble character of the Gold elemental/colloid particle that results in its opsonization and splenic accumulation after co-inhalation exposure.

5.4 Iron oxide nanoparticles

The iron oxide NPs (IONPs) have supraparamagnetic water proton-relaxation properties and include the ore-derived Hematite (Fe_2O_3; Fe^{III}, 2), and also the chemically-synthesized crystal Magnetite (Fe_3O_4; Fe^{II}, 1: Fe^{III}, 2) form by basic pH ultrasound-assisted co-precipitation of ferric and ferrous chlorides [83], with both iron forms containing unpaired electrons reduced by mono-Oxygen. The various IONPs are (size range): SPIONs (50–180 nm), ultrasmall PIONs (USPIONs, 10–50 nm) and the very small SPIONs (VS-SPIONs <10 nm) (**Table 2**) The IONPs are r_2-molar relaxivity contrast agents with above paramagnetic effect relaxational rate properties (R_2) that result in T_2-weighted contrast enhancement with percussion frequency around transverse magnetic plane as compared to the Gd^{3+} and Mn^{2+} transitional metals with T_1 shortening, and longitudinal relaxation rate (R_1) increasing magnetic properties per concentration of agent per time $(M \cdot msec)^{-1}$ with a reciprocal size-dependent enhancement effect high-molar relaxivity (r^2) paramagnetic effect contrast enhancement for long T_R, long T_E sequence MRI (T_2W-seq) in the inverse correlation IONP size-to-relaxation rate effect [84], and a direct relationship between loading amount and magnetic strength [26]. Silica-coated Magnetite particles also developed with nanoporous silica exterior shell with an aminosilane organic coat for PEGylation and multimodality applicability [7] of either a low or high contrast enhancement molar relaxivity ratio (r_2/r_1) [88, 89] (**Table 2**); there also exists a class of cross-linked iron oxides (CLIO) in core for improved stability of exterior surface for amination $(CLIO\text{-}NH_2)_n$ for additional covalency and functionalization [90].

Iron oxide NPs of spherical shape by SEM, and mean size 60 nm by TEM, result in lymphocyte cell viability decrease in an inverse relation to ROS and toxicity upon accumulation, which is responsive to applied Thymoquinone (TQ, *ox*) [95], a Phase II detoxification enzyme (NQO1) substrate that metabolizes to cyclohexa-2,5-diene (methyl, isopropyl)-yl-1,4-diol, with maximum reduction in ROS at mid- concentration in a low, high concentration parabolic relationship of ROS production, and alike to the toxicity type that results in A549 adenocarcinoma cells upon exposure to ZnO NP (NM110) in which oxidative toxicity is more than genotoxicity [94]. Although the potential for interaction with the nuclear DNA exists as determined by UV (200–350 nm)-visible spectra- scopy, particle toxicity for such less soluble crystalline particulate matter is primary non-genomic extranuclear toxicity. For bioengineered iron oxide NP infusions, the *in vivo* pharmacokinetics and biodistribution is measured by QAR following ^{59}Fe radiolabeled-AMI SPION NP infusion, which results in an increase in transverse tissue relaxivity (R_2) with T_2W-negative contrast enhancement in liver over the spleen or lung organs [97] and is by hepatic macrophage-reticuloendothelial system cell internalization.

Endogenous iron in its free oxidized form (Fe^{3+}) readily reduces, and in its hydroxylate ferric form bonds into Apoferritin (18 nm), a variable diameter multi-α-helix 24mer subunit cage (MW 474 kDa) that can hold 3.5–4 Fe (II)/O_2 with between 220 and 1900 (– 2220) molecules of >50% oxidized ferrous iron (Fe^{3+}) [98] as Ferrihydrite, $Fe^{3+}O_2H \cdot nH_2O$ [99]. Accrual of iron oxide densities is seen on TEM of pollution-exposed population sample specimens with inelastic scattering EELS spectra showing the iron oxide-containing Hematite, Magnetite and Goethite nanoparticle forms (30–50 nm) [100], some of which, could be Ferritin protein-assoc. forms such as 2 L-Ferrihydrite/ABACA, and Maghemite-like that are neurotoxic and also visible by nano-diffraction TEM of small volume samples [101]. Inaddition to epithelium- and olfactory nerve ending- internalized exogenous particles, the toxicity of the oxide particulates is also from extracellular bioaccumulation potential due to microglia-macrophage overload and deposition as it is in the case of Ferritin adsorption on inosilicates and formed asbestos bodies in tissue interstitium.

5.5 Cesium-, zinc- and cobalt- oxides

Inaddition to the effects of AgNP-20 nm and AgNP-70 nm particles (Sub-section B), the effects of other major oxide forms of the other transitional metals have also been studied in the human eosinophil cell culture model, and include the rutile and crystalline forms, TiO_2 (anatase crystal oxide), CeO_2 (crystalline) and ZnO (crystalline) NPs particles have been also studied in the human eosinophil cell culture model [39]. The respective particle aggregate populations follow a multimodal distribution, but could also fit a *log* normalized distribution as a single population per DLS after particles dispersion in cell culture medium such as RPM1 1640 with particle size measured in-solution by dynamic light scattering, and in the examples of CeO_2, ZnO and TiO_2 in which the *log*-normalized CeO_2 particle distribution is left-shifted as compared to for the other two with a smaller particle size distribution; and as to the cell viability and gene expression effects, the elemental oxides with favorable electro-potentials to maintain in dry/TEM particle size (TiO_2-NM101), or decrease in size upon de-aggregation (CeO_2) have less potential for negative effects on cell viability and toxicity [94], and of these three oxides, the effect of the largest dehydrated particulate of the NM series, ZnO-NM110 (132 nm, TEM) is towards ROS generation [94] and apoptosis [39]. The molecular gene expression changes that occur as a result of cell internalization include the activation of IL-8 and MIP-1α/−1β (ZnO, AgNP $_{20\ nm}$), RANTES (ZnO) genes with non-change in *CXCL9* (*alias* MIG) gene and cytokine levels gene expression [39].

The genomic effects following the local intratacheal instillation of Cobalt NP (-Nano-Co, 50 μg/mouse; μ = 20 nm, TEM) and Nano-TiO2 (μ = 28 nm) have been studied in the mutant guanine phosphoribosyltransferase (*gpt*) assay for selection of positive *Cre*-activated cell clones with integrated transfections by application of 6-thioguanine to cells in culture, and crossed transgenic mice study endpoints [40]. Based on such study data, it can be specified that: i) there is an increase in local protein concentration due to increased permeability of capillaries, and decreased neutrophil number is due to cell diapedesis inaddition to an increase in local inflammatory cytokine (*CXCL1*) concentration; ii) there is an increase in genomic DNA base transversions (G- > T) with Nano-Co colloidal Cobalt suspension exposure, but not with control or Nano-TiO_2 exposure as determined by 8-OHdG levels and non-WT *gpt* gene single-pair base sequence mutation frequency analysis; and iii) there is cell cycle progression upon exposure to Nano-Co that surface oxidizes (Co^{2+}) with the potential for assoc. anionic serum/plasma constituents, and can be seen in lung parenchymal cells by IHC staining for *PCNA* and Ki67 (*MKI67*) positivity as compared to Titania (TiO_2) that is reduced and in oxide form. There is a probability that there is correlation to bond dissolution rates as the water solubility (mol_{metal}/L) of titanium- (Ti_2O_3, TiO, n/a; TiO_2, 10^{-9} M) and iron- oxides (FeO, Fe_2O_3, $10^{-10} - 10^{-12}$ M) is low [37], the oxide-types being Ferritin cage substrates; and the potential for inflammatory/ROS stress, DNA alterations and cell proliferation/transformation exists with high concentration exposure to transition metals of oxides that have higher solubility.

5.6 Titanium oxides

Cell membrane (CM)-associated cellular esterase-mediated dichlorofluorescein diacetate (H_2DCFDA) acetate cleavage and reduction (H_2DCF) assay is coupled to the with DCF oxidation fluorescence assay for ROS detection; by the cell culture-based ROS generation assay, amorphous and rutile TiO_2 particle matter reactivity is determined by size and crystal phase [41]. TiO_2 samples are prepared by aerosol reactors, particle size distributions differences are generated by spectrometry (SMPS), and morphologic characterization of the size distribution is by filter paper electron microscopy (SEM, TEM) and by x-ray diffraction (XRD) for particle size characterization of particle distribution phases. There are monodisperse distributions of amorphous phase TiO_2 particles within the >30 ≤ 53 nm size interval are most reactive in peroxide generation (H_2O_2) per particle surface area (S.A., μmol/m^2). Particle number and concentration are the variables for 3 nm-sized particle-mediated ROS, while 41–53 nm particle size is favorable to cell ROS generation in the order of amorphous > para-crystalline (anatase) > rutile for a 34–102 nm range particle size distribution. In this study design, particle number and concentration are the additional variables to particle size with increased ROS stress determined for the amorphous particles in a single time point experimental data acquired at 15 min. in which case solubility differences would be negligible between the particle types.

The dose-dependent effects of the anatase and crystalline forms of TiO_2 particulates are assessed in the 3 T3 fibroblast cell culture model exposure model by nuclear division index (NDI) and the Cytochalasin B F-actin inhibitor binucleate micronucleation chromosomal damage (BNMN) assay with non-aggregated mean particle hydrodynamic size ranges (DLS) determined in cell culture medium for nano-sized anatase (An-10; 26 nm), bulk anatase (B-An; 260 nm), nano-sized rutile (Ru-10; 82 nm) and bulk rutile (B-Ru; 755 nm) [42]. There is maintained cell internalization to a minimum of 28 nm (An-10; > 28 ≤ 53 nm) for anatase forms (TiO_2) with no change either assay (BNMN, NDI), however this is not the case for the rutile forms, which have a right-shifted bimodal size distribution, aggregate, and are present in the media supernatant rather than within cells at 24 hrs; the B-Ru particles either endocytose less, or endocytose initially. The 82 nm-mean size Ru-10 crystalline particles are more oblong in shape (TEM) with a one-order more negative zeta (ζ)-potential (DLS) than the egg-shaped/spherical An-10 particles.

5.7 Quantum dot nanoparticles

The intermediate group metal-metalloid QDs with semiconductor properties have been applied for biochemical luminescence detection [85, 86]. The quantum dots are comprised of a transition metal (Group 12; i.e. Cd, Zn) and metalloid (Group 16; Se, S) bonded atoms as core (CdSe) and shell (i.e. ZnS) particles with conduction to valence band orbital fluorescence emission ($E_{m\ \lambda}$), **Table 2**. Being of monodisperse nanometer scale size distributions, the particles have applications in electronics including for LED photoluminescence (PL). QDs with Tungsten Sulfide (WS_2) core have a disperse distribution with one modal interval in-between 4 and 6 nm, and absorbance (λ_{abs}) wavelengths at 246, 278, 333, 365 nm and $E_{m\ \lambda}$ in-between 370 and 500 nm wavelengths with this range being conducive to wide-range of emissions color application. As to the shell and organic coating, there is a concentration-dependent cation interaction-mediated decrease in PL intensity applicable to the detection of a number of milieu transition metals (Pb, Cd, Hg, Fe^{3+} (Fe^{2+}), Cu) [86], and also in other functionalized organic molecule-coated QDs such as Molybdenum Selenium ($MoSe_2/COO^-$, NH^{3+}, SH) consistent with ionic chelate neutralization that results in an increase in emission intensity (i.e. Cu^{2+}/COO^-) [85], which also affords efficiency in detection by threshold-based binomially. PEGylation- and peptide-incorporated quantum dot-based NPs with larger size (> 12 nm) with increased blood circulatory $t_{1/2}$ time and Stokes shift emission for PL at $\lambda_{800\ nm}$ (NIR) [8] with applicability for effect on only irradiated tumor tissue angiogenic capillary-interstitium barrier pore size that can also be studied by other small NPs with monodisperse distributions in the lower nano-size range NPs [102]; and the initial feasibility of targeting to vascular malformations can be studied by endothelium-targeted QDs nanoparticles (i.e. $\alpha_V \beta_3$ integrin, [69]).

The size distributions of Cys-QDs have been shown to be narrow per QD color (E_{m} $_{\lambda}$, 515–584 nm) and within the 2.85–4.31 nm range by TEM for dry particle size, and within the 4.36 (GFC), 4.64 nm - 7.22 (GFC), 8.65 nm (DLS) range by size exclusion chromatography and dynamic light scattering in solution {Choi, 2007 #9} [9]. There is a difference in aggregation potential for particles with non-neutral exterior properties, but indifference in size for zwitterion ion organic molecule-coated shell wet QDs inaddition to those in solution with DHLA-PEG coating and near neutral exteriors. Cys-QDs show renal elimination clearance after direct infusion with threshold limits determinable by bladder intravital fluorescence microscopy (IFM) intensity; there is an increase in blood half-life ($t_{1/2}$) beginning at a hydrodynamic diameter (HD) of 4.99 nm [9], which is consistent with the pore size limit to peritubular renal capillary secretion. The pharmacodynamics of tissue distribution for neutral charge QDs with larger diameters (8.65 nm) and zwitterionic surface is hepatic > lung (> spleen), and that of smaller particles (4.36 nm) is a slight shift towards hepatic elimination due to the renal peritubular capillary reabsorption pore size threshold; splenic RES cell clearance is for non- cationic, −anionic or -cationoneutral micro-sized particles.

5.8 Pharmacokinetic models and hyper-permeability analyses

The circulatory transcapillary transfer potentials (P_c, P_i, π_c, π_i), reflection coefficient (σ) and permeability coefficient (L_p) are the three variables in the relationship for hydraulic flux calculation; first the forward transfer water flux per unit area (J_v/A) is determined, and then the ratio of the rate of solute flux-to-rate applied of water flux ratio (J_s/J_v; a.u.) is determined [103]. Since hydraulic or water conductivity coefficient (L_p)-adjusted flux is affected by the presence of circulatory macromolecular particulate matter, there is the Peclet variable adjustment in the modified Starling relationship [104]; Appendix I. Microvascular fluid exchange and pharmacokinetic modeling); and the relationship is also applicable for determining of hydraulic flux of water (or solute in Ref. to water) in the presence of pharmacologically-applied macromolecule interactions in modeling secondary to exterior vdW hydrophilicity interactions of capillary wall permeable macromolecules with water; there are two relationships, one that is of hydrostatic pressure favoring filtration when there is a lower Peclet coefficient, and the other of the effect of the permeable fraction of the macromolecules and increased oncotic effect on water filtration. The overall model for transcapillary permeation is a three layer plus one layer barrier [105] in which there are three capillary layers (EGL, endothelium, basement membrane) with the fourth layer being the interstitial matrix, and at which vdW interactions occur.

Endothelial barrier hyper-permeability and the altered pharmacodynamics of affected tissues can be also be studied *in vivo* by real-time quantitative MRI (qDCE-MRI) at high resolution and analyzed by contrast enhancement-based multivariable compartmental modeling with several parameters, C_t (t), C_p (t), V_p, K_{ep} and V_e and the time (t), time constant (τ)-dependent fractional solutions for the forward transfer constant, K_{trans}, are determined (Appendix I. Microvascular fluid exchange and pharmacokinetic modeling). The pharmacokinetic toxicity assessment-applicable models have the common assumption that the transfer between compartments is linear although plasma concentration-coupled tissue clearance ($K_{ep,}$ per min) is exponential decay ($1/e^{Kep \cdot t}$) for (small) molecules that are renally-cleared and less toxic. The common relationship for the DCE-MRI pharmacokinetic models are based on the earlier Crone indicator dilution method [106] in which tissue extraction (E, 1 – $e^{-P^{*}SA/F}$; Appendix I) is related to experimentally-determined tissue capillary permeability-surface area product ($P^{*}S.A.$, a.k.a. K_{trans}) normalized to tissue blood flow (F); tissue extraction is also related to clearance (efflux), $e^{-Kep/F}$.

There is also a model relationship for the non-linear (integrated) increase in tissue volume of distribution (V_D') over the baseline (V_D, fractional y-intercept) over experimental efflux time (minutes) as the forward transfer rate per min (K_{trans}, min^{-1}) in model selection (Model 1, 2 or 3; [107]). The generalized kinetic model (GKM) is based on Model 3 as the three-parameter model with outward efflux (K_{trans}) inoto extracellular tissue space (EES, V_D) blood plasma inward rate constant (K_{ep}, K_b) on the Patlak x-axis plot; the y- intercept V_D parameter is sometimes required for the model [107]; and there is a strong correlation between the ^{14}C-AIB QAR K_i forward transfer constant (K_{trans}, K_i) and the Gd-DTPA DCE-MRI GKM model K_{trans} parameter [108]. DCE-MRI-based bi-compartmental modeling is in T_1-weighted concentration space (mM) with a linearized signal intensity (SI) to DTPA/DOTA (MW 404 Da)-chelated Gadolinium (Gd, MW 157 Da) concentration determination by the product of the molar relaxivity (r_1^{-1}), the longitudinal relaxivity (R_1), and dynamic imaging SI/SI_0 ratio of the low T_R/T_E dual -flip angle (12°, 3°) FFE sequence-based imaging [64]; and the effective blood-brain/tumor barrier tumor interstitium permeability limit to macromolecular therapeutics is 12 nm as determined by serial DCE-MRI of Gd-DTPA-G1 through -G8 PAMAM dendrimer generations (2–14 nm).

Several studies have compared the GKM kinetic modeling parameters for the detection of the degree of hyperpermeability across tumor types to tumor tissue histopathology, and show high correlation between K_{trans} or V_e and lesion grade (r, 0.72–0.78) along high probability for dissimilarity with high discriminatory AUC between Grade II and Grade IV lesions (Sens: 100%, Spec: 93.3%) [109]. Other MRI-based parameters include the blood oxygenation level dependent (BOLD) susceptibility weighted imaging (GE SWI) gradient-echo (GE) Δ R_2*plateau tumor signal parameter in the presence of tortuous vascular density (V_D) [110], which shows better correlation with microvascular vascular density (MVD, 5–10 μm vessels) and vascular density than the GKM model-derived parameter, and also sensitivity and specificity for distinguishing molecular gene expression marker differences in high-grade glioma (HGG) [111] and could be applicable for other tumor types (i.e. exposure-related). Thus, there is applicability of pharmacokinetic model-based parameter determination for differentiation of earlier grade and higher grade lesions within the range for lower permeability brain neoplasms with DCE imaging (Appendix I), and also could have positive predictive value (PPV) for early detection of non-nervous system peripheral solid tumors in worker populations with comorbid risk factors and higher probability for malignancy.

These further developed models are also applicable for the study of circulatory effects of experimental high-molecular weight, density and size or aspected-size therapeutic agents with an increased probability for risk due to non-selectively and extended half-life-related systemic toxicity. Thus, since bioengineered particulate matter toxicity occurs to tissue capillary walls, and the lining endothelium with secondary alterations to the endothelium epicalyx/glycocalyx layer (EGL) and the sub-endothelium basement membrane collagen subunits, which are endothelial cell-secreted deposition, these methods can be utilized for initial and serial determinations of toxicity-related alterations at *in vitro*, *in situ* and *in vivo* temporal resolution.

6. Geologic detritus particulates

6.1 Inosilicates

Asbestos is the double-chain silica tetrahedral-based inosilicate with sharing of two or three atoms each and is comprised of aspected particulate fibers called amphibole fibers with: i) Non-serpentine subtypes Actinolite, Grunerite, Anthophyllite, Crocidolite and Tremolite (ρ = 2.58–2.83 g·mL^{-1}) (**Table 1**), these being the primary and law regulated asbestos fiber subtypes; and ii) the serpentine asbestos form is Chrysotile (ρ = 2.53 g·mL^{-1}) within the density interval of the silicates 2.196 g·mL^{-1} (amorphous) - 2.648 g·mL^{-1} (α-quartz). The amphibole asbestos subtypes are defined short fiber (SAF) of less than 5 μm in length, and the long fiber (LAF) asbestos ≥5 μm based on study of Crocidolite, Tremolite, Dawsonite and Wollastonite [31].

Asbestosis disease/mesothelioma, and related lung carcinoma result is an increase in mortality (%) based on an early epidemiologic cohort study of disease-prevalence associated mortality (1950s–70s) in which 17,800 U.S. and Canada trade workers (i.e. welders, shipfitters) were the 10- year prospective cohort for recording of Asbestosis onset radiographic findings in asbestos insulators [32]; i) mesothelioma is associated with a significant increase in mortality over expected in persons with plaques as compared to controls (Ref. [32] Table 8 (df = 6), post-hoc X^2 p-value (Pr_{sig}), 0.001; Appendix II. Post-analyses: Chi Square; **Table 6**, under respective table), and there is also a low probability for lung carcinoma-associated increase in mortality that could be significant (Ref. [32] Table 8, X^2 p-value, 0.15, α = 0.05); ii) there is an increase in lowest grade abnormalities on X-ray over expected with asbestosis exposure in the 0–9 year and 10–19 year duration of contact Ref. [32] **Table 2** (df = 4), post-hoc X^2 p-values, 0.002, 1.65E-08; **Table 4**), and the 40+ duration of exposure groups post-analysis confirms an decrease in observed Grade I asbestosis X-ray abnormalities with an increase in Grade 2 asbestosis-like abnormalities over expected; **Table 4**, post-hoc X^2 p-value, 7.455E-12); and iii) there is shift from fewer observed abnormal X-ray findings to normal in-between 1956 and 1960 and 1961 Ref. [32] Table 10 (df = 3), post-hoc X^2 p-value, 0.0055; **Table** 7). An increase in mortality percent (%; Ref. [32] **Table 5**; data not post-hoc analysed) from lung (bronchogenic) carcinoma is maximal between the 25–29 and 35–39 years of employment exposure strata, and the increase from mesothelioma between the 30–34 and 40–44 years of the same with the skew of the distribution, and the majority in Shipwrights and Boilermakers. The higher mortality from pleural mesothelioma is in the 35–39 year exposure interval, and the higher proportion of peritoneal mesothelioma is over a 45+ period due to metastatic pleural disease. Thus, asbestos exposure results in disease with latency, increased bronchogenic carcinoma earlier and mesothelioma later, in addition to there being the non-linear increase in risk of lung carcinoma with cigarette smoke co-exposure in workers with asbestosis as reported by the group.

There is an increase in lung tumor and mesothelioma with exposure to >8 μm length particles that have narrower diameters with the highest correlation to tumor for particles with a > 32- to 16- fold aspect ratio, i.e. in length: width strata, > 8 μm ≤ 0.25 μm (r = 0.80), and > 4–8 μm ≤ 0.25 μm (r = 0.63) [31]; and there is the particle aspect-tumor initiation relationship after deposition by inhalation and trans-mesothelial peritoneal metastases, or after direct intraperitoneal inoculation in animal models. The other determinant of toxicity appears to be atomic constituency, for example, of some more recently identified asbestos forms (IARC, 2012; [30]), Winchite [$(Ca^{2+}, Na^{1+})\ Mg^{2+}_4\ (Al^{3+}, Fe^{3+}, Mn^{2+})\ (Si_8O_{22})\ (OH)_2$] and Richterite [$Na^{1+}_2, Ca^{2+})\ Mg^{2+}_5\ (Si_8O_{22})(OH)_2$], in addition to Crocidolite [$Na^{1+}_2, Fe^{2+}_3, Fe^{3+}_2, Si_8O_{22}\ (OH)_2$] containing Riebeckite [$Na^{1+}_2Si_8O_{22}(OH)_2$] as the classic common asbestos type as an example of ionic ferrous and/or ferric iron in a polyhedral lattice of silicon oxide. In addition to the presence of covalently bond SiO_2 lattice in asbestos fibers, these are comprised of intermolecular ionic bond element interactions (Fe^{2+}, Fe^{3+}), and ferruginous bodies marked by Perl's oxidized iron stain (Fe^{3+}) within tissue interstitium has also been studied in a cumulative exposure cross-sectional study of synchrotron-based x-ray radio-opacity imaging of asbestos bodies with associated Ferritin (variable size), and X-ray fluorescence microscopy (μ-XRF) for elemental composition analysis, which are intracellularly localizing in macrophage-type cells and also present extracellularly (interstitial); and the asbestos body fluorescence analysis reveals high amounts of K > Mg > Na > Fe, and energy absorption analysis (micro-XANES) reveals presence of ferruginous asbestos fiber body constituents, ferrihydrite-containing protein Ferritin (61.7%), and Hematite (22.1%) and Crocidolite (16.2%) [33].

Also, the U.S. Geological Survey asbestos fiber mixtures have now been characterized by aerodynamic equivalent diameter (D_{ae}), equivalent diameter (D_{eq}) by TEM, and settling velocity (V_t) measurements at elutriator flow funnel settings to capture asbestos fibers ≤2.5 microns, and K-factor atomic number (Z)-corrected element composition analysis [36]. Thirteen percent (%) of the mixture is respirable fraction fibers by elutriator recovery at expected respiratory flow (see Appendix I, RPM sampling efficiency); and by STEM mode energy-dispersive x-ray (EDX) spectrum analysis, elemental composition correlations are determined [36]: Fe to Ca or Mg, and Ca to Na ratios are negatively correlated for co-presence, while Mg and Ca, Fe and Na (r = 0.573), and K and Na, show positive correlations. The asbestos particle aspect and composition toxicity results in different histopathological subtypes of mesotheliomas includes a breadth of transformed cell types including fusiform (fibrogenic, osteogenic, giant cells) and pleiomorph (medullar, tubulopapillar).

6.2 Silica and inosilicates

The least soluble particulates include SiO_2 and TiO_2, and aluminum (III) oxide (Al_2O_3) is less, and CuO and non-amorphous carbon black (Printex 90) are least, gene expression effects of which have been correlated with solubility; the least soluble result in greater magnitude gene expression effects upon internalization with maintained positive correlation for presence of tissue neutrophils, Saa-1 (liver) and Saa-3 (lung) mRNA [112]. In addition to gene and protein expression responses to bioengineered particle exposure, differences in gene expression between inosilicates, rutile Cristobalite crystalline silica (non-quartz) and amorphous silica concentrations of crystalline silica and Crocidolite asbestos, are studied in comparative cell lines, primary (normal) human bronchial epithelial cell (NHBE) and BEAS 2B on the SV40 virus-integrated epithelial cell line by cDNA hybridization microarray of larger sets of genes and/or qRT-PCR of certain gene sets [34, 38].

The effects of mined Vermiculite deposits (Libby amphibole, LA) composed of a mixture of Winchite (83%), Richterite (11%) and Tremolite (6%) asbestos with TEM-based length: width on primary human airway epithelial cells (HAECs) are studied with qRT-PCR measurement of inflammatory marker gene expression in normalized dose-response effect (lL-8, IL-6, COX-2, TNF) [35]; and the geometric mean (μ) lengths of the fiber sets are: RTI Amosite, 10.4 μm (n = 359), LA2000, 3.34 μm (n = 433, width > 1), LA2007 2.47 μm (n = 268) and UICC Amosite, 2.07 μm (n = 222) [35]. There is a higher potency effect and increase in chemokine IL-8 mRNA (*CXCL8*) levels in response to RTI Amosite particle number dose/cm^2 cell for equivalent concentrations of ionizable iron (inorganic) and chelatable iron (Fe^{3+}) present in both RTI Amosite and UICC Amosite samples as can be determined by ICP-optical emission spectroscopy. The difference in the inflammation-mediated toxicity potential of the fiber sets is due to the prolonged residence time of aspected iron-containing asbestos fiber types inaddition to the probability of the adsorption of oxidized iron forms.

There is methodological validity in same group studies. The findings of two comparative studies with normal human bronchial epithelial cell (NHBE)- and BEAS-2B immortalized normal epithelial cell- types exposed to low- or high-dose crystalline silica, and at 15 and 75 × 10^6 $\mu m^2/cm^2$, or low dose inosilicate (iron-containing asbestos). There are no cell viability differences for low- or high-dose silica-exposed BEAS 2B cells, however there are differences in gene expression by cDNA microarray with shift towards *FOS* (cFOS) and IL-6 related gene expression with Cristobalite silica exposure ($p < 0.05$/cut off ≥2.0-fold) with reduced false positive risk (FDR 5%) [38], see **Table 1**. The rank order of gene changes between control and respective exposure comparisons is high dose Cristobalite > low dose > Amorphous with little difference in cell adhesion pathway gene activation between the two silica types by gene ontology (GO) molecular pathways correlation analysis; and there are differences in gene expression between exposure between normal and virally-transformed cell types. Low dose Cristobalite silica results in minimal differential gene expression in NHBE cells (3 genes; **Table 1**) [34]; the Asbestos exposure pathway is more towards *JUN*, IL-8 and MMP-1 metalloprotease gene expression by qPCR for specific gene expression, and common silica and asbestos pathway activation includes glucocorticoid (GR), whereas divergent pathways are extracellular matrix synthesis proteoglycan with Crocidolite asbestos exposure, and the Aryl hydrocarbon receptor (*AHR*) pathway with crystalline silica exposure.

6.3 Lead

Lead is found in steel at 0.15–0.35% concentration in addition to carbon, and is released during the abrasive blasting process, and poses a risk, as there is little positive or negative correlation and independence to air flow velocity and there is an increase in exposure-related Lead concentration in blood in both blaster and vacuumer [113]. There is the effect of exposure during the ore lead extraction smelting process and to combusted lead particulates over prolonged duration during which dissolution favors ionic Lead (Pb^{2+}) release. There is an increase in the stratified years interval SMR (1–5 yr., 5–20 yr., 20+ yr) for renal carcinoma and non-malignant respiratory disease (emphysema, pneumoconiosis) [46]: this is in employees with BLL at 56.3 μg/dL (2.73 μM) and 0.366 ppm average airborne Lead concentration exposure (3.1 mg/m^3; **Table 3**, Appendix III. Particulates toxicity industrial hygiene), but is with low probability for anemia (10%), in a low co-exposure to Arsenic (0.0133 ppm) and Cadmium (0.0016 ppm) population of workers with unknown smoking history. Inaddition to accumulation levels in long bone parts (i.e. tibia, 99.4 μM), there is a negative x- (patella Pb, μg/g), y- (LINE-1 methylation) variable regression relationship for increasing Lead levels to 40 μg/g bone concentration (0.194 mM) that is consistent with decreased inactive LINE-1 DNA and RT DNA methylation with increased Pb concentration; this evidence comes from a cross-sectional study (2010) of global DNA CpG island methylation of retrotransposon viral origin human DNA integrated long repetitive base sequence elements (LINE-1) and reverse transcriptase (RT)-encoding elements [114].

There are also increases in inflammatory marker expression (IL-6) inaddition to of both vascular permeability factor gene (*VEGFA*) and decoy receptor (sVegfr1) gene expression as per another more recent single timepoint study (2017) enrolled Lead-Zinc workers with elevated BLL (37 μg/dL) and ZPP levels as compared to controls with unknown alcohol use history [47]. Additional support for the Lead transition metal-type toxicity effect comes from cell culture exposure with applied Lead Acetate (100 nM – 1 μM; **Table 1**) on transiently-transfected chondrocytes with measured reporter luciferase (LUC) activity of AP-1 and Nf-KB plasmids in response to Pb $(C_2H_3O_2)_2$ alone or with applied PTHrP [50], in which its application alone results in a decrease in cell survival factor gene Nf-KB without change in AP-1 factor subunit genes (*FOS*, *JUN*) activity, but an increase in AP-1 activity with PTHrP as effect modifier. The effect of released ionic Pb is condition-dependent and can in certain instances shift the cell response towards pro-angiogenic progression. The effect of released ionic Lead (Pb^{2+}) is co-exposure condition-dependent, and can in certain cell types and instances shift the cell response towards pro-angiogenic phenotype, inaddition to its known negative effects on neuronal cell growth in association with Casp8 gene promoter hypomethylation (CpG) [48] and over activation resultant neurotoxicity.

6.4 Manganese

Manganese has paramagnetic properties, and is present in various oxidation states, Mn^{2+} ($Mn(OH)_2$, basic), Mn^{3+} (Mn_2O_3, $Mn(OH)_3$) or Mn^{4+} (MnO_2) in the presence of H_2O_2 reactive Oxygen species (ROS) with co-product hydroxides formed (OH^-, OH·) as per the Pourbaix electrical potential to pH relationship [115]. It is reactive with Transferrin as Mn^{3+} in the presence of base (HCO_3^-) similar to that of ferric iron, but with less affinity than ionic Chromium (Cr^{3+}) due to a greater ratio of bicarbonate to metal as determined by optical absorbance [116], and as shown by the Cr $(OH)_x$-Transferrin peak on EPR spectroscopy [116]. An example of exposure to particulate matter metal mixture is in welding with fume generation, and dissimilar solid or flux-cored pre-weld composition-containing wire-dependent secondary aerosolizing particulates of differing solubilities and toxicity.

These hazardous by-products of welding have been characterized [117]: i) the size distribution of particles generated is between 3 to 180 nm (SEM, TEM); ii) the mixture in a) solid wire gas metal arc welding (GMAW; S_1, S_2) is a combination of oxides, Fe_xO_y, Mn_xO_y, CrO_4^{2-} (Cr^{VI}), Cr_2O_3 ($Cr^{III/IV}$), SiO_2 and BiO_2, and in b) flux-cored wire arc welding (FCAW; F_1, F_2), this includes the additional composition of ionically-bonded Alkalis (Na^+, K^+; Group 1, Period 3, 4) and Halogen (F^-; Group 17, Period 2) elements (XPS); and iii) the PBS soluble metal is a small proportion in solid wire welding, while in flux wire welding, it is most of the (by-) product comprised of Mn, Cr^{III} and Cr^{VI}, predominantly CrO_4^{2-} (Cr^{VI}). The weld by the FCAW process has flux by-product with increased solubility and the weld proportion of metal product content improved in purity with less metal oxide, but there is a large proportion percentage (%) with higher *Pr* for sig. For concentration-dependent cytotoxicity at 1 day and genotoxicity as early as 3 hrs at 10^{-4} M concentration in the flux FCAW F_1

and F_2 process type welding [117]. Thus, any combustion by-product composition has Manganese in its ionic (Mn^{2+}) form, which is the energetically-favorable form at pH < 8, in both sub-type processes (S_2, F_2), which is supported by the experimental data showing solubility decreases with addition of increasing concentration of Fluorine (F^-).

As to the Manganese effects on genomics, DNA re-sequencing results shows Manganese efflux transporter gene *SLC30A10* and *SlC39A14* polymorphism ([118]; **Table 1**), and changes result in other related gene expression such as *RBFOX1* [119], *TXN8* [48] and MPP+ responsive influx transporter gene *DMT1* [120], for example bisulfite Cytosine (– > Uridine) substitution DNA epigenomics shows methylation of the splicing regulator gene *RBFOX1* (A2BP1) at position cg02042823, which has been studied by qT-PCR and RNA-seq transcriptomics. The double-allele mutant *lox*ed gene (*RBFOX1*$^{-/-}$) murines show a decrease in exon inclusion for certain channel genes such *GABRG2* and *GRIN1* [119] with the net effect being lower stimulus intensity thresholds required to evoke EPSPs and less inhibitory ion flux as per potential recordings showing this effect.

In addition to study of alterations in gene methylation of the *RBFOX1* gene, study of cell morphology shows altered gender-dimorphic dendritic spine count with spatial frequency lower in females (10 μm^{-1}), and higher in males, after subcutaneous exposure to repeated dosing of ionic Manganese as $MnCl_2$ [121] such that this would result in Dopaminergic with a sex-specific dimorphic effect on neurite bleb budding in association with dendrite density. Similar local concentrations achieved in the striatum in both and the inverse inhalation effect of Maghemite (Fe_2O_3) in the Welder workers [122] with the clinical sign of cocked-gait as sign of basal ganglia and midbrain nuclear group toxicity with internal capsule upper motor neuron disinhibition effects due to exposure intensity toxicity early in Manganism, and then exposure duration-related toxicity in Manganese-induced Parkinsonism).

The experimental data on the transfected overexpression of the divalent influx transporter gene, *DMT1*, that transports Mn^{2+} present at increased local concentration in SH-SY5Y pre-differentiated neuroblastoma neuronal cell line show the effects of Manganese (Mn^{2+}) in comparison to ferrous Iron (Fe^{2+}) with more of an initial shift in cell compliance (P_{eff} C, a.u.) to the phosphorylation of the c-JUN AP1 transcription factor terminal kinases [123], and towards the resultant p53-mediated R-shift for activation of the *Bax*-, *Bad*- and *Bim*- apoptosis cascade, and/or DNA repair and cell cycle changes. For Manganese, i) the negative electrical potential and neutral pH favors its ionic and soluble form; and ii) it is the presence of efflux channel and certain cell surface internalization receptors that results in its initial effect followed by its later apoptosis-related toxicity to basal ganglia cell types.

6.5 Copper ore arsenate

Either Arsenic or Stibium, Group 15 (semimetals), are bound to Sulfur (Group 16, nonmetal), and there is 91% dissolution of Tennantite at high temperature (212° F) over a 2 hour period in basic electrolyte solution (NaOH) containing Na_2S [124]. The mined copper ore contains transition metals, Copper, Arsenic and Iron in polyhedral/hextetrahedral configuration with some forms containing ionic free forms of Iron ($Fe^{3+/2+}$) and Zinc (Zn^{2+}) (**Table 1**). Copper in oxidation states 1^+ or 2^+ results in interactions with metallothioneins with overload effect on DNA, resulting in conformational changes [125]. Certain studies show the effect of ionic Copper in form of Cu (II) acetate ($((CH_3)_2COO\text{-})_2$ on cells that when applied results in overexpression of the

NFκB as per plasmid transfectant fluorescence and of pathway and related genes as determined by qRT-PCR and confirmed by siRNA data [43]. Copper (Cupric) exposure to cells (i.e. Hepatoma G2) results in a time-dependent increase in NFκB2 target gene *CXCL8* (IL-8) and *IER3* gene expression, inaddition to overexpression of *NFE2L2* (NRF-2) and Phase II detoxification pathway genes, *HMOX1* (HO-1) and *GCLC*. In desaturation conditions (Section IV), the dissolution kinetics favor conversion of As_2S_3 to H_3ASO_3 and hydrogen sulfide gas (H_2S); Arsenate $(AsO_4)^{3-}$ and Arsenite $(AsO_3)^{3-}$ in respective trivalent and pentavalent oxidation states are slow dissolution molecules with the potential to transform cells in low-chronic exposure conditions with ROS pathway and related gene activation [126], and hypoxia-associated genes (i.e. HIF-1α) in transformed cells [44]. In the Chile Copper ore sub-regions (various forms, see **Table 1**) with mining area-specific differences in exposure risk [127], there is increased gender-specific mortality in a low-smoke exposure prevalence population of men and women ≥30 years of age as analysis by the age-weighted standard mortality ratio (SMR; [14]) [45] supports increased mortality due to carcinomas of the bladder (SMR 6.0, M; 8.2, F), kidney (SMR, 1.6, M; 2.7, F), pulmonary (SMR, 3.8, M; 3.1) and dermatologic (SMR, 7.7, M; 3.2, F), and the population attributable risk (PAR) in that mining region is 9.7% (M) and 4.9% (F) for ore exposure-related carcinoma. Leaching into drinking water reservoirs with peak exposure concentration at 870 ug/dL (116 μM) results in an increase in mortality due to lung cancer is attributable to increased ore particulates air concentration. The presence of additional transitional metals, Iron, Stibium and Zinc present in Copper ore particulates could result in effect modification.

7. Exposure prevention and surveillance guidelines

In the various sectors there are increased levels of process-generated smoke particulates or aerosol mists and related hazards that require the implementation of the hierarchy of controls and access to online information repositories [128]. Surveillance of occupationally-associated promontory habits prevalence has shown co-existence of comorbidity such as in the asbestos abatement workers [129]. Chest X-ray radio-opacities are present in individuals not exposed to dusts in an intercontinental population with gender- and age- specific differences with low prevalence at ≤50 yoa, in females (F, 0.4%) or blue collar workers working in non-exposure conditions (0.21–0.25%) [130]. Exposure surveillance comprises spirometry and CXR for monitoring of both PMF and simple CWP (under MSHA (30 USC 801–962) with a PEL of 2 mg/m^3 (REL 1 mg/m^3) or reduced PEL for silica content (30 CFR 70.101, see **Table 4**, Appendix III), and the requirement of a Black Lung (B) read (42 CFR Part 37) as per ILO classification for radiographs (1980) [131], **Table 1**. In the particulates exposure-monitored individuals with FEV_1/FVC ratio increases due to silica or asbestos exposure, there is a higher probability for parenchymal abnormality by age and Laborer/Cleaners class (Model I), but not in an un-conditional Model (II) without inclusion of asbestos exposure in logit probability [132].

Surveillance measures can be proposed for higher risk work groups in similar exposure conditions, and at the action limit there is the requirement for substance-specific NIOSH-approved respiratory protective device use under the OSHA Respiratory Protection Standard (29 CFR 1910.134). There is environmental risk for carcinogenesis with air exposure concentration at $8.40x10^{-2}$ (Cr^{VI}) and majority of years of life lost due to water exposure to Free Cr^{6+} ($2.09x10^{-5}$; YLL_{water}, 92%) [133] [REF] with the effective cancer risk at two-orders higher concentration for the particulate, aerosol and water partitioned agents tested (i.e. Hg^{II}, Cd^{II}; Lindane, DDT). There is increased risk for lung carcinoma in certain worker sub-groups such as Crane, Derrick and Hoist workers (OR_{adj} 14.4; 3.36% PAR), and with ≥10 years exposure to Coal Dust (IARC 1) adjusted for other variables shows greater odds of lung cancer in cases over non-cancer (OR_{adj} 2.0) and cancer (OR_{adj} 1.5) control group populations [134], as per a 1992 two-control group case-control study applying multi-coefficient variable probability regression (unconditional) modeling; as per the study, any duration of cigarette smoke exposure ≥20/day results in increased risk of lung carcinoma (OR_{adj} 2.1) in Asbestos–exposed persons.

Area sampling of air for aerosolized viruses transmitted by speak, cough or sneeze [54] and particulates is by general elutriation devices [16], carbon-filled adsorption capillary tubes for lipophilic gaseous vapors, or personal dosimeter devices with pore size threshold filters (**Table 5**, Appendix III), to adhere to certain exposure threshold monitoring requirements at the action level (AL) to begin biomonitoring. The OSHA air concentration medical action level for Lead (Pb^{II}, Pb^{2+}, IARC 2A; MW = 206.2 Da) is 0.03 mg/m^3 (10^{-10} M) with a PEL (ceiling limit) of 50 ug/m^3 under 29 CFR 1910.1025 (https://www.osha.gov/laws-regs), and Manganese (Mn^{II}) has a threshold limit value (TLV, respiratory limit)-8 hr. TWA of 0.02 mg/m^3 as recommended by the American Conference of Governmental Industrial Hygienists (ACGIH) [15]. Occupational surveillance of inhalable dust exposure includes for IARC Group I carcinogens, Asbestos (29 CFR 1910.1001, Industry), Silica (29 CFR 1926.1153, Construction), Chromium (CrO_3, Cr^{VI}; 1910.1026, GI), and also Arsenic (Ar^{V}, inorganic; 29 CFR 1910.1018) for which there exists a permissible exposure limit (PEL) of 10 ug/m^3 with an AL that is one-half of this limit.

Traditional X-ray diffraction has sufficient sensitivity for detection of high-Hounsfield unit pathology in linear scale with exposure monitoring reads as per the ILO reads classification system [131], and has high specificity to detect early metastatic foci (91%), but low sensitivity (41%) [135]. Non-contrast chest CT can be utilized for study temporal tracking of pulmonary lesions at high-resolution as it has improved sensitivity (74%). T_1-weighted MRI at early time point after injection early contrast enhancement (ECE; 4.5 min) can be utilized for signal intensity (SI)-based hyper-permeability detection of mesothelioma foci by region of interest (ROI) as it has high sensitivity (91%) and 15% false positive rate with a high-correlation κ statistic (0.77) [136]; and ^{18}FDG PET has utility for quantitative detection of solid tumor tissue hypermetabolism and metastatic disease with a high true positive rate (92%) and high specificity (93%) [135]. The diagnostic determination of occupationally-related neoplasia can include the application of macromolecular contrast enhancement agents for sensitive and specific detection of the hyperpermeable pathology of solid tumors on CT and on MRI during surveillance or treatment course.

8. Conclusions

There are several mineral ores containing transitional atoms with polyhedral bond structure; there is the utilization of minerals of mined mineral ores in the various biotechnology sectors; there is the development of bioengineered nanoparticles for enhanced accumulation and retention effects; there is the bioaccumulation of human-made synthetic materials including plastic particulates with very slow dissolution rates and potential for particulate matter environmental toxicity that alter the normal cell molecular mechanisms.

Certain permissible threshold limits are established by the U.S. regulatory agencies for minimizing environmental or occupational exposure to individual mineral constituents measured by mass spectroscopy or by fluoroscopy or to whole particulate matter. Causality for certain mineral exposure-associated toxicity and disease is established by the combined evaluation of *in vitro* test results, animal study experimental data and human epidemiologic findings by probability of significance. Methods that are applied in study design include exposure air sampling by elutriator air flow, characterization of particulate dimensions and shapes by electron microscopy or particle light scattering, measurement of particulate matter dissolution, and biomonitoring of biochemical metabolites in workplace surveillance, gene expression level alterations by cDNA microarray or by RNA sequencing with pre-amplification by qPCR, small animal subject and human blood plasma and tissue imaging with pharmacokinetic modeling, descriptive statistics and epidemiologic -based risk delineation by alternative hypothesis and p-value thresholds.

The reasons for particulate matter toxicity can be understood with knowledge of the diversity of topics covered in this chapter on: i) aerosol and particulate matter classes, ii) natural and bioengineered particle types, iii) respiratory tree deposition mechanisms, iv) imperfect particle properties and flow regimes, v) solvation variables, vi) compartmental modeling parameters, and vii) workplace exposure limits and guidelines. The material covered in this chapter will be pertinent to the determination of causal and synergistic relationships for new toxicants and mixtures, and industrial, mining and construction sector exposure air sampling, high-resolution tissue bioimaging and monitoring of general and specific external exposomic effects on DNA and polymorphism and for the further understanding of mechanisms of particulate matter toxicity, and disease initiating and promoting variables.

Abbreviations

MS	mass spectroscopy
ICP-MS	Inductively coupled plasma-MS
ICP-OES	ICP-optical emission spectroscopy
EF-TEM	energy filtered-TEM
EEL-TEM	electron energy loss spectroscopy
SANS, SAXS	small angle neutron/x-ray scattering
SEM	scanning emission microscopy
XANES	X-ray absorption near-edge structure (element specific)
XRD, XRF	X-ray diffraction or fluorescence
DSL	dynamic light scattering
EELS	electron energy loss spectroscopy
DSC	differential scanning calorimetry
FTIR	Fourier transform infrared resonance
PID	photoionization detection
XPS	X-ray photoelectron spectroscopy
GFC	Gel-filtration chromatography
LINE-1	long interspersed nuclear elements
cDNA micro-array	differential RNA expression (DE) by cDNA hybridization library

qRT-PCR	quantitative RNA reverse transcription DNA polymerase chain reaction
RNA-seq	RNA sequencing
gRNA-seq	guide RNA sequencing
FPG	formamidopyrimidine glycosylase
SOD	superoxide dismutase
UHCL	unsupervised/unlabeled data hierarchical clustering analysis
GO	gene ontology DE pathway analysis
IPA	ingenuity pathway analysis
IPRL	isolated perfused rat lung
QAR	quantitative autoradiography
DCE-MRI	dynamic contrast-enhanced 1H-nuclear magnetic imaging
FFE	fast-field echo, gradient-recall (GE) echo
SE	spin-echo, fast-spin echo
GKM	generalized kinetic model
PAR	population attributable risk
RR	relative risk
OR	odds ratio
SMR	standardized mortality ratio
HR	hazards ratio
Logit	logistic probability
HA	alternate hypothesis
HO	null hypothesis
□2	Chi-Square non-parametric
FDR	false discovery rate
REL	recommended exposure limit
PEL	permissible exposure limit
TLV	threshold limit value
AL	action level
TWA	time weighted average
PPE	personal protective equipment
IARC A1	Human carcinogen
A2	Suspected human carcinogen
A3	Confirmed animal carcinogen
A4	Not classifiable as a human carcinogen
A5	Not suspected as a human carcinogen
ILO	International Labor Organization
OSHA-DOL	Occupational Safety and Health Administration
NIOSH-CDC	National Institute of Occupational Safety and Health
ACGIH	American College of Governmental Industrial Hygienists

Conflict of interest

None.

Declarations and acknowledgements

The writing of this chapter is supported by the National Institute of Biomedical Imaging and Bioengineering, National Institutes of Health with also feedback on microscopy methods by Richard Leapman and colleagues (Maria Aronova), and the Department of Environmental Medicine and Public Health, Mount Sinai School of Medicine, USA with feedback in the system by John Meyer on respiratory particulates, and James Godbold in Biostatistics on post-analyses of Ref. 31 study data. Additional feedback was provided by Behzad Ghanbarian on solution percolation, Mark Goldberg and Jack Caravanos on the air sampling and industrial hygiene, Phil Landrigan on the initial Lead smelter study and exposure thresholds, Roberto Lucchini on the effects of Manganese exposure-related toxicity to susceptible populations, Paul Tofts on the vascular plasma term modification in the general kinetic model, Charles Michel on the convective effects of macromolecules, and Gunter Oberdorster on the inhalation toxicology aspects. All previous co-authors including John Butman, Steve Fung, Ariel Kanevsky, Colin Wilson, Haitao Wu, Alioscka Sousa and Matthew Hall have reviewed the pertinent contents.

A. Appendices

A.1 Microvascular fluid exchange and pharmacokinetic modeling

Unidirectional three-parameter transcapillary permeability model issue extraction by indicator dilution

$$E_{\mathrm{F}} = 1 - \frac{1}{e^{\frac{P \cdot S}{CBF}}} \tag{1}$$

where E_F, brain tissue indicator extraction fraction, a.u.

P, permeability, $x10^{+1}$ mm/sec

S, surface area, $x10^{+2}$ mm^2

Q, blood flow (i.e. CBF), microvascular capillary flow per unit mass of tissue per min, ml/60 sec · g tissue, mm^3/sec

if P·S product - > ∞, *then* $E_f \approx 1$ and all of the indicator is extracted from blood plasma into the EES and is a small molecule permeable across the blood-CSF barrier and the blood-brain barrier and is a barrier permeability indicator

if P·S product - > 0, *then* $E_f \approx 0$ and none of the indicator is extracted and the indicator is a macromolecular contrast agent and blood volume indicator

Bidirectional flux two-parameter transcapillary permeability model for hydraulic conductivity with Peclet variable for capillary connective flow

$$\frac{J_{\mathrm{V}}}{A} = Lp\left[\Delta P - \sigma^2\, \pi_c\, \frac{\left(1 - \frac{1}{e^{Pe}}\right)}{\left(1 - \sigma\frac{1}{e^{Pe}}\right)}\right] \tag{2}$$

where J_v, effective hydraulic flux into tissue in 10^{-6} cm per sec

A, surface area for transcapillary exchange in cm-squared

L_p, hydraulic conductivity coefficient of water permeability in cm per sec · cm H_2O

σ, reflection coefficient to transcapillary macromolecule transfer across capillary wall, a.u.

π_c, capillary osmotic pressure in mm Hg or cm H_2O

P_e, Peclet variable ratio for macromolecule convective flow through capillary, a.u.

Bidirectional three-parameter generalized kinetic model (GKM) pharmacodynamic model with vascular plasma volume adjustment term (as per section Refs.)

$$K\text{trans} = \frac{C_t(t) - V_p * C_t(t)}{\int_{\text{Ʈ}}^{t} \frac{C_p(t)}{e^{Kep\ (t-\text{Ʈ})}} dt} \tag{3}$$

where C_t (t), concentration in tissue at time t in minutes

C_p (t), concentration in plasma at time t

Ʈ, imaging time at initial sequence contrast enhancement

V_p, plasma vascular volume

V_e, extracellular extravascular volume, tissue interstitium volume

K_{trans}, K_i, positive slope forward (outward) transfer rate constant per minute (or apparent tissue V_D (V_D')) per minute)

K_{ep}, K_b, negative slope reverse (inward) transfer rate per minute

K_{trans} range: indication.

Upper tier: high-hyperpermeability lesion, i.e. inflammation

Mid tier: mid-hyperpermeability lesion, i.e. high grade tumor

Lower tier: low-hyperpermeability lesion, i.e. low grade tumor

A.2 Chi Square post-analyses

	Exposure duration (yrs)				
X-ray Grade (of abnormality)	*0–9*	*10–19*	20–29	30–39	*40+*
Asbestosis Grade 1	Exp, 24 (Obs, *36*)	Exp, 113 (Obs, *158*)	Exp, 38 (Obs, 35)	Exp, 114 (Obs, 102)	Exp, *77* (Obs, 35)
Asbestosis Grade 2	Exp, 8 (Obs, *0*)	Exp, 39 (Obs, *9*)	Exp, 13 (Obs, 17)	Exp, 39 (Obs, *49**)	Exp, 27 (Obs, *51*)
Asbestosis Grade 3	Exp, 3 (Obs, *0*)	Exp, 15 (Obs, *0*)	Exp, 5 (Obs, 4)	Exp, 16 (Obs, 18)	Exp, 11 (Obs, *28*)
Chi Test *p*-value	*0.00017*	*5.507E-13*	0.4294	0.1313*	*6.303E-17*

**trend for difference between groups (Grade 1, −2 and − 3).*

Table 3.
Table 2 post-analysis of grade 1, grade 2 and grade 3 X-ray abnormalities (of ref. [32]).

	Exposure duration (yrs)				
X-ray Grade (of abnormality)	*0–9*	*10–19*	20–29	30–39	*40+*
Asbestosis Grade 1	Exp, 27 (Obs, *36*)	Exp, 124 (Obs, *158*)	Exp, 39 (Obs, 35)	Exp, 112 (Obs, 102)	Exp, 64 (Obs, 35)

	Exposure duration (yrs)				
X-ray Grade (of abnormality)	***0 9***	***10–19***	**20–29**	**30–39**	***40+***
Asbestosis Grade 2	Exp, 9 (Obs, *0*)	Exp, 43 (Obs, *9*)	Exp, 13 (Obs, 17)	Exp, 39 (Obs, 49)	Exp, 22 (Obs, *51*)
Chi Test *p*-value	*0.002*	*1.649E-08*	0.5043	0.1565	*7.455E-12*

Table 4.
Table 1 (a) post-analysis of grade 1 and grade 2 X-ray abnormalities (of ref. [32]).

	Exposure duration (yrs)				
X-ray Grade (of abnormality)	**0–9**	**10–19**	**20–29**	**30–39**	**40+**
Asbestosis Grade 2	Exp, 0 (Obs, 0)	Exp, 6 (Obs, 9)	Exp, 15 (Obs, 17)	Exp, 112 (Obs, 102)	Exp, 64 (Obs, *35*)
Asbestosis Grade 3	Exp, 0 (Obs, 0)	Exp, 3 (Obs, 0)	Exp, 6 (Obs, 4)	Exp, 39 (Obs, 49)	Exp, 22 (Obs, *51*)
Chi Test *p*-value	Und.	0.1677	0.6361	0.9615	0.5655

Table 5.
Table 1 (a) post-analysis of grade 2 and grade 3 X-ray abnormalities (of ref. [32]).

	Exposure duration (yrs)						
	Lung cancer	***Mesothelioma***	**G.I. cancer**	**All other cancer**	**Ischemic heart disease**	**Chronic bronchitis**	**All other causes**
Plaques	Exp, 15 (Obs, 19)	Exp, 15 (Obs, 23)	Exp, 7 (Obs, 7)	Exp, 13 (Obs, 7)	Exp, 40 (Obs, 34)	Exp, 11 (Obs, 9)	Exp, 27 (Obs, 22)
Controls	Exp, 8 (Obs, 4)	Exp, 8 (Obs, 0)	Exp, 4 (Obs, 4)	Exp, 7 (Obs, 7)	Exp, 23 (Obs, 29)	Exp, 7 (Obs, 9)	Exp, 16 (Obs, 21)
Chi Test *p*-value	0.1548	*0.0012*	0.9995	0.9859	0.3166	0.5104	0.263

Table 6.
Table 8 post-analysis of X-ray asbestosis contact-related pleural plaques in exposed and non-exposed groups (as per ref. [32]).

	Onset of work (interval)			
X-ray Grade (of abnormality)	**< 1950**	**1951–1955**	**1956–1950**	**> *1961***
Normal	Exp, 90 (Obs, 81)	Exp, 107 (Obs, 98)	Exp, 180 (Obs, 174)	Exp, 164 (Obs, *188*)
Abnormal	Exp, 76 (Obs, *85*)	Exp, 91 (Obs, 100)	Exp, 153 (Obs, 159)	Exp, 139 (Obs, 115)
Chi Test *p*-value	0.1702*	0.1935	0.4986	*0.0055*

**trend for difference between groups (Normal, Abnormal).*

Table 7.
Table 10 post-analysis of abnormal and normal X-ray findings for beginning work year intervals (of ref. [32]).

A.3 Particulates toxicity industrial hygiene (adapted from Ref. [15])

I. Threshold limit value

$$TLV\text{ppm} = \frac{(TLV)\ (\text{Air vol})}{(MW)}; \tag{4}$$

$$TLV\text{ppm} = \frac{(TLV)\ (\text{Air vol})}{(MW\text{conv})} \tag{5}$$

where 1 $^{(a)}$, MW, g/mol, or 1 $^{(b)}$, MW (conv., mg/mol), MW x 1000 mg/mol.

1 $^{(a)}$ TLV, g/1 x 10^6 L, or 1 $^{(b)}$ mg/m^3

1 $^{(a)}$ Air vol, 24.45 L/mol, or 1 $^{(b)}$, 0.02445 m^3/mol

II. Time-weighted mean

$$TWA\text{m}\,[\,] = \frac{\sum(n_1 * [\,] + \ ...\ + n_m * [\,])}{\sum(n_1 + \ ...\ + n_m)} \tag{6}$$

where TWA_m, Time-weighted average concentration.

[], concentration

n, hours of exposure at TLV concentration

m, hours in shift

III. Fraction of TLV exposure

$$\text{TLVfract} = \sum\left(\frac{[\,]_1}{TLV_1} + \ ...\ + \frac{[\,]_z}{TLV_z}\right) \tag{7}$$

where TLV_{fract}, overall fraction of threshold limit value exposure (> 1, exposure limit has been exceeded).

[], concentration

z, TLV for specific chemicals beginning with first exposure []

IV. Sampler efficiency

A. IPM, Inhalable particle matter size range, 50% cumulative reference level

$$\text{Finh}\ (D\text{a}) = -11.8\ (ln\ \ (D\text{a})) + \beta 0 \tag{8}$$

Finh (D_{ae}), IPM efficiency.

β_0 = 101.

(, non-inclusion x variable.

], inclusion x variables.

where particle aerodynamic diameter D_a, (0 μm; 1 μm, 2 μm, 5 μm, 10 μm, 20 μm, 30 μm, 40 μm, 50 μm, 100 μm], R^2 = 0.97.

where 50% efficiency is at standard particle D_a = 100 μm, and 97% efficiency is at std. particle D_a = 1 μm (actual values)

B. RPM, Respirable particle matter size range, 100% cumulative reference level

$$\text{Fresp }(Da) = -3.79(Da)2 + 2.84(Da) + \beta 0 \tag{9}$$

Fresp (D_{ae}), RPM efficiency (small particle).

β_0 = 99.4.

where particle aerodynamic diameter D_a, [0 μm, 1 μm, 2 μm, 3 μm, 4 μm], R^2 = 0.998

$$\text{Fresp }(Da) = \beta 0\, e^{-0.648\ (D_a)} \tag{10}$$

F_{resp} (D_{ae}), RPM efficiency (large particle).

β_0 = 767.

where particle aerodynamic diameter D_a, [4 μm, 5 μm, 10 μm, 6 μm, 7 μm, 8 μm, 10 μm], R^2 = 0.992.

A.4 End of chapter educational objectives exercise – Particulate matter toxicity

i. What are the bioengineered macro-particulate classes for extended release effects (**Table 2**)?

- Colloid (1–1 μm),
- Suspension (> 1 μm)
- Emulsion (H_2O-in-oil (*w*/*o*), oil-in-H_2O (*o*/w)) ± inclusion
- Micelle (10–200 nm)
- Liposome (100–580 nm)

ii. How are bioengineered particles categorized (**Table 2**)?

- Soft nanoparticles ($3^x + 2$, x = C-C bond)
- Hard nanoparticles ($n^x + m$, x = metal oxide bond)

iii. Geological detritus is (**Table 1**):

- Transition metal oxide/polyhedral bond structure particulates

iv. What are the particle properties for elimination or uptake?

- Charge/exterior oxidation state (neutral, poly-neutral, anionic, cationic, cationo-neutral)
- Size (1.65–2.09 nm, renal filtration; 4.75 nm, renal reabsorption)
- Aspect (w, l)

v. What are the terms and relationship for inhalable particle diameters?

- Aerodynamic diameter (D_{ae})
- Settling velocity diameter (D_a)
- Volume equivalent diameter (D_{ve})
- Mass equivalent diameter (D_{me})

vi. What are the determinants in the inhalable particulate matter flow regime?

- Effective particle density (${\rho_{eff}}^{II}$)
- Particle density (ρ_p)
- Dynamic shape factor (χ)
- Knudsen path length (K_n)
- Velocity-slip correction parameter (C)

vii. What are mechanisms of smelter particle formation?

- Condensation for collision product molecules
- Coagulation for particles

viii. What are the variables of dissolution?

- Particle type (hard, soft)
- Dreiding forcefield energy (D_E)
- Free energy threshold ($G_{critical}$)
- Surface energy barrier to dissolution (H)

- Defusion entropy (S_f)
- Partition coefficient (P)
- Solution electrolyte saturation ($1 - \alpha$)
- Volume per molecule (ω)
- Temperature (T)

ix. What are the variables of percolation?

- Matrix porosity (φ), pore size (r_m), pore number (r_n)
- Volumetric water content (θ)
- Percolation (p)
- Critical threshold for percolation (p_C)
- Cross-over threshold (p_x)
- Power variable (t, q)
- Molecule size (a)
- Macro-diffusivity (D)
- Diffusivity, unbounded solvent (D_∞)

x. What are the epidemiologic/statistical methods?

- Categorical
  - Relative risk (RR), Odds ratio (OR), Chi Square (χ^2; Appendix II), HR: Logistic regression
  - Sensitivity and specificity
- Numeric
  - Parametric methods, SMR

xi. What are the primary variables for water and solute permeability modeling (Appendix I)

- Starling model, 2-compartment, bidirectional
- Oncotic pressure, π_c; π_i, π_g
- Hydrostatic pressure, p_c, p_i

- Reflection coefficient, σ (a.u.)
- Hydraulic permeability coefficient, L_p
- Fluid flux parameter, $J_v/A/\Delta P$
- Solute flux parameter, $J_s/A/\Delta C$
- Peclet convection diffusion ratio, P_e (a.u.)

xii. What are the primary variables for solute pharmacokinetic modeling (Appendix I)

- Universal diffusional concentration variables, C_b, $C_{t\ \text{intracellular}}$
- QAR model, 2-compartment, uni-directional (integrated/non-linear)
- Influx parameter, K_i
- GKM model, 2-compartment, bi-directional (integrated decay)
- Transfer parameters, K_{trans}, K_{ep}
- Fractional variables, v_p, v_e

xiii. What are US agency threshold limits or groups for toxic exposure surveillance for respirable size range particulates aerosol (Appendix III, **Table 3** footnotes)?

- Threshold limit value, $(\text{shift, TLV})^{-1}$
- Permissible exposure limit (PEL)
- Recommended exposure limit (REL)
- Short-term exposure limit (STEL)
- Action level (AL)

xiv. What is the industrial hygiene approach?

- Air sampling/personal dosimetry
    - Particulates, flow rate-calibrated sampling pumps with size-specific filters
    - Vapors, monitoring badges and/or solid sorbent tubes
- Hierarchy of controls implementation

xv. What are the methods for analysis of cell macromolecular changes in response to exposure?

- Fluorescence absorbance/emission (*Abs*, *Em* SI, λ)
- X-ray diffraction (SI, 2-θ°)
- FTIR reflectance or transmission spectroscopy (SI, λ^{-1})
- Mass/ionization spectroscopy (SI, m/z)
- ^{1}H NMR spectroscopy (SI, ppm)
- TEM: $STEM_{2.6E2\ e-/nm2}$, $EFTEM_{10E6\ e-/nm2}$ (SI, σ)

xvi. Methods for analysis for DNA polymorphism and transcription factor/complex binding, RNA expression, methylation and conformational changes include:

- DNA segments sequencing (reads mapping, assembly; ATAC, CHiP)
- RNA-sequencing
- Guide RNA-sequencing/Cas9-CRISPR
- qRT-PCR with primers
- cDNA microarray library with FDR *Q*-value correction
- Bisulfite C - > U conversion
- Circular dichroism (CD) spectroscopy (θ°, λ)

xvii. What are the determinants of particulate matter cell genomic and epigenomic toxicity?

- Particulate matter exterior properties aspect ratios
- Transition metal dissolution rates and exterior properties/oxidation states (cell surface interaction, nuclei acid binding/conformational changes)
- AhR and related pathway agonism
- Efflux transporter and related gene expression polymorphisms
- Gene promoter or body methylation changes

Author details

Hemant Sarin
National Institute of Biomedical Imaging and Bioengineering, National Institutes of Health, Bethesda, Maryland, USA

*Address all correspondence to: hemantsarin74@gmail.com

References

[1] Albers P, Maier M, Reisinger M, Hannebauer B, Weinand R. Physical boundaries within aggregates – Differences between amorphous, Para-crystalline, and crystalline structures. Crystal Research and Technology [Internet]. 2015;**50**(11):846-865. Available from: https://onlinelibrary.wiley.com/doi/abs/10.1002/crat.201500040

[2] Dove PM, Han N, Wallace AF, De Yoreo JJ. Kinetics of amorphous silica dissolution and the paradox of the silica polymorphs. Proceedings of the National Academy of Sciences of the United States of America. 2008;**105**(29): 9903-9908

[3] Yang W, Wang L, Mettenbrink EM, DeAngelis PL, Wilhelm S. Nanoparticle Toxicology. Annual Review of Pharmacology and Toxicology. 2021; **61**(1):269-289 [Internet].Available from: https://www.annualreviews.org/doi/abs/10.1146/annurev-pharmtox-032320-110338

[4] Buseck PR, Adachi K, Gelencsér A, Tompa É, Pósfai M. Are black carbon and soot the same? Atmospheric Chemistry and Physics Discussions. 2012;**2012**:24821-24846 [Internet]. Available from: https://acp.copernicus.org/preprints/12/24821/2012/

[5] Landrigan PJ, Lucchini RG, Kotelchuck D, Grandjean P. Chapter 29 - principles for prevention of the toxic effects of metals. In: Nordberg GF, Costa M, editors. Handbook on the Toxicology of Metals (Fifth Edition) [Internet]. Amsterdam: Academic Press, Elsevier B.V.; 2022. pp. 685-703. Available from: https://www.sciencedirect.com/science/article/pii/B9780128232927000267

[6] Dykman LA, Khlebtsov NG. Methods for chemical synthesis of colloidal gold. Russian Chemical Reviews [Internet]. 2019;**88**(3):229. DOI: 10.1070/RCR4843

[7] Bruce IJ, Sen T. Surface modification of magnetic nanoparticles with alkoxysilanes and their application in magnetic bioseparations. Langmuir. 2005;**21**(15):7029-7035

[8] Schipper ML, Iyer G, Koh AL, Cheng Z, Ebenstein Y, Aharoni A, et al. Particle size, surface coating, and PEGylation influence the biodistribution of quantum dots in living mice. Small [Internet]. 2009;**5**(1):126-134. Available from: http://www.ncbi.nlm.nih.gov/pubmed/19051182

[9] Choi HS, Liu W, Misra P, Tanaka E, Zimmer JP, Itty Ipe B, et al. Renal clearance of quantum dots. Nature Biotechnology. 2007;**25**(10):1165-1170

[10] Arnida J-AMM, Ray A, Peterson CM, Ghandehari H. Geometry and surface characteristics of gold nanoparticles influence their biodistribution and uptake by macrophages. European Journal of Pharmaceutics and Biopharmaceutics [Internet]. 2011;**77**(3): 417-423. Available from: https://www.sciencedirect.com/science/article/pii/S0939641110002997

[11] Missaoui WN, Arnold RD, Cummings BS. Toxicological status of nanoparticles: What we know and what we don't know. Chemico-Biological Interactions. 2018;**295**:1-12

[12] Rahman H, Harbison R. Benign dusts (nuisance dusts). In: Hamilton & Hardy's Industrial Toxicology. Wiley; 2015. pp. 931-934

[13] Pott P. Chirurgical Observations Relative to the Cataract, the Polypus of

the Nose, the Cancer of the Scrotum, the Different Kinds of Ruptures, and the Mortification of the Toes and Feet. London: T. J. Carnegy; 1775. p. 208

[14] Silva dos S. Cancer Epidemiology: Principles and Methods. Lyon, France: IARC; 1999

[15] American Conference of Governmental Industrial Hygienists. 2021 TLVs and BEIs: Based on the Documentation of the Threshold Limit Values for Chemical Substances and Physical Agents & Biological Exposure Indices [Internet]. Cincinnati, OH: ACGIH; 2021. Available from: https://portal.acgih.org/s/store#/store/browse/detail/a154W00000BOag7QAD

[16] DeCarlo PF, Slowik JG, Worsnop DR, Davidovits P, Jimenez JL. Particle morphology and density characterization by combined mobility and aerodynamic diameter measurements. Part 1: Theory. Aerosol Science and Technology [Internet]. 2004;**38**(12):1185-1205. DOI: 10.1080/027868290903907

[17] Jackson CL, Chanzy HD, Booy FP, Drake BJ, Tomalia DA, Bauer BJ, et al. Visualization of dendrimer molecules by transmission electron microscopy (TEM): Staining methods and Cryo-TEM of vitrified solutions. Macromolecules [Internet]. 1998;**31**(18):6259-6265. DOI: 10.1021/ma9806155

[18] Stetefeld J, McKenna SA, Patel TR. Dynamic light scattering: A practical guide and applications in biomedical sciences. Biophysical Reviews. 2016; **8**(4):409-427

[19] Neu-Baker NM, Dozier AK, Eastlake AC, Brenner SA. Evaluation of enhanced darkfield microscopy and hyperspectral imaging for rapid screening of TiO(2) and SiO(2) nanoscale particles captured on filter media. Microscopy Research and Technique. 2021;**84**(12):2968-2976

[20] Valsesia A, Parot J, Ponti J, Mehn D, Marino R, Melillo D, et al. Detection, counting and characterization of nanoplastics in marine bioindicators: A proof of principle study. Microplastics and Nanoplastics [Internet]. 2021; **1**(1):5. DOI: 10.1186/s43591-021-00005-z

[21] Deng J, Shoemaker R, Xie B, Gore A, LeProust EM, Antosiewicz-Bourget J, et al. Targeted bisulfite sequencing reveals changes in DNA methylation associated with nuclear reprogramming. Nature Biotechnology. 2009;**27**(4): 353-360

[22] Flowers P, Theopold K, Langley R, Robinson WR. Chemistry. Second ed. OpenStax; 2019. p. 1331

[23] Park JH, Mudunkotuwa IA, Kim JS, Stanam A, Thorne PS, Grassian VH, et al. Physicochemical characterization of simulated welding fume from a spark discharge system. Aerosol Science and Technology. 2014;**47**(7):768-776

[24] Cotta MA. Quantum dots and their applications: What lies ahead? ACS Applications Nano Materials [Internet]. 2020;**3**(6):4920-4924. DOI: 10.1021/acsanm.0c01386

[25] Nabavinia M, Beltran-Huarac J. Recent Progress in iron oxide nanoparticles as therapeutic magnetic agents for cancer treatment and tissue engineering. ACS Applied Bio Materials. 2020;**3**(12):8172-8187

[26] Bulte JW, Duncan ID, Frank JA. In vivo magnetic resonance tracking of magnetically labeled cells after transplantation. Journal of Cerebral

Blood Flow and Metabolism. 2002;**22**(8): 899-907

[27] Rhaman MM, Islam MR, Akash S, Mim M, Noor Alam M, Nepovimova E, et al. Exploring the role of nanomedicines for the therapeutic approach of central nervous system dysfunction: At a glance. Frontiers in Cell and Development Biology. 2022;**10**: 989471

[28] Dobrovolskaia MA, Aggarwal P, Hall JB, McNeil SE. Preclinical studies to understand nanoparticle interaction with the immune system and its potential effects on nanoparticle biodistribution. Molecular Pharmaceutics. 2008;**5**(4): 487-495

[29] Brown JS, Gordon T, Price O, Asgharian B. Thoracic and respirable particle definitions for human health risk assessment. Particle and Fibre Toxicology. 2013;**10**:12 [Internet]. Available from: http://www.ncbi.nlm.nih.gov/pubmed/23575443

[30] International Agency for Research on Cancer. Arsenic, metals, fibres, and dusts: A review of human carcinogen. In: IARC Monographs on the Evaluation of Carcinogenic Risks to Humans. Vol. 100 C. Lyon, France: IARC; 2012

[31] Boulanger G, Andujar P, Pairon JC, Billon-Galland MA, Dion C, Dumortier P, et al. Quantification of short and long asbestos fibers to assess asbestos exposure: A review of fiber size toxicity. Environmental Health. 2014;**13**:59

[32] Selikoff IJ, Hammond EC. Asbestos-associated disease in United States shipyards. CA: a Cancer Journal for Clinicians. 1978;**28**(2):87-99 [Internet]. Available from: https://acsjournals.onlinelibrary.wiley.com/doi/abs/10.3322/canjclin.28.2.87

[33] Pascolo L, Gianoncelli A, Schneider G, Salomé M, Schneider M, Calligaro C, et al. The interaction of asbestos and iron in lung tissue revealed by synchrotron-based scanning X-ray microscopy. Scientific Reports. 2013;**3**: 1123

[34] Perkins TN, Peeters PM, Shukla A, Arijs I, Dragon J, Wouters EF, et al. Indications for distinct pathogenic mechanisms of asbestos and silica through gene expression profiling of the response of lung epithelial cells. Human Molecular Genetics. 2015;**24**(5): 1374-1389

[35] Duncan KE, Cook PM, Gavett SH, Dailey LA, Mahoney RK, Ghio AJ, et al. In vitro determinants of asbestos fiber toxicity: Effect on the relative toxicity of Libby amphibole in primary human airway epithelial cells. Particle and Fibre Toxicology. 2014;**11**:2

[36] Webber JS, Blake DJ, Ward TJ, Pfau JC. Separation and characterization of respirable amphibole fibers from Libby, Montana. Inhalation Toxicology. 2008;**20**(8):733-740

[37] Meijerink MJ, de Jong KP, Zečević J. Assessment of oxide nanoparticle stability in liquid phase transmission electron microscopy. Nano Research [Internet]. 2019;**12**(9):2355-2363. DOI: 10.1007/s12274-019-2419-3

[38] Perkins TN, Shukla A, Peeters PM, Steinbacher JL, Landry CC, Lathrop SA, et al. Differences in gene expression and cytokine production by crystalline vs. amorphous silica in human lung epithelial cells. Particle and Fibre Toxicology. 2012;**9**(1):6

[39] Vallières F, Simard JC, Noël C, Murphy-Marion M, Lavastre V, Girard D. Activation of human AML14.3D10 eosinophils by

nanoparticles: Modulatory activity on apoptosis and cytokine production. Journal of Immunotoxicology. 2016; **13**(6):817-826

[40] Wan R, Mo Y, Zhang Z, Jiang M, Tang S, Zhang Q. Cobalt nanoparticles induce lung injury, DNA damage and mutations in mice. Particle and Fibre Toxicology. 2017;**14**(1):38

[41] Jiang J, Oberdörster G, Elder A, Gelein R, Mercer P, Biswas P. Does nanoparticle activity depend upon size and crystal phase? Nanotoxicology. 2008;**2**(1):33-42

[42] Uboldi C, Urbán P, Gilliland D, Bajak E, Valsami-Jones E, Ponti J, et al. Role of the crystalline form of titanium dioxide nanoparticles: Rutile, and not anatase, induces toxic effects in Balb/3T3 mouse fibroblasts. Toxicology In Vitro. 2016;**31**:137-145 [Internet]. Available from: https://www.sciencedirect.com/science/article/pii/S0887233315300060

[43] McElwee MK, Song MO, Freedman JH. Copper activation of NF-kappaB signaling in HepG2 cells. Journal of Molecular Biology. 2009;**393**(5): 1013-1021

[44] He J, Wang M, Jiang Y, Chen Q, Xu S, Xu Q, et al. Chronic arsenic exposure and angiogenesis in human bronchial epithelial cells via the ROS/miR-199a-5p/HIF-1α/COX-2 pathway. Environmental Health Perspectives. 2014;**122**(3):255-261

[45] Smith AH, Goycolea M, Haque R, Biggs ML. Marked increase in bladder and lung cancer mortality in a region of northern Chile due to arsenic in drinking water. American Journal of Epidemiology. 1998;**147**(7):660-669

[46] Steenland K, Selevan S, Landrigan P. The mortality of lead smelter workers: An update. American Journal Public Health [Internet]. 1992;**82**(12): 1641-1644. Available from: http://www.ncbi.nlm.nih.gov/pubmed/1456339

[47] Machoń-Grecka A, Dobrakowski M, Boroń M, Lisowska G, Kasperczyk A, Kasperczyk S. The influence of occupational chronic lead exposure on the levels of selected pro-inflammatory cytokines and angiogenic factors. Human & Experimental Toxicology. 2017;**36**(5):467-473

[48] Bozack AK, Rifas-Shiman SL, Coull BA, Baccarelli AA, Wright RO, Amarasiriwardena C, et al. Prenatal metal exposure, cord blood DNA methylation and persistence in childhood: An epigenome-wide association study of 12 metals. Clinical Epigenetics [Internet]. 2021;**13**(1):208. DOI: 10.1186/s13148-021-01198-z

[49] Balachandran RC, Mukhopadhyay S, McBride D, Veevers J, Harrison FE, Aschner M, et al. Brain manganese and the balance between essential roles and neurotoxicity. Journal of Biological Chemistry [Internet]. 2020;**295**(19): 6312-6329. Available from: https://www.sciencedirect.com/science/article/pii/S0021925817484985

[50] Zuscik MJ, Pateder DB, Edward Puzas J, Schwarz EM, Rosier RN, O'Keefe RJ. Lead alters parathyroid hormone-related peptide and transforming growth factor-β1 effects and AP-1 and NF-κKB signaling in chondrocytes. Journal of Orthopaedic Research [Internet]. 2002;**20**(4): 811-818. Available from: https://onlinelibrary.wiley.com/doi/abs/10.1016/S0736-0266%2802%2900007-4

[51] Miller FJ, Gardner DE, Graham JA, Lee RE, Wilson WE, Bachmann JD. Size considerations for establishing a

standard for inhalable particles. Journal of the Air Pollution Control Association [Internet]. 1979;**29**(6):610-615. DOI: 10.1080/00022470.1979.10470831

[52] Timbrell V. THE INHALATION OF FIBROUS DUSTS. Annals of the New York Academy of Sciences. 1965;**132**(1): 255-273 [Internet]. Available from: https://nyaspubs.onlinelibrary.wiley.com/doi/abs/10.1111/j.1749-6632.1965.tb41107.x

[53] Allen MD, Raabe OG. Slip correction measurements of spherical solid aerosol particles in an improved Millikan apparatus. Aerosol Science and Technology [Internet]. 1985;**4**(3): 269-286. DOI: 10.1080/02786828508959055

[54] Pöhlker ML, Krüger OO, Förster JD, Berkemeier T, Elbert W, Fröhlich-Nowoisky J, et al. Respiratory aerosols and droplets in the transmission of infectious diseases. Physics Medicine. arXiv. 2021:1-50. eprint={2103.01188v4}

[55] Yalkowsky SH, Valvani SC. Solubility and partitioning I: Solubility of nonelectrolytes in water. Journal of Pharmaceutical Sciences. 1980;**69**(8): 912-922 [Internet]. Available from: https://www.sciencedirect.com/science/article/pii/S0022354915433568

[56] Sahimi M, Jue VL. Diffusion of large molecules in porous media. Physical Review Letters. 1989;**62**(6):629-632

[57] Ghanbarian B, Daigle H, Hunt AG, Ewing RP, Sahimi M. Gas and solute diffusion in partially saturated porous media: Percolation theory and effective medium approximation compared with lattice Boltzmann simulations. Journal of Geophysical Research - Solid Earth. 2015;**120**(1):182-190 [Internet]. Available from: https://agupubs.onlinelibrary.wiley.com/doi/abs/10.1002/2014JB011645

[58] Bonny JD, Leuenberger H. Matrix type controlled release systems: I. effect of percolation on drug dissolution kinetics. Pharmaceutica Acta Helvetiae. 1991;**66**(5–6):160-164

[59] Pillai G, Ceballos-Coronel ML. Science and technology of the emerging nanomedicines in cancer therapy: A primer for physicians and pharmacists. SAGE Open Medicine. 2013;**1**: 2050312113513759

[60] Hobbs SK, Monsky WL, Yuan F, Roberts WG, Griffith L, Torchilin VP, et al. Regulation of transport pathways in tumor vessels: Role of tumor type and microenvironment. Proceedings of the National Academy of Sciences of the United States of America. 1998;**95**(8): 4607-4612

[61] Tomalia DA, Reyna LA, Svenson S. Dendrimers as multi-purpose nanodevices for oncology drug delivery and diagnostic imaging. Biochemical Society Transactions. 2007;**35**(Pt 1):61-67

[62] Rupp R, Rosenthal SL, Stanberry LR. VivaGel™ (SPL7013 gel): A candidate dendrimer – Microbicide for the prevention of HIV and HSV infection. International Journal of Nanomedicine. 2007;**2**(4):561-566 [Internet]. Available from: https://www.tandfonline.com/doi/abs/10.2147/IJN.S2.4.561

[63] Wang J, Li B, Qiu L, Qiao X, Yang H. Dendrimer-based drug delivery systems: History, challenges, and latest developments. Journal of Biological Engineering [Internet]. 2022;**16**(1):18. DOI: 10.1186/s13036-022-00298-5

[64] Sarin H, Fung SH, Kanevsky AS, Wu H, Wilson CM, Vo H, et al. Quantitative gadolinium chelate-

enhanced magnetic resonance imaging of normal endothelial barrier disruption from nanoparticle biophilicity interactions. Materials Today: Proceedings [Internet]. 2021;**45**: 3795-3799. Available from: https://www.sciencedirect.com/science/article/pii/S2214785321006283

[65] Kaminskas LM, Boyd BJ, Porter CJH. Dendrimer pharmacokinetics: The effect of size, structure and surface characteristics on ADME properties. Nanomedicine [Internet]. 2011;**6**(6): 1063-1084. Available from: https://www.futuremedicine.com/doi/abs/10.2217/nnm.11.67

[66] Olson LB, Hunter NI, Rempel RE, Yu H, Spencer DM, Sullenger CZ, et al. Mixed-surface polyamidoamine polymer variants retain nucleic acid-scavenger ability with reduced toxicity. iScience [Internet]. 2022;**25**(12):105542. Available from: https://www.sciencedirect.com/science/article/pii/S2589004222018144

[67] Sousa AA, Aronova MA, Wu H, Sarin H, Griffiths G, Leapman RD. Determining molecular mass distributions and compositions of functionalized dendrimer nanoparticles. Nanomedicine (London, England). 2009;**4**(4):391-399

[68] Kobayashi H, Kawamoto S, Choyke PL, Sato N, Knopp MV, Star RA, et al. Comparison of dendrimer-based macromolecular contrast agents for dynamic micro-magnetic resonance lymphangiography. Magnetic Resonance in Medicine. 2003;**50**(4):758-766 [Internet]. Available from: https://onlinelibrary.wiley.com/doi/abs/10.1002/mrm.10583

[69] Banerjee D, Harfouche R, Sengupta S. Nanotechnology-mediated targeting of tumor angiogenesis. Vascular Cell [Internet]. 2011;**3**(1):3. DOI: 10.1186/2045-824X-3-3

[70] Prabhakar U, Maeda H, Jain RK, Sevick-Muraca EM, Zamboni W, Farokhzad OC, et al. Challenges and key considerations of the enhanced permeability and retention effect for nanomedicine drug delivery in oncology. Cancer Research. 2013;**73**(8):2412-2417

[71] Danhier F, Ansorena E, Silva JM, Coco R, Le Breton A, Préat V. PLGA-based nanoparticles: An overview of biomedical applications. Journal of Controlled Release. 2012;**161**(2):505-522

[72] Makadia HK, Siegel SJ. Poly lactic-co-glycolic acid (PLGA) as biodegradable controlled drug delivery carrier. Polymers (Basel). 2011;**3**(3): 1377-1397

[73] Fasehee H, Dinarvand R, Ghavamzadeh A, Esfandyari-Manesh M, Moradian H, Faghihi S, et al. Delivery of disulfiram into breast cancer cells using folate-receptor-targeted PLGA-PEG nanoparticles: In vitro and in vivo investigations. Journal of Nanobiotechnology. 2016;**14**:32

[74] Liu Z, Davis C, Cai W, He L, Chen X, Dai H. Circulation and long-term fate of functionalized, biocompatible single-walled carbon nanotubes in mice probed by Raman spectroscopy. Proceedings of the National Academy of Sciences of the United States of America. 2008;**105**(5): 1410-1415

[75] Singh R, Pantarotto D, Lacerda L, Pastorin G, Klumpp C, Prato M, et al. Tissue biodistribution and blood clearance rates of intravenously administered carbon nanotube radiotracers. Proceedings of the National Academy of Sciences of the United

States of America. 2006;**103**(9): 3357-3362

[76] Kotagiri N, Kim JW. Stealth nanotubes: Strategies of shielding carbon nanotubes to evade opsonization and improve biodistribution. International Journal of Nanomedicine. 2014;**9 Suppl 1**(Suppl. 1):85-105

[77] Åslund AKO, Vandebriel RJ, Caputo F, de Jong WH, Delmaar C, Hyldbakk A, et al. A comparative biodistribution study of polymeric and lipid-based nanoparticles. Drug Delivery and Translational Research. 2022;**12**(9): 2114-2131

[78] Tsuchiya K, Uchida T, Kobayashi M, Maeda H, Konno T, Yamanaka H. Tumor-targeted chemotherapy with SMANCS in lipiodol for renal cell carcinoma: Longer survival with larger size tumors. Urology [Internet]. 2000; **55**(4):495-500. Available from: http://europepmc.org/abstract/MED/10736490

[79] Deschamps F, Moine L, Isoardo T, Tselikas L, Paci A, Mir LM, et al. Parameters for stable water-in-oil lipiodol emulsion for liver trans-arterial chemo-Eembolization. Cardiovascular and Interventional Radiology. 2017; **40**(12):1927-1932

[80] Ahnfelt E, Degerstedt O, Lilienberg E, Sjögren E, Hansson P, Lennernäs H. Lipiodol-based emulsions used for transarterial chemoembolization and drug delivery: Effects of composition on stability and product quality. Journal of Diabetes Science and Technology. 2019;**53**:101143 [Internet]. Available from: https://www.sciencedirect.com/science/article/pii/S1773224719303922

[81] He P, Ren E, Chen B, Chen H, Cheng H, Gao X, et al. A super-stable homogeneous Lipiodol-hydrophilic chemodrug formulation for treatment of hepatocellular carcinoma. Theranostics. 2022;**12**(4):1769-1782 [Internet]. Available from: http://www.ncbi.nlm.nih.gov/pubmed/35198072

[82] Kinbara K. Monodisperse engineered PEGs for bio-related applications. Polym Journal [Internet]. 2018;**50**(8):689-697. DOI: 10.1038/s41428-018-0074-2

[83] Wu W, Wu Z, Yu T, Jiang C, Kim WS. Recent progress on magnetic iron oxide nanoparticles: Synthesis, surface functional strategies and biomedical applications. Science and Technology of Advanced Materials. 2015;**16**(2):23501

[84] Jeon M, Halbert MV, Stephen ZR, Zhang M. Iron oxide nanoparticles as T (1) contrast agents for magnetic resonance imaging: Fundamentals, challenges, applications, and Prospectives. Advanced Materials. 2021; **33**(23):e1906539 [Internet]. Available from: http://www.ncbi.nlm.nih.gov/pubmed/32495404

[85] Dhenadhayalan N, Lin TW, Lee HL, Lin KC. Multisensing capability of MoSe2 quantum dots by tuning surface functional groups. ACS Applied Nano Materials [Internet]. 2018;**1**(7): 3453-3463. DOI: 10.1021/acsanm.8b00634

[86] Singh VK, Mishra H, Ali R, Umrao S, Srivastava R, Abraham S, et al. In situ functionalized fluorescent WS2-QDs as sensitive and selective probe for Fe^{3+} and a detailed study of its fluorescence quenching. ACS Applied Nano Materials [Internet]. 2019;**2**(1):566-576. DOI: 10.1021/acsanm.8b02162

[87] Bumb A, Brechbiel MW, Choyke PL, Fugger L, Eggeman A, Prabhakaran D, et al. Synthesis and characterization of

ultra-small superparamagnetic iron oxide nanoparticles thinly coated with silica. Nanotechnology [Internet]. 2008; **19**(33):335601. DOI: 10.1088/0957-4484/19/33/335601

[88] Mathieu P, Coppel Y, Respaud M, Nguyen QT, Boutry S, Laurent S, et al. Silica coated iron/iron oxide nanoparticles as a nano-platform for T (2) weighted magnetic resonance imaging. Molecules. 2019;**24**(24):4629. DOI: 10.3390/molecules24244629

[89] Iqbal MZ, Ma X, Chen T, Zhang L, Ren W, Xiang L, et al. Silica-coated super-paramagnetic iron oxide nanoparticles (SPIONPs): A new type contrast agent of T1 magnetic resonance imaging (MRI). Journal of Materials Chemistry B [Internet]. 2015;**3**(26): 5172-5181. DOI: 10.1039/C5TB00300H

[90] Wunderbaldinger P, Josephson L, Weissleder R. Crosslinked iron oxides (CLIO): A new platform for the development of targeted MR contrast agents. Academic Radiology. 2002;**9** (Suppl 2):S304-S306

[91] Pandya AD, Iversen TG, Moestue S, Grinde MT, Mørch Ý, Snipstad S, et al. Biodistribution of poly(alkyl cyanoacrylate) nanoparticles in mice and effect on tumor infiltration of macrophages into a patient-derived breast cancer xenograft. Nanomaterials (Basel). 2021;**11**(5):1140. DOI: 10.3390/nano11051140

[92] Baig N, Kammakakam I, Falath W. Nanomaterials: A review of synthesis methods, properties, recent progress, and challenges. Materials Advances [Internet]. 2021;**2**(6):1821-1871. DOI: 10.1039/D0MA00807A

[93] Liu P, Chen W, Liu C, Tian M, Liu P. A novel poly (vinyl alcohol)/poly (ethylene glycol) scaffold for tissue engineering with a unique bimodal open-celled structure fabricated using supercritical fluid foaming. Scientific Reports. 2019;**9**(1):9534

[94] García-Rodríguez A, Rubio L, Vila L, Xamena N, Velázquez A, Marcos R, et al. The comet assay as a tool to detect the genotoxic potential of nanomaterials. Nanomaterials (Basel). 2019;**9**(10):1385. DOI: 10.3390/nano9101385

[95] Ansari MO, Parveen N, Ahmad MF, Wani AL, Afrin S, Rahman Y, et al. Evaluation of DNA interaction, genotoxicity and oxidative stress induced by iron oxide nanoparticles both in vitro and in vivo: Attenuation by thymoquinone. Scientific Reports [Internet]. 2019;**9**(1):6912. DOI: 10.1038/s41598-019-43188-5

[96] Lee P, Kim JK, Jo MS, Kim HP, Ahn K, Park JD, et al. Biokinetics of subacutely co-inhaled same size gold and silver nanoparticles. Part Fibre Toxicology [Internet]. 2023;**20**(1):9. DOI: 10.1186/s12989-023-00515-z

[97] Shen T, Weissleder R, Papisov M, Bogdanov A Jr, Brady TJ. Monocrystalline iron oxide nanocompounds (MION): Physicochemical properties. Magnetic Resonance in Medicine. 1993;**29**(5): 599-604 [Internet]. Available from: https://onlinelibrary.wiley.com/doi/abs/10.1002/mrm.1910290504

[98] Xu B, Chasteen ND. Iron oxidation chemistry in ferritin. Increasing Fe/O2 stoichiometry during core formation. The Journal of Biological Chemistry. 1991;**266**(30):19965-19970 [Internet]. Available from: http://www.ncbi.nlm.nih.gov/pubmed/1939058

[99] Walls MG, Cao C, Yu-Zhang K, Li J, Che R, Pan Y. Identification of ferrous-

ferric Fe3O4 nanoparticles in recombinant human ferritin cages. Microscopy and Microanalysis. 2013; **19**(4):835-841 [Internet]. Available from: http://www.ncbi.nlm.nih.gov/pubmed/23800760

[100] Maher BA, Ahmed IA, Karloukovski V, MacLaren DA, Foulds PG, Allsop D, et al. Magnetite pollution nanoparticles in the human brain. Proceedings of the National Academy of Sciences of the United States of America. 2016;**113**(39): 10797-10801

[101] Cowley JM, Janney DE, Gerkin RC, Buseck PR. The structure of ferritin cores determined by electron nanodiffraction. Journal of Structural Biology. 2000;**131**(3):210-216 [Internet]. Available from: http://www.ncbi.nlm.nih.gov/pubmed/11052893

[102] Kobayashi H, Reijnders K, English S, Yordanov AT, Milenic DE, Sowers AL, et al. Application of a macromolecular contrast agent for detection of alterations of tumor vessel permeability induced by radiation. Clinical Cancer Research. 2004;**10**(22): 7712-7720

[103] Michel CC, Curry FE. Microvascular permeability. Physiological Reviews. 1999;**79**(3): 703-761

[104] Levick JR, Michel CC. Microvascular fluid exchange and the revised Starling principle. Cardiovascular Research [Internet]. 2010;**87**(2):198-210. DOI: 10.1093/cvr/cvq062

[105] Sarin H. Physiologic upper limits of pore size of different blood capillary types and another perspective on the dual pore theory of microvascular permeability. Journal of Angiogenesis Research. 2010;**2**:14

[106] Crone C. The permeability of capillaries in various organs as determined by use of the indicator diffusion method. Acta Physiologica Scandinavica. 1963;**58**: 292-305

[107] Ewing JR, Brown SL, Lu M, Panda S, Ding G, Knight RA, et al. Model selection in magnetic resonance imaging measurements of vascular permeability: Gadomer in a 9L model of rat cerebral tumor. Journal of Cerebral Blood Flow and Metabolism. 2006;**26**(3): 310-320

[108] Ferrier MC, Sarin H, Fung SH, Schatlo B, Pluta RM, Gupta SN, et al. Validation of dynamic contrast-enhanced magnetic resonance imaging-derived vascular permeability measurements using quantitative autoradiography in the RG2 rat brain tumor model. Neoplasia. 2007;**9**(7):546-555

[109] Li X, Zhu Y, Kang H, Zhang Y, Liang H, Wang S, et al. Glioma grading by microvascular permeability parameters derived from dynamic contrast-enhanced MRI and intratumoral susceptibility signal on susceptibility weighted imaging. Cancer Imaging [Internet]. 2015;**15**(1):4. DOI: 10.1186/s40644-015-0039-z

[110] Pathak AP, Penet MF, Bhujwalla ZM. MR molecular imaging of tumor vasculature and vascular targets. Advances in Genetics. 2010;**69**:1-30

[111] Kong LW, Chen J, Zhao H, Yao K, Fang SY, Wang Z, et al. Intratumoral susceptibility signals reflect biomarker status in gliomas. Scientific Reports [Internet]. 2019;**9**(1):17080. DOI: 10.1038/s41598-019-53629-w

[112] Gutierrez CT, Loizides C, Hafez I, Brostrøm A, Wolff H, Szarek J, et al.

Acute phase response following pulmonary exposure to soluble and insoluble metal oxide nanomaterials in mice. Particle and Fibre Toxicology [Internet]. 2023;**20**(1):4. DOI: 10.1186/s12989-023-00514-0

[113] Guth K, Bourgeois M, Harbison R. Assessment of Lead exposures during abrasive blasting and vacuuming in ventilated field containments: A case study. Occupational Diseases and Environmental Medicine. 2022;**10**: 116-131

[114] Wright RO, Schwartz J, Wright RJ, Bollati V, Tarantini L, Park SK, et al. Biomarkers of lead exposure and DNA methylation within retrotransposons. Environmental Health Perspectives. 2010;**118**(6):790-795

[115] Boytsova OV, Shekunova TO, Baranchikov AE. Nanocrystalline manganese dioxide synthesis by microwave-hydrothermal treatment. Russian Journal of Inorganic Chemistry [Internet]. 2015;**60**(5):546-551. DOI: 10.1134/S0036023615050022

[116] Aisen P, Aasa R, Redfield AG. The chromium, manganese, and cobalt complexes of transferrin. Journal of Biological Chemistry [Internet]. 1969; **244**(17):4628-4633. Available from: https://www.sciencedirect.com/science/article/pii/S0021925818936707

[117] Hedberg YS, Wei Z, McCarrick S, Romanovski V, Theodore J, Westin EM, et al. Welding fume nanoparticles from solid and flux-cored wires: Solubility, toxicity, and role of fluorides. Journal of Hazardous Materials [Internet]. 2021; **413**:125273. Available from: https://www.sciencedirect.com/science/article/pii/S0304389421002363

[118] Lindner S, Lucchini R, Broberg K. Genetics and epigenetics of manganese toxicity. Current Environmental Health Reports. 2022;**9**(4):697-713

[119] Gehman LT, Stoilov P, Maguire J, Damianov A, Lin CH, Shiue L, et al. The splicing regulator Rbfox1 (A2BP1) controls neuronal excitation in the mammalian brain. Nature Genetics. 2011;**43**(7):706-711

[120] Salazar J, Mena N, Hunot S, Prigent A, Alvarez-Fischer D, Arredondo M, et al. Divalent metal transporter 1 (DMT1) contributes to neurodegeneration in animal models of Parkinson's disease. Proceedings of the National Academy of Sciences of the United States of America. 2008;**105**(47): 18578-18583

[121] Madison JL, Wegrzynowicz M, Aschner M, Bowman AB. Gender and manganese exposure interactions on mouse striatal neuron morphology. Neurotoxicology [Internet]. 2011;**32**(6): 896-906. Available from: https://www.sciencedirect.com/science/article/pii/S0161813X11000969

[122] Antonini JM, Santamaria AB, Jenkins NT, Albini E, Lucchini R. Fate of manganese associated with the inhalation of welding fumes: Potential neurological effects. Neurotoxicology [Internet]. 2006;**27**(3):304-310. Available from: https://www.sciencedirect.com/science/article/pii/S0161813X05001415

[123] Sarin H. Pressure regulated basis for gene transcription by delta-cell micro-compliance modeled in silico: Biphenyl, bisphenol and small molecule ligand models of cell contraction-expansion. PLoS One. 2020;**15**(10):1-66

[124] Cuevas J, Bruckard WJ, Pownceby MI, Sparrow GJ, Torpy A. Alkaline sulphide leaching of tennantite in copper flotation concentrates to

selectively dissolve arsenic. Mineral Processing and Extractive Metallurgy [Internet]. 2022;**131**(3):229-238. DOI: 10.1080/25726641.2021.1948319

[125] Govindaraju M, Shekar HS, Sateesha SB, Vasudeva Raju P, Sambasiva Rao KR, Rao KSJ, et al. Copper interactions with DNA of chromatin and its role in neurodegenerative disorders. Journal of Pharmaceutical Analysis. 2013;**3**(5): 354-359

[126] Li G, Lee LS, Li M, Tsao SW, Chiu JF. Molecular changes during arsenic-induced cell transformation. Journal of Cellular Physiology. 2011; **226**(12):3225-3232

[127] Onder M, Onder S. Evaluation of occupational exposures to respirable dust in underground coal mines. Industrial Health. 2009;**47**(1):43-49

[128] Gorman T, Dropkin J, Kamen J, Nimbalkar S, Zuckerman N, Lowe T, et al. Controlling health hazards to hospital workers: A reference guide. NEW SOLUTIONS: A Journal of Environmental and Occupational Health Policy [Internet]. 2014;**23**(1_suppl): 1-169. Available from: https://journals.sagepub.com/doi/abs/10.2190/NS.23.Suppl

[129] Osinubi OY, Afilaka AA, Doucette J, Golden A, Soriano T, Rovner E, et al. Study of smoking behavior in asbestos workers. American Journal of Industrial Medicine. 2002; **41**(1):62-69

[130] Meyer JD, Islam SS, Ducatman AM, McCunney RJ. Prevalence of small lung opacities in populations unexposed to dusts. A literature analysis. Chest. 1997; **111**(2):404-410

[131] Office IL. Guidelines for the Use of the ILO International Classification of Radiographs of Pneumoconiosis. Geneva: International Labor Office; 2011

[132] de la Hoz RE, Weber J, Xu D, Doucette JT, Liu X, Carson DA, et al. Chest CT scan findings in world trade Center workers. Archives of Environmental & Occupational Health. 2019;**74**(5):263-270

[133] Chatham-Stephens K, Caravanos J, Ericson B, Sunga-Amparo J, Susilorini B, Sharma P, et al. Burden of disease from toxic waste sites in India, Indonesia, and the Philippines in 2010. Environmental Health Perspectives. 2013;**121**(7):791-796 [Internet]. Available from: https://ehp.niehs.nih.gov/doi/abs/10.1289/ehp.1206127

[134] Morabia A, Markowitz S, Garibaldi K, Wynder EL. Lung cancer and occupation: Results of a multicentre case-control study. British Journal of Industrial Medicine. 1992;**49**(10): 721-727

[135] Krabbe CA, Pruim J, van der Laan BF, Rödiger LA, Roodenburg JL. FDG-PET and detection of distant metastases and simultaneous tumors in head and neck squamous cell carcinoma: A comparison with chest radiography and chest CT. Oral Oncology. 2009; **45**(3):234-240

[136] Armato SG 3rd, Blyth KG, Keating JJ, Katz S, Tsim S, Coolen J, et al. Imaging in pleural mesothelioma: A review of the 13th international conference of the international mesothelioma interest group. Lung Cancer. 2016;**101**:48-58